T0313383

Productivity Improvement in Manufacturing SMEs

Application of Work Study Techniques

Productivity Improvement in Manufacturing SMEs

Application of Work Study Techniques

By

Thomas Munyai, Boysana Mbonyane, and Charles Mbohwa

CRC Press is an imprint of the
Taylor & Francis Group, an **informa** business
A PRODUCTIVITY PRESS BOOK

CRC Press
Taylor & Francis Group
6000 Broken Sound Parkway NW, Suite 300
Boca Raton, FL 33487-2742

CRC Press is an imprint of Taylor & Francis Group, an Informa business

Printed on acid-free paper

International Standard Book Number-13: 978-1-138-74711-1 (Hardback)

Visit the Taylor & Francis Web site at
http://www.taylorandfrancis.com

and the CRC Press Web site at
http://www.crcpress.com

Contents

Preface

This is a research book on productivity improvement in manufacturing small and medium-sized enterprises (SMEs). It focuses on work study as a technique to improve the productivity of manufacturing SMEs in South Africa and the rest of the world, including African developing industries. This research monograph presents an in-depth, reader-friendly exposition of current challenges that hamper the productivity of organizations.

Various concepts affecting the productivity of manufacturing industries and SMEs are presented, such as manufacturing, productivity, productivity measurement, productivity improvement, layout, location, efficiency, and ergonomic issues. Alternative techniques are introduced and applied and formally reported addressing the current challenges facing manufacturing SMEs. These techniques include qualitative and quantitative improvement tools such as method study, work measurement, business reengineering, operational research, and supplementary organizational effectiveness–related techniques. Work study as a tool is addressed in terms of how it impacts physical capital, technological capital, and management in improving the productivity of manufacturing SMEs. Work study is a stimulating area of operations management, industrial engineering, and management sciences that has a considerable effect on the productivity of manufacturing industries, both large and small. This tool can be applied in other businesses, not only manufacturing but also services businesses. This research book will focus on presenting a comprehensive overview of the field of manufacturing, taking a rational and realistic approach.

Work study as a discipline includes a merging of topics from manufacturing, industrial engineering, and management sciences. Even though our readers are not work study specialists but specialize in finance, marketing, safety management, project management, and information technology across the globe, they will find the book fascinating and beneficial as we develop a fundamental knowledge of manufacturing. This research book also presents the reality of manufacturing in the cases in point of Companies A, B, C, and D in all chapters.

This monograph encompasses 12 chapters. Chapter 1 presents the main research results of work done on the background of manufacturing small and medium enterprises (SMEs) in South Africa. Chapter 2 reviews work study and productivity theory to identify research gaps and provide the scope of the study that informed the rest of the work and of the monograph. Grounded theory, addressing the historical background of work study, is discussed, and its paradigm presented from its origins to its current status. This chapter also reviews productivity, productivity measurement, and productivity improvement in manufacturing SMEs with a view to identifying their suitability and adaptability to the South African environment. Chapter 3 addresses issues of effectiveness versus efficiency in manufacturing SMEs in order to develop new ideas and perspectives that could be applied to the study. Chapter 4 identifies factors influencing productivity in manufacturing SMEs as well as providing a detailed understanding of physical capital factors, technological capital factors, and management challenges influencing the productivity of manufacturing SMEs, based on a thorough review of the literature.

Chapter 5 focuses on the environment for manufacturing SMEs, building on a critical assessment of previous studies in the area. This chapter identifies the selection of environmental factors hindering the efficient operation of the manufacturing process, followed by the role of standard working procedures in ensuring a safe and healthy environment in manufacturing SMEs. These procedures involve ergonomics and good housekeeping.

Chapter 6 reviews the literature on work study, in particular the technique of method study, to examine its relevancy and applicability to modern productivity improvement tools. Work study characteristics based on method study are also discussed in manufacturing SMEs.

Chapter 7 provides a literature background on the second technique of work study, that is, work measurement. The review builds a case for the need for work measurement in modern industry. Work study characteristics based on work measurement are regarded as similar to those of method study, but they depend on the nature of the work being done.

Chapter 8 introduces studies done regarding the impact of work study on physical capital in relation to the productivity of manufacturing SMEs. Chapter 9 presents studies done involving the impact of work study on technological capital in relation to the productivity of manufacturing SMEs. Chapter 10 reports on studies done concerning the impact of work study on management in relation to the productivity of manufacturing SMEs.

Chapter 11 presents research results on presentation and report writing in manufacturing SMEs. The last chapter considers contributions that result from this work in an integrated manner, reflecting on research applications and directions for future work.

Acknowledgment

We would like to acknowledge our colleague Mr. Boitumelo Ramatsetse for his contributions to this first edition. Mr. Boitumelo Ramatsetse is a lecturer in the Department of Industrial Engineering at Tshwane University of Technology, and worked diligently together with the authors to redesign most of the diagrams used in this research book.

About the Authors

Thomas Munyai is lecturer and head of department in Operations Management at Tshwane University of Technology, Pretoria, South Africa. Currently, he is a doctoral student in Engineering Management at the University of Johannesburg with a main focus on the development of a model to empower small businesses to enhance productivity in Gauteng, South Africa. He holds an MBA from the University of South Africa (UNISA), a BSc (hons) in Mathematical Statistics (Wits University), and a BSc Cum Laude in Mathematics and Statistics (University of Limpopo). He has 20 years' experience in the manufacturing and food processing industry, as a factory manager, operations manager, product and sales manager, and procurement manager. His professional memberships include the South African Production and Inventory Control Society (SAPICS). His fields of interest are optimization through Lean Six Sigma, work study, productivity, linear and non-linear modeling, statistics, mathematics, simulation, process mapping modeling, and manufacturing technology.

Boysana Mbonyane is a senior lecturer in the Department of Operations Management at the University of South Africa, Pretoria, South Africa. Currently, he is a doctoral student in Engineering Management at the University of Johannesburg with a main focus on the development of a framework for efficiency in relation to physical and technological capital to yield continuous growth in small business in Gauteng, South Africa. He completed his master's in Business Leadership in 2014 and a master of Technology degree in Business Administration in 2007 at the University of South Africa (UNISA). He also completed his BCom (hons) in Industrial Psychology at the University of Johannesburg in 2006 and a BTech in Business Administration in 1999 at the University of South Africa. His undergraduate qualifications are a national diploma in Organization and Work Study, completed in 1995 at Technikon Witwatersrand. He spent 10 years in the mining industry and 10 years as a lecturer at the University of Johannesburg. Boysana joined UNISA in February

2010. He completed some of the post-graduate modules in Entrepreneurship and Small Business Management at the University of Pretoria. His professional memberships include the Southern Africa Institute for Management Services (SAIMAS). His research area of interest is in work study/organizational effectiveness; productivity; ergonomics; operations management; entrepreneurship and small businesses.

Charles Mbohwa is a professor in the Faculty of Engineering and the Built Environment. He is an established researcher and professor in the field of sustainability engineering and energy. He was chairman and head of the Department of Mechanical Engineering at the University of Zimbabwe from 1994 to 1997 and was vice-dean of postgraduate studies research and innovation in the Faculty of Engineering and the Built Environment at the University of Johannesburg from 2014 to 2017. He has published more than 350 papers in peer-reviewed journals and conferences, 10 book chapters, and 3 books. He has a Scopus h-index of 11 and Google Scholar h-index of 14. Upon graduating with a BSc honors in Mechanical Engineering from the University of Zimbabwe in 1986, he was employed as a mechanical engineer by the National Railways of Zimbabwe. He holds a masters in Operations Management and Manufacturing Systems from the University of Nottingham and completed his doctoral studies at Tokyo Metropolitan Institute of Technology in Japan. He was a Fulbright Scholar visiting the Supply Chain and Logistics Institute at the School of Industrial and Systems Engineering, Georgia Institute of Technology; and a Japanese Foundation fellow. Dr. Mbohwa is a fellow of the Zimbabwean Institution of Engineers and is a registered mechanical engineer with the Engineering Council of Zimbabwe. He has been a collaborator on projects of the United Nations Environment Program. He has also visited many countries on research and training engagements, including the United Kingdom, Japan, Germany, France, the United States, Brazil, Sweden, Ghana, Nigeria, Kenya, Tanzania, Malawi, Mauritius, Austria, the Netherlands, Uganda, Namibia, and Australia. He has received several awards including a British Council scholarship, a Japanese Foundation fellowship, a Kubota Foundation fellowship, and a Fulbright fellowship.

Chapter 1

Research Approaches and Main Results on Manufacturing SMEs in South Africa

1.1 Introduction

This research book focuses on the application of work study in the productivity of manufacturing small and medium enterprises (SMEs) in South Africa. Work study is defined as "a discipline advising management in managing manufacturing operations through proper use of resources such as physical capital, technological capital and management commitment in order to improve productivity in manufacturing SMEs" and is outlined in Figure 1.1. This diagram presents all the various areas of the work study and the manufacturing concepts impacting on manufacturing SMEs as discussed throughout this research book. On the contrary, productivity focuses on the relationship between the products produced and the resources used in manufacturing SMEs (Gobinath, Elangovan, & Dharmalingam, 2015:46–9). Based on the introduction, this book provides a breakdown of manufacturing SMEs' productivity problems based on research trends and challenges facing the country; research methodology; research findings; background of manufacturing SMEs and its definitions; and production systems, as well as value-adding drivers for the manufacturing SMEs.

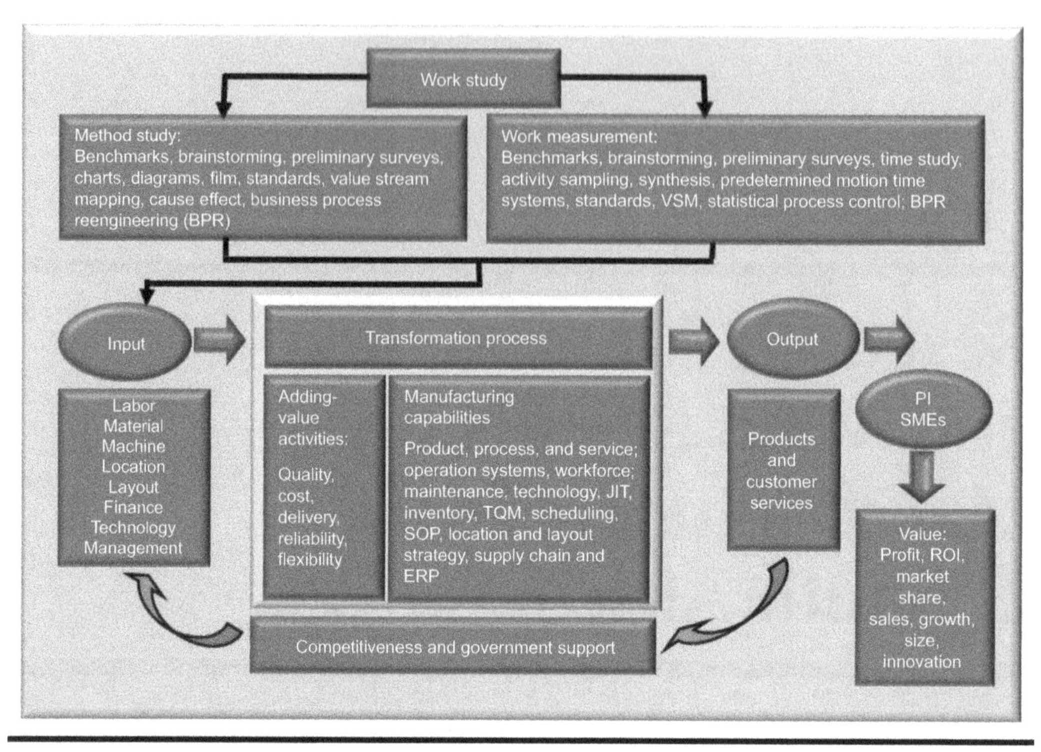

Figure 1.1 Productivity improvement in manufacturing SMEs: Application of work study techniques. (From Author, 2017.)

Manufacturing SMEs productivity problems: research trends and challenges is the next section to be discussed.

1.1.1 Manufacturing SMEs Productivity Problems: Research Trends and Challenges

Low productivity is currently a serious challenge facing manufacturing SMEs. These SMEs are functioning under the expected manufacturing output levels (Coka, 2014:1; ILO, 2015:25; Statistics SA, 2015:7). According to Statistics SA (2015:7), the production output ratio decreased from 4.6 in 2010 to −0.1 in 2014 in terms of gross domestic product (GDP). See production output ratio in Table 1.1.

Various manufacturing capital resource elements, such as physical capital and technological capital, are currently leading to a decline in the productivity of manufacturing SMEs countrywide and globally (Habidin & Yusof, 2013:61; Jain, Bhatti, & Singh, 2014:299; de Carvalho, Ho, & Pinto, 2014:612; Soda, Sachdeva, & Garg, 2015:890; Schwab, 2015:3). Firstly, regarding physical capital, poor physical infrastructure, human resources problems (less efficient in labor), material, and finance (overhead cost) and energy represent a serious problem facing

Table 1.1 Annual Percentage Change in the Total Index Showing Decrease of the Physical Volume of Manufacturing Production From 4.6 in 2010 to –0.1 in 2014

Month	*2010*	*2011*	*2012*	*2013*	*2014*	*2015*
January	2.2	2.6	3.0	3.4	2.5	–2.3
February	1.2	6.7	4.4	–1.9	1.6	—
March	6.4	5.3	–2.5	–1.7	1.0	—
April	7.9	0.0	1.8	7.3	–1.9	—
May	6.8	1.0	5.4	2.3	–3.9	—
June	8.9	1.3	0.7	0.7	0.2	—
July	7.2	–6.1	7.0	5.8	–8.2	—
August	5.5	6.1	2.2	–0.5	–0.6	—
September	2.3	8.0	–2.2	–3.8	8.9	—
October	1.8	2.4	3.3	2.1	2.0	—
November	4.0	3.9	3.8	0.1	–2.1	—
December	2.0	2.7	1.1	3.0	0.9	—
Years	**4.6**	**2.8**	**2.3**	**1.2**	**–0.1**	—

Source: Statistics, S.A., Quarterly Labour Force Survey: Quarter 2: 2015. Available at: http://www.statssa.gov.za/publications/P0211/P02112ndQuarter2015.pdf, 2015.

Note: Bold text indicates production output decreased from 4.6 in 2010 to –0.1 in 2014.

manufacturing industries among SMEs in developing countries, in particular in South Africa, where low productivity will continue to occur and SMEs will fail to secure the market share globally (Schwab, 2015:39). Secondly, with regard to technological capital, manufacturing industries are experiencing inefficiencies in their businesses due to lack of innovation among employees, since employees do not have the educational background that would assist them in using their creativity to operate machines (ILO, 2015:61).

Based on the foregoing statement with regard to physical and technological capital, competition continues to increase in the market worldwide, which emphasizes the value of productivity pertaining to quality, price, and delivery in manufacturing SMEs. These SMEs are under growing pressure to keep their positions in the market. Manufacturing SMEs need to consider innovation, reducing lead time and bottlenecks, and faster product quality market delivery with the aim of continuous measurement and productivity improvement (Hilmola et al., 2015).

Small Enterprise and Development Agencies (SEDA) and Productivity South Africa (SA) are authorized by the South African government to support manufacturing SMEs in enhancing productivity and developing the economy in their businesses (Coka, 2014:1; SEDA, 2014:31). Furthermore, the manufacturing engineering sector's education and training authority (MERSETA) provides manufacturing SMEs with training and skills to accelerate the growth of business productivity levels (Manufacturing Indaba, 2014:3; ILO, 2015:25; Statistics SA, 2015:7). Although SEDA, Productivity SA, and MERSETA are delegated by the government for financial support as well as training and skills provision, productivity continues to fall off in manufacturing SMEs (Coka, 2014:1; Herrington, Kew, & Kew, 2014:45; Manufacturing Indaba, 2014:1; SEDA, 2014:31; ILO, 2015:25; Schwab, 2015:39; Statistics SA, 2015).

The role of the government in addressing the business policies, infrastructure, and regulations on the productivity of manufacturing SMEs requires underpinning. In the meantime, the government has introduced a 10% import tariff for the World Trade Organization in order to protect local manufacturing industries, in particular SMEs, in the country, whereby manufacturing SMEs are part of this tariff. On the other hand, China is still sinking the market with low-priced commodities to accommodate the South African manufacturing SMEs (Ojediran, Wintoki, & Odumade, 2016:6–7). Since the growth of productivity is considered as the key to improving GDP per capita and hence the standard of living of people in the country, if productivity decreases, this becomes an economic problem, and GDP suffers (Gibb & Luiz, 2011:309; South African Reserve Bank, 2013:5).

Currently, the evolution of the literature on productivity improvement indicates that there are a large number of cases of the problem of productivity decline globally among manufacturing SMEs (Debnath & Sebastian, 2014:6–7; Singh & Ahuja, 2015:125; Vinodh et al., 2015:395; Sanjog et al., 2016:34–44; Hooi & Leong, 2017:2–13). So, the improvement of productivity among manufacturing SMEs, as indicated by various academic scholars along with SA agencies, increases employment and economic activity in the country (Herrington et al., 2014:43; Schwab, 2015:42; SEDA, 2012:63; Coka, 2014:26–31). On the contrary, low productivity results in a falling off of manufacturing SMEs' businesses and the GDP of the country dwindling (Herrington et al., 2014:45). When GDP decreases, unemployment increases, and this becomes an economic crisis in the country (ILO, 2015:12). By the same token, Urban and Naidoo (2012:147); Bogue (2014:117–21); Hamilton, Nickerson, and Owan (2015:117); Huggins, Morgan, and Williams (2015:474–81); and Ozigbo (2015:1–5) emphasize the importance of productivity, which results in job creation by manufacturing SMEs and, ultimately, improved GDP for economic growth.

Given the low productivity, it becomes vital that research be done on factors that are required to enable manufacturing SMEs to improve their productivity. Problems encountered by manufacturing SMEs are numerous and can be described, among other things, as being operational and environmental in nature. However, much more is needed among manufacturing industries, in particular manufacturing SMEs, to improve their productivity levels. The challenge is to encourage the use of work study in physical capital, technological capital, and management in relation to the productivity of manufacturing SMEs in South Africa.

Even though the literature sources used are sufficient enough on work study as the application tool to improve the productivity of manufacturing SMEs, studies pinpointing the work study as the key tool were embarked on to improve the productivity of manufacturing SMEs in developed countries worldwide (Habidin & Yusof, 2013:61; Jain et al., 2014:289; Soda et al., 2015:893–4). The outcome of the results in the literature were then realized to be successful in research (Habidin & Yusof, 2013:61; de Carvalho et al., 2014:612; Soda et al., 2015:890). So, not enough scientific literature prevails for work study on improving the productivity of manufacturing SMEs in South Africa.

The purpose of this research book is to introduce work study tools that will enable manufacturing SMEs improve the productivity rate of their businesses in South Africa. Based on the problems that manufacturing SMEs are facing in the country, there is a lack of understanding of the scientific literature on the application of work study through the use of physical capital and technological capital to improve the productivity of manufacturing SMEs in South Africa. As reported by Gashi, Hashi, and Pugh (2014:408), insufficient attention has been paid to how work study influences physical capital, technological capital, and management in relation to the productivity of manufacturing SMEs in South Africa. Hence, the objective of the study is to investigate the application of work study in physical and technological capital utilization on the productivity of manufacturing SMEs in South Africa. The next section outlined focuses on the literature survey of primary data, secondary data, and case studies used as theoretical and practical tools in the research methodology:

- With data gathered from government agencies, the research attempts to identify productivity problems facing manufacturing SMEs in South Africa.

- Secondary data also is read to compare various literature sources supporting the use of work study to investigate physical capital and technological capital in relation to the productivity of manufacturing SMEs.
- It also aims to evaluate the case studies that will expose shortcomings and recommend possible solutions for productivity improvement in South Africa.

1.1.2 Research Methodology

In order to acquire a deeper and more comprehensive understanding of the nature of the problems facing manufacturing SMEs in terms of their productivity level, various literature sources were selected from the database in the form of journal articles and information from government agencies such as Productivity SA and SEDA, including case studies of Companies A, B, C, and D in South Africa. These journals involve Science Direct; ProQuest; Emerald; Taylor & Francis; Elsevier; Inderscience Enterprises Ltd.; Springer Science, and other secondary databases, to mention a few, which are related to manufacturing SMEs, and they were utilized to gather secondary data for productivity measurement among manufacturing SMEs.

Key words used to gather secondary data were *work study*, *productivity*, *physical* and *technological capital*, *management*, *manufacturing processes and systems*, *operational capabilities*, and *adding-value activities*; interviews and observations from companies around the Gauteng region of South Africa

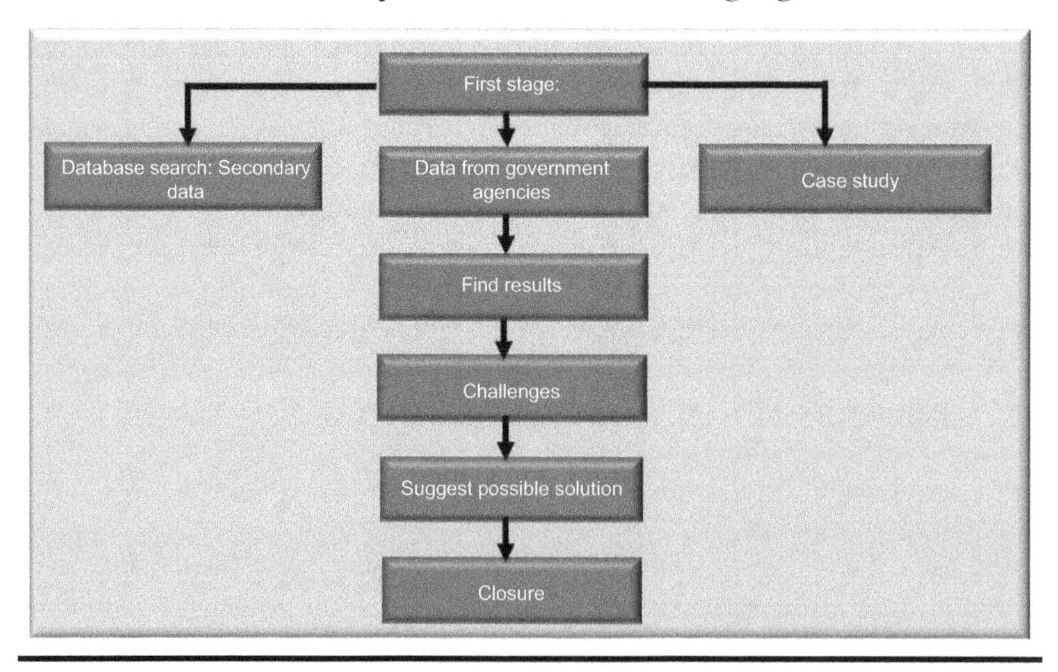

Figure 1.2 Research methodology procedure. (From Author, 2017.)

were used, as well as documents from the agencies. The aim of using key words was to restructure and eliminate the studies whose major focus was not related to work study. This research book provides a procedure of the research methodology in Figure 1.2.

The findings are the next section to be presented.

1.1.3 Research Findings

The emphasis of the exploratory survey was on current literature from 2010 to 2017; data collected from government agencies from 2010 to 2017, and real-time cases studied through interviews conducted and observations made in 2017. The information gathered was through the operation of manufacturing SMEs locally and worldwide. Endeavors were made regarding these three research methods to gather the information through the impact of work study on physical and technological capital as well as management in relation to the productivity of manufacturing SMEs. The aim of the research book was to support the validity of the conclusions drawn. A summary of the results of the literature, government agencies in South Africa, and case studies of Companies A, B, C, and D is presented in respective subheadings in Table 1.2.

Subsequent to the analysis of the literature review, primary data collected through government agencies, and use of case studies, 1000 journal articles, 10 government documents, and 4 case studies were selected to justify the findings of the research. In terms of the literature, the research studies conducted were from developed countries such as the United States, the United Kingdom, and France and from developing countries such as Brazil, Russia, India, China, South Africa, and Turkey with the aim of overviewing challenges facing manufacturing SMEs globally. Looking at various methods used in collecting data, there is a lack of application of work study in improving the productivity of manufacturing SMEs in South Africa.

To enable further understanding of the challenges facing manufacturing SMEs locally and in the rest of the world and the research methodology and findings addressed in this chapter, the first stage to be presented as an overview in manufacturing is the background of manufacturing SMEs.

1.2 The Background of Manufacturing SMEs

In this section, manufacturing SMEs are studied and expressed to provide a detailed understanding of what these SMEs encompass. The background of manufacturing in manufacturing SMEs in this study involves the definitions

Table 1.2 Case Studies of Manufacturing Companies A, B, C, and D in Gauteng, South Africa

Challenges Facing Manufacturing SMEs	*Recommended Approach*	*Academic Scholars*
1. Literature Review on the Impact of Work Study on Physical and Technological Capital in Relation to the Productivity of Manufacturing SMEs		
Physical capital and productivity of manufacturing SMEs	Need for application of work study tools such as method study and work measurement in improving productivity of manufacturing SMEs	Pandey et al. (2014:113–4); Chompu-inwai, Jaimjit, and Premsuriyanunt (2015:1359); Sharma and Shah (2016:582–99); Ohu et al. (2016:48–9); Hooi and Leong (2017:2–13)
Technological capital and productivity of manufacturing SMEs	Need for application of work study tools such as method study and work measurement in improving productivity of manufacturing SMEs	Gupta, Acharya, and Patwardhan (2013:635–45); Sylla et al. (2014:475–7); Silva et al. (2015:176); Prasad, Khanduja, and Sharma (2016:409–424)
Management commitment and productivity of manufacturing SMEs	Need for application of work study tools such as method study and work measurement in improving productivity of manufacturing SMEs	Chompu-inwai et al. (2015: 1353–9); Duran, Cetindere, and Aksu (2015:10–11); Małachoski and Korytkowski (2016:166–70); Pedersen and Slepniov (2016:45–49); Sanjog et al. (2016:34–44); Srinivasan et al. (2016:364–76).

(*Continued*)

Table 1.2 (Continued) Case Studies of Manufacturing Companies A, B, C, and D in Gauteng, South Africa

Challenges Facing Manufacturing SMEs	*Recommended Approach*	*Academic Scholars*
2. Primary Data on Low Productivity Collected from Companies: Small and Medium Enterprises through Government Agencies in South Africa		
The production output ratio decreased from 4.6 in 2010 to –0.1 in 2014 in South Africa	Need for research work study tools such as method study and work measurement in improving productivity of manufacturing SMEs	Statistics SA (2015:7)
Most government agencies indicate that productivity is a serious dilemma facing manufacturing SMEs in South Africa	Need for research work study tools such as method study and work measurement in improving productivity of manufacturing SMEs as well as company visits to validate the results with regard to the productivity of manufacturing SMEs	South African Reserve Bank (2013:5); Coka (2014:1); Herrington et al. (2014:45); Manufacturing Indaba (2014:3); Ojediran et al. (2016:6–7); ILO (2015:61); Schwab (2015:39); SEDA (2014:31); Statistics SA (2015:7)
Management commitment and productivity of manufacturing SMEs	Need for research work study tools to advise management in growing their manufacturing SMEs through Productivity SA in South Africa	Coka (2014:1)

(*Continued*)

Table 1.2 (Continued) Case Studies of Manufacturing Companies A, B, C, and D in Gauteng, South Africa

Challenges Facing Manufacturing SMEs	*Recommended Approach*	*Academic Scholars*
3. Productivity Challenges Facing Companies A, B, C, and D: Small and Medium Enterprises in Gauteng, South Africa		
Company A: **Poor supply chain** Late deliveries, incorrect sizes, stock not available in other instances **Human element** Employee resistance to change; uncertainty; lack of incentives No compliance with standard operating procedures **Machine** Aging machines **Inventory** Poor inventory control, delays, and waste **Environment/layout** Haphazard areas; insufficient space for working and storage of material and products for delivery **Technology** No technology **Transport** 35% of delivery is carried out inefficiently, whereby products are delivered late **Management shortcomings** Poor inventory control; poor planning and poor employee safety control; poor sharing of responsibility	Needs research and development to grow the business Organize communication-training workshop Need for formal training Compliance with standards based on ISO 14001 and ISO 18001 Certification needed by the company for ISO 14001 and ISO 18001 Elimination of unnecessary transport and movement of employees New technology is also needed to grow the business Need job enrichment training to empower employees Improvement of layout through signage and appropriate space Reinforcement of safety measures by management Procuring what is necessary for inventory control and quality improvement	Collaboration of universities, government agencies, industries, and other stakeholders

(*Continued*)

Table 1.2 (Continued) Case Studies of Manufacturing Companies A, B, C, and D in Gauteng, South Africa

Challenges Facing Manufacturing SMEs	*Recommended Approach*	*Academic Scholars*
Company B: **Poor supply chain** Lead times very low, dependent on suppliers for quantity of material required, and not enough suppliers **Human element** Most aging employees, maternity leave for most young women, heavy box handling, no incentives, and poor communication by management **Machine** Aging machines **Inventory** Poor inventory control, delays, and waste **Environment/layout** Insufficient space for working and storage of material and products for delivery **Technology** No technology **Transport** 20% late delivery **Management shortcomings** Poor planning	Needs research and development to grow the business Revisit formal organizational structure to improve duties of employees from the top to the floor Organize communication-training workshop Need for formal training Compliance with standards based on ISO 9001, ISO 14001, and ISO 18001 Certification needed by the company for ISO 9001, ISO 14001, and ISO 18001 New technology is also needed to grow the business Need job enrichment training to empower employees Improvement of layout through signage and appropriate space Reinforcement of safety measures by management Liaise with Productivity SA to improve on productivity performance	Collaboration of universities, government agencies, industries, and other stakeholders

(*Continued*)

Table 1.2 (Continued) Case Studies of Manufacturing Companies A, B, C, and D in Gauteng, South Africa

Challenges Facing Manufacturing SMEs	*Recommended Approach*	*Academic Scholars*
Company C: **Suppliers** Late delivery and unavailability of stock **Employee** Employee resistance and no incentives, high employee turnover **Material** Too much line-based material flow and waste **Machine** Tooling problems **Environment** Haphazard areas, insufficient space, poor safety control and poor standard operating procedures, and no standards in place **Machine** Old machines are utilized, and as a result delays are encountered by Company C as a manufacturing business Competition is now becoming a serious dilemma for companies to survive and grow in the market **Inventory** The types of material affected in Company C include bad quality material, cracks, and hardness. These problem areas are picked up late, at the completion of the final product, which is costly for the company. As a result, the productivity of Company C is compromised. **Management** Poor planning, poor forecast on buying, poor inventory control	Needs research and development to grow the business Organize communication training workshop Revisit formal organizational structure to improve duties of employees from the top to the floor Encourage incentives Need for formal training Compliance with standards based on ISO 14001 and ISO 18001 Certification needed by the company for ISO 18001 New technology is also needed to grow the business Need job enrichment training to empower employees Improvement of layout through signage and appropriate space Reinforcement of safety measures by management	Collaboration of universities, government agencies, industries, and other stakeholders

(*Continued*)

Table 1.2 (Continued) Case Studies of Manufacturing Companies A, B, C, and D in Gauteng, South Africa

Challenges Facing Manufacturing SMEs	*Recommended Approach*	*Academic Scholars*
Company D: **Supplier** Major change with international supply, poor supply chain **Employees** Severe ill-health, absenteeism **Material** Waste and damage are incurred due to poor quality systems and poor material handling **Machinery** Maintenance on a monthly basis **Technology management** Poor planning **Transport** 30% late delivery	Needs research and development to grow the business Organize communication training workshop Revisit formal organizational structure to improve duties of employees from the top to the floor Encourage incentives Need for formal training Compliance with standards based on ISO 14001 and ISO 18001 Certification needed by the company for ISO 18001 New technology is also needed to grow the business Need job enrichment training to empower employees Improvement of layout through signage and appropriate space Reinforcement of safety measures by management Liaise with Productivity SA to improve on productivity performance	Collaboration of universities, government agencies, industries, and other stakeholders

Source: Mbohwa, C., Letter of permission to conduct an interview: Productivity improvement in manufacturing SMEs: Application of work study techniques, University of Johannesburg, Johannesburg, South Africa, 2017.

of manufacturing SMEs, the various manufacturing systems in which manufacturing SMEs are operating, the manufacturing process, and manufacturing performance objectives.

In the manufacturing process, various types of manufacturing systems are introduced with the aim of making manufacturing SMEs add value in their business (De Snoo, Van Wezel, & Wortmann, 2011:1352; Brown, 2012:378–9).

Singh and Singh (2014:184) assert that since the manufacturing process is regarded as the backbone of any industrialized nation, its importance is emphasized by the fact that it improves and ensures economic activity. A country's level of manufacturing activity is directly related to its economic health. Manufacturing technologies have continually gone through gradual but revolutionary changes. These advancements in manufacturing technologies, like the continuous improvement approach, have brought about a metamorphosis in the world's manufacturing industrial scene.

1.2.1 The Definitions of Manufacturing SMEs

The concept of SMEs differs in the manner in which they are defined from one country to the next. For the purpose of this book, this research will focus on the structure of the manufacturing SME in developed countries versus developing countries. The reason for this is that this book must be able to be read and understood by local and global SMEs. Manufacturing SMEs are defined in the form of qualitative and quantitative information. For instance, Stockes and Wilson (2010:4) define SMEs, in terms of qualitative information, as "a business firm managed by its owners in a personalized manner; occupying a relatively small share of the market in economic terms, and an independent business that does not form part of a larger enterprise." See Table 1.3 for the quantitative definition of SMEs in a developed region such as Europe.

Table 1.3 The Quantitative Definition in Europe

EU SME Definition				
Enterprise Category	*Headcount*	*Turnover*	*Or*	*Balance Sheet Total*
Medium sized	<250	≤€50 million		≤€43 million
Small	<50	≤€10 million		≤€10 million

Source: Stockes, D., and Wilson, N., *Small Business Management and Entrepreneurship,* South-Western Cencage Learning, Andover, 2010.

The next definition is provided in a South African context, since the monograph focuses on productivity challenges facing manufacturing SMEs in South Africa, This definition is presented under the National Small Business Act 26 of 2003 (Republic of South Africa, 2003). The qualitative definition is described as "a separate and distinct business entity, including cooperative enterprises and non-governmental organisations, managed by one owner or more which, including its branches or subsidiaries, if any, is predominantly carried on in any sector or sub-sector of the economy and which can be classified as a small and a medium enterprise (SME)."

The quantitative grouping of SMEs for Table 1.4 is defined next as the definition of SMEs in a developing country such as South Africa.

National Credit Regulator (NCR) (2011:23) report the quantitative definition of Brazil, Russia, India, and China, referred to as BRIC, which are all believed to be at a similar stage of newly advanced economic development. The annual report of the National Credit Regulator (2011:63) introduces South Africa into the BRIC economy, which is thereby expanded into BRICS. The synopsis of the BRIC economy consists of number of employees and turnover.

BRIC does not show the asset value for manufacturing SMEs in the definition as developed regions like Europe and the United States do, including developing countries like South Africa. When South Africa was introduced into the BRIC economies, the acronym was expanded to BRICS. The reason for showing the differences of quantitative definition for developed regions like Europe and the United States as compared to BRICS is that the study intends to show how manufacturing SMEs in developed countries differ from developing countries in terms of size, turnover, and assets. Furthermore, only manufacturing SMEs for BRICS are defined because South Africa is mainly part of the BRICS and is where trade and competiveness largely take place

Table 1.4 The Quantitative Definition in South Africa

Sector or Sub-Sectors in Accordance with the Standard Industrial Classification	*Size or Class*	*Total Full-Time Equivalent of Paid Employees Less Than*	*Total Annual Turnover Less Than*	*Total Gross Asset Value (Fixed Property Excl) Less Than*
Manufacturing	Medium	200	R40.00 m	R15.00 m
	Small	50	R10.00 m	R3.75 m

Source: Republic of South Africa, *National Small Business Act 26 of 2003*. Government Printer, Pretoria, 2003.

Table 1.5 United States, Brazil, Russia, India, and China (BRIC) SME Definitions

Enterprise Category	*USA*	*Brazil*	*Russia*	*India*	*China*
Number of Employees					
Medium sized	<500	<100–499	≤101–250	0	<300–2000
Small	<100	<29	≤15–100	0	<300
Micro	<0	<19	≤0	0	0
Turnover					
Medium sized	0	0	R60–99 m	1 B RUB max	Y30–Y300 m
Small	0	0	R50–60 m	400 m RUB max	<Y30
Micro	0	0	<R50 m	0	0

Source: National Credit Regulator (NCR). 2011. Literature Review on Small and Medium Enterprises' Access to Credit and Support in South Africa. Pretoria, South Africa. Available from: http://www.ncr.org.za/pdfs/Literature%20Review%20on%20SME%20Access%20to%20Credit%20in%20South%20Africa_Final%20Report_NCR_Dec%202011.pdf. [Accessed on 2015-05-20.]

in terms of export and import. See Tables 1.5 and 1.6 for Europe, the United States, BRIC, and South Africa, respectively.

For the purpose of the study, research will focus on the definition of the National Small Business Act 26 of 2003 (Republic of South Africa, 2003), which refers to "a separate and distinct business entity, including cooperative enterprises and non-governmental organisations, managed by one owner or more which, including its branches or subsidiaries, if any, is predominantly carried on in any sector or sub-sector of the economy and which can be classified as

Table 1.6 RSA SMEs Definitions

Sector or Sub-Sectors in Accordance with the Standard Industrial Classification	*Enterprise Category*	*Headcount Less Than*	*Turnover Less Than*	*Balance Sheet Total Less Than*
Manufacturing	Medium	200	R40.00 m	R15.00 m
	Small	50	R10.00 m	R3.75 m
	Very small	20	R4.00 m	R1.50 m
	Micro	5	R0.15 m	R0.10 m

Source: Republic of South Africa, *National Small Business Act 26 of 2003*. Government Printer, Pretoria, 2003.

a small and a medium enterprise (SME)." Whereas with quantitative classification, the information in Table 1.6 defines SMEs in terms of three main criteria: employment, turnover, and asset value.

1.2.2 Manufacturing Systems

Grütter (2010:2) explains many ways in which one has to look at an operation to manage it well, and this can be done from a systems point of view. Findings from the authors define a system as "a collection of parts that work together to function as a whole." Furthermore, characteristics of systems are addressed in various ways. Systems have inputs that make up the system and inputs that are processed by a system; they consist of dynamically linked, interdependent elements; and they transform (change) inputs into outputs by performing certain activities on some of the inputs. System boundaries depend on one's point of view, whereby the output of one system can be the input of another system. A system has a purpose and runs better if there is feedback from the output side to the input side to monitor whether a system is working as it should be.

This section introduces different kinds of systems exercised in manufacturing SMEs, and these systems include the assembly (line) system (Choudhari, Adil, & Ananthakumar, 2012:1338) and mass, continuous, batch production as well as project systems (Choudhari et al., 2012:1338; Ozturk et al., 2015:120–21)

As described by Choudhari et al. (2012:1353), an assembly system is considered to be a system used in arranging a sequence of work activities, a series of stages and components utilized in the operation with the intention of putting together parts or components to complete a product.

According to Hu et al. (2015:982) a mass-production system involves producing products in batches of large quantities that are yielded by a nonstop manufacturing operation required by the customers.

The next system to be described in manufacturing SMEs is a continuous manufacturing system, which focuses on products generated as nonstop manufactured products with the intention of meeting the customers' requirements. A batch manufacturing system includes products of similar features generated by a machine under the same conditions (Ozturk et al., 2015:120–21).

Aagaard, Eskerod, and Madsen (2015:222–25) categorize project systems into two different types of manufacturing system operating within manufacturing SMEs. The first one is a formal project whereby the plant and equipment infrastructure is built by manufacturing SMEs in such a way that they are continuously operating efficiently in accordance with market demands. For example, a project can be done by a team of technical specialists operating from different operational sections, consulting with manufacturing SMEs with the aim of executing multifaceted and expert-based work activities within a limited period. Another project is informal, whereby manufacturing SMEs involve a group of people working in a situation in which a once-off job is done on a small scale.

The background of the manufacturing process among manufacturing SMEs is the next aspect presented in this chapter.

1.2.3 Manufacturing Process

In this section, the background of the manufacturing process is examined on the basis of numerous literature sources and is discussed with the aim of providing a detailed insight into this concept. Zhang and Gregory (2011:744) and Singh and Bakshi (2014:173) also refer to this process as the *input-transformation-output model.* Inputs indicated in manufacturing SMEs include employee resources (Helkiö & Tenhiäla, 2013:234); capital (Zhang & Gregory, 2011:740; Helkiö & Tenhiäla, 2013:234); material, machinery, and technology (Choudhari et al., 2012:3700–2; Dora et al., 2016:1–18); as well as energy (Dora, Kumar, & Gellynck, 2016:18).

As noted by Lee et al. (2013:38), the manufacturing process is described as a process centering on the utilization of input resource factors such as material, machinery, methods, and time measurements for generating products in manufacturing SMEs.

As indicated by Singh and Bakshi (2014:169–90); Youn et al. (2014:376); and Huang, Tan, and Ding (2015:80–9), the transformation process in manufacturing SMEs is all about the gathering of work activities assigned to employees exposed to machinery and tools, during which the parts used are converted to a final product for the customers. According to Zhang and Gregory (2011:744) and Singh and Bakshi (2014:173), outputs in manufacturing SMEs are displayed in the form of products generated to meet customer requirements.

As pointed out by Robinson, Sanders, and Mazharsolook (2015:234), the manufacturing process for manufacturing SMEs involves production planning and control and efficient use of the environment to transfigure inputs that include employees' capability and involvement, material usage, and machine and equipment utilization, along with efficient time, into the final product in order to reach the customer's needs.

According to Bi et al. (2015:265–70), the manufacturing process for manufacturing SMEs not only focuses on management capabilities such as production planning and control and efficient use of the environment but also on other capabilities such as both manual and machine operation, the efficient use of inventory and technology to transform inputs that include employees' capability and involvement; material usage; and machine and equipment utilization through the support of efficient time, cost, quality, flexibility, and fast delivery to avoid waste and to meet or exceed the customers' requirements.

Manufacturing process refers to a process that involves the substitution of input materials generated as waste for reuse and recycling through the management of the environment improved by using technology in manufacturing SMEs (Zhou & Zhao, 2016:420; Lentes et al., 2017, 480–1). At the completion of the manufacturing process, this section is followed by competitive priorities exercised in manufacturing SMEs, which are discussed next.

1.2.4 Competitive Priorities as Operational Capabilities and Value-Adding Drivers for the Manufacturing SMEs

In this section, competitive priorities as operational capabilities and value-adding drivers for the manufacturing SMEs are explained and described in order to provide a detailed understanding of concepts regarding these capabilities and adding-value drivers in manufacturing SMEs. The first area of competitive priorities addressed is operational capabilities.

1.2.4.1 Operational Capabilities

This section identifies various operational capabilities used in manufacturing SMEs. These capabilities involve product, process, and service design and operation systems (Lee et al., 2013:38–41; Mellor, Hao, & Zhang, 2014:195–99; Prasad, Jha, & Prakash, 2015:271–83; Ayeni, Ball, & Baines, 2016:38–41); employees' ability and involvement (Mellor et al., 2014:199; Ayeni et al., 2016:50); total productive maintenance (Jain et al., 2014:299; Dora et al., 2016:9); technology (Mellor et al., 2014:195; Ayeni et al., 2016:41); just-in-time delivery (JIT) (Panizzolo et al., 2012:769–85); inventory management (Panizzolo et al., 2012:771; Ayeni et al., 2016:52); total quality management (TQM) (Bhamu & Sangwan, 2014:881); scheduling (Ayeni et al., 2016:39); standard operating procedure (SOP) (Gupta & Jain, 2015:80); location and layout (Ali & Suleiman, 2016:308–12); supply chain management (SCM) (Mellor et al., 2014:195); enterprise resource planning (ERP) (Huang & Handfield, 2015:2–3); and capacity (Ayeni et al., 2016:41).

According to Lee et al. (2013:38); Mellor et al. (2014:196); Prasad et al. (2015:283); and Ayeni et al. (2016:39), product or goods design in manufacturing SMEs focuses on building up a detailed specification of a finished product manufactured a standardized structure in accordance with the customers' requirements.

Process design in manufacturing SMEs involves planning and arranging the manufacturing process by management with the aim of attaining an organized set of operations in order to improve product and service quality for the customer (Lee et al., 2013:39; Mellor et al., 2014:197; Prasad et al., 2015:271; Ayeni et al., 2016:41; Dora et al., 2016:9).

As described by Lee et al. (2013:39); Mellor et al. (2014:198); Prasad et al. (2015:282); and Ayeni et al. (2016:41), service design in manufacturing SMEs is a devised method of service, which is provided to customers with the aim of retaining customers as well as building and maintaining sustained relationships with customers.

Employees' ability and involvement focuses on their training, knowledge, and the skills involved in work activities carried out to ensure that knowledgeable and multiskilled employees engage their motivation and effort toward performing well at their workstations in order to meet or exceed customers' requirements, which gives credit to manufacturing SMEs (Mellor et al., 2014:199; Ayeni et al., 2016:42). Jain et al. (2014:299) and Dora et al. (2016:9) refer to total productive maintenance as a process of introducing a scheduled plan of machine service or repairs in order to avoid or reduce machine downtime.

As noted by Mellor et al. (2014:198) and Ayeni et al. (2016:41), technology involves features of effective communication, modern machinery or equipment efficiently used for appropriate handling of material, and an efficient operational process achieved through the latest automation technology, as well as a healthy and safe environment to provide quality service delivery. Quality service delivery then contributes to sustainable competitiveness in manufacturing SMEs locally and worldwide. Panizzolo et al. (2012:784–5) state that JIT delivery encompassed deliveries supported by the use of SCM and or ERP system software that will assist in downsizing inventory levels and better interdepartmental coordination, so that delivery schedules are improved to avoid bottlenecks during the manufacturing process in manufacturing SMEs.

As reported by Panizzolo et al. (2012:771) and Ayeni et al. (2016:52), inventory management in manufacturing SMEs entails material requirement planning (MRP) and use of employee capacity by preparing appropriate space allocation for the storage of material as well as calculating the amount of material required to suit the space available and to control orders made in order to ensure on-time or early-time deliveries for customers. Capacity involves optimum utilization of employee resources such as ordering to manage customer demand to avoid backlogs in manufacturing SMEs. Bhamu and Sangwan (2014:901–2) explain that TQM in manufacturing SMEs focuses on a management approach to long-term sustainability involving all members of the organization to engage in improving the businesses' processes, products, and services and instilling an organizational culture geared toward customer satisfaction. According to Ayeni et al. (2016:39), scheduling is prioritizing through planning by management on the control of received orders according to delivery period, shortest processing times, or first-come-first-served principles in manufacturing SMEs. As reported by Gupta and Jain (2015:80), SOP involves documents comprising instructions and steps to be followed, with descriptions of each step indicating how a task must be performed accordingly. These steps assist those involved in the work activities to work out the simplest way of doing the job in order to achieve quality requirements as well as safety requirements without any risk or injury.

Ali and Suleiman (2016:308–12) distinguish location strategy from layout strategy in manufacturing SMEs. With location strategy, premises are arranged to facilitate cleaning in the workplace and provide proper supervision of tools for easy accessibility and product flow in order to avoid inventory damages, whereas with layout strategy, the manufacturing process is physically arranged for appropriate process flow, proper employee adjustment to the environment, and good health and safety practices, including protection against any unhygienic and unsafe working conditions. SCM is the adequate movement of material and range of activities carried out to manufacture a quality product to be delivered on time or earlier to the customer as per the requirement (Mellor et al., 2014:195). Huang and Handfield (2015:2–3) refer to ERP as a system that standardizes business processes and supports decisions concurrent with the planning and managing of businesses through supplementing the need for an automatic interface between manufacturing activities and corresponding transactions for effective communication in manufacturing SMEs.

The subsequent section to be presented is value-adding activities.

1.2.4.2 Value-Adding Activities

In this section, value-adding activities are studied and discussed from various literature sources with the aim of providing a breakdown of the structure and a detailed background of these activities as regards the productivity progress of manufacturing SMEs. As described by Mellor et al. (2014:196–9), internal value activities comprise quality, cost, shorter delivery cycles, flexibility, and speed is exercised in the manufacturing processes and services rendered to the customer by manufacturing SMEs. Quality involves material consumed and equipment used by the manufacturing SME to generate the product that will meet or exceed the customers' requirements (Oakland, 2011:4; O'Neill, Sohal, & Teng, 2016:381).

As indicated by Shavarini et al. (2013:1109) and Hung, Hung, and Lin (2015:198), *cost* in manufacturing SMEs is expenditure used by operations managers on employee effort, machine utilization, and material used in operations taking place in the manufacturing process. Delivery is the transportation of goods produced by manufacturing SMEs to the customers (Mellor et al., 2014:195–7).

Mellor et al. (2014:195–7) and Mishra, Pundir, and Ganapathy (2016:382) refer to *flexibility* in manufacturing SMEs as the ability to adapt to changes in product design as well as dealing with changes on the product mix. The concept *speed* means how rapid operations are as a result of the excellence of employees' and or machine operational practice in manufacturing SMEs (Wu, Melnyk, & Swink, 2012:140).

Adding-value activities that are exercised in the manufacturing process and impact on the effective running of manufacturing SMEs and contribute to their productivity level are referred to as quality, cost, product price, speed, flexibility, reliability, dependability, and delivery (Liu & Liang, 2015:1021–5). Product price focuses on money charged on the product sold by the manufacturing SMEs to the customer and for competitiveness with other SMEs locally and worldwide (Liu & Liang, 2015:1022; Piya, Khadem, & Shamsuzzoha, 2016:445), whereas reliability is the process of sustaining strategic partnerships between manufacturers and suppliers as well as between manufacturers and customers to maintain product quality and price and sustain delivery for customers (Hudnurkar, Rathod, & Jakhar, 2016:633). Liu and Liang (2015:1024) and Ağan et al. (2016:1874) explain that dependability involves the process of making manufacturing SMEs successful through building relationships with suppliers and customers to reach the goals of the business. A background of case study results concerning Companies A, B, C, and D is the subsequent section to be addressed.

1.3 Case Study Results: Companies A, B, C, and D

1.3.1 The Reality of Manufacturing: How Company A Is Dealing with Factors Influencing Productivity in Its Business

1.3.1.1 Manufacturing at Company A

Company A has been in existence since the early twentieth century, its emphasis being on effectively manufacturing turning parts and components according to customer requirements and outlines. This company is located in Gauteng, South Africa. Operations managers in Company A generate products locally on a day-to-day basis in order to meet people's needs. These products are produced in the form of a variety of steel components. These components are produced in large quantities on an assembly line at a rate of 250 components per hour.

1.3.1.2 Vision

The vision of Company A is to manufacture its product to comply with the ISO 9001 standard as well as maintain its business relationship with its customers.

1.3.1.3 Mission

The mission of the company is to become one of the most competitive companies in the manufacturing industrial sector around the Gauteng region.

1.3.1.4 Background of the Company

During the year 2000, Company A opened its manufacturing plant in the west of Gauteng, and it continued to grow in 2017 with eight employees in manufacturing, overseen by three supervisors; two quality control staff; three administrative staff; and two finance staff, headed by the operations managing director and the operations manager. The company consists of four departments, namely, administrative and finance; manufacturing Sections 1, 2, and 3; and administrative staff headed by two levels, the managing director and the operations manager. The operations managing director delegates and ensures that the operations manager directs supervisors in their section to make sure that the employees achieve the results requested by the company. The manufacturing structure of Company A is depicted in Figure 1.3.

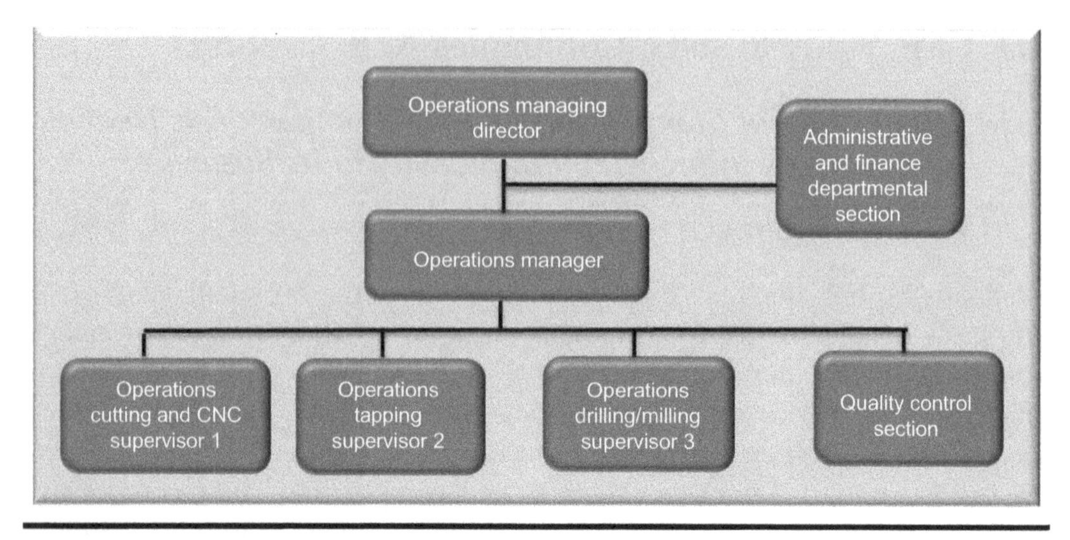

Figure 1.3 Organogram of Company A. (From Author, 2017.)

For Company A to be competitive, it must provide services to customers involved in mining, gas, and electrical businesses as well as households. These customers are situated locally and in other countries in sub-Saharan Africa. The objectives and targets of Company A are indicated in Table 1.7 with the intention of measuring the actual results against the targeted results.

In terms of Table 1.7, injuries and damages are stringently avoided. In addition, inappropriate suppliers' delivery, customer returns, and complaints are

Table 1.7 Actual Results versus Planned Target

No.	*Objective*	*Planned Targeted*	*Actual Results*
1	Poor product quality	10	Less than 10
2	Customer returns (rework)	≤10	Less than 5
3	Customer complaints	≤5	4
4	Lost time injuries	≤6	0
5	Minor injuries (no lost time)	≤12	Less than 10
6	Supplier poor material quality	≤5	2
7	Stock accuracy	≥85%	Less than 85%
8	Components produced per shift	10,000	6000
9	Number of operators	8	8

also eliminated to a certain extent. The serious challenge facing Company A is SOP in terms ISO 14001 whereby employees are exposed to inappropriate use of the workplace in terms of cleaning and good housekeeping as well as damage to inventory. The second challenge is SOP in terms of ISO 18001 whereby employees do not comply with occupational health and safety (OHS) standards, which are measures aiding the employee to avoid serious injuries, ill health, and accidents. Finally, stock accuracy is a serious situation facing Company A, which may result in high costs being incurred. Company A competes with other component engineering manufacturing companies in terms of pricing, flexibility, product quality, and delivery. Despite its competitors, Company A is still going strong in attempting to improve the productivity of its business locally and internationally.

The types of resources used by the company to prepare the final product are as follows:

- Three supervisors and eight operators with a support function comprising two finance staff and five administrative staff are employed to make the manufacturing process a success for Company A. Operations managers recruit experienced people from other companies and even new hires, providing them with on-the-job training in their workplace. These managers also arrange an effective employee schedule and manufacturing design layout of the products that need to be produced for customers.
- The materials used include mild steel, nylon, brass, stainless steel, and PVC.
- The types of machines used are for turning, drilling, tapping, plating, and packaging.
- The building used uses the same door for the incoming material and the outgoing final product. In addition, safety is taken into consideration so that employees do not get injured or become ill.
- The warehouse only caters for inventory that is stored temporarily (the order is made, and when the product is ready, the customer immediately comes to collect it or it is delivered on time to avoid delays).
- The storage of tools is controlled by the operations manager to ensure that operators return tools to where they belong after use.

For Company A to be competitive, it must provide services to customers involved in mining, gas, and electrical businesses as well as households. The intentions and target goals of Company A were to measure the actual results against the targeted results in order to ensure productivity progress in the company. The plan target was for 10,000 components to be produced by eight employees in an 8-hour shift, but the challenge facing Company A was that the actual results were less than the expected results. The actual results were 6000 components, whereby each operator produced 750 instead of 1250 components per shift as planned by the company.

Company A only utilizes technology when focusing on intangible components. This company liaises with its customers using emails and the telephone to order steel components. Secondly, printers are also used to print the documents required for records of ordered and manufactured products.

1.3.1.5 Background of the Manufacturing Process in Company A

A diagram of the manufacturing process for Company A is presented in Figure 1.4, showing how resources are converted into products for customer satisfaction.

There are a number of capabilities that the operations managing director and operations manager in Company A utilize in order to ensure that the company continues to grow in productivity. The capabilities provided are employee involvement and abilities; product, process, and service; SOPs in terms of product and service quality; layout; location (storeroom); inventory management; machine maintenance; scheduling; JIT technology; and SCM.

The operations manager in Company A prepares product, process, and service design in the form ordering by the customers. This manager writes the quotation for the customer and sends feedback via email with the approved price. When the customer is happy with the sample and quotation, a sample is then prepared by the quality section through the delegation of the operations manager. The administrative staff will prepare a job card indicating a sample of the product already designed for the customer. The signature will be attached by the operations manager and the time scheduling will be prepared as to how and when the product is going to be produced. At the completion of planning, the operations manager will instruct the manufacturing section to go ahead with the generation of the product following the product specification and the drawing. The manufacturing section will check the material specification to ensure that it is in line with the manufacturing process for the final product. The operators are always reminded to follow the SOP to produce the product required in terms of quality standards. Maintenance in Company A is done by operators on a daily basis following the checklist outlining the grease, oil, and lubricant levels. The second

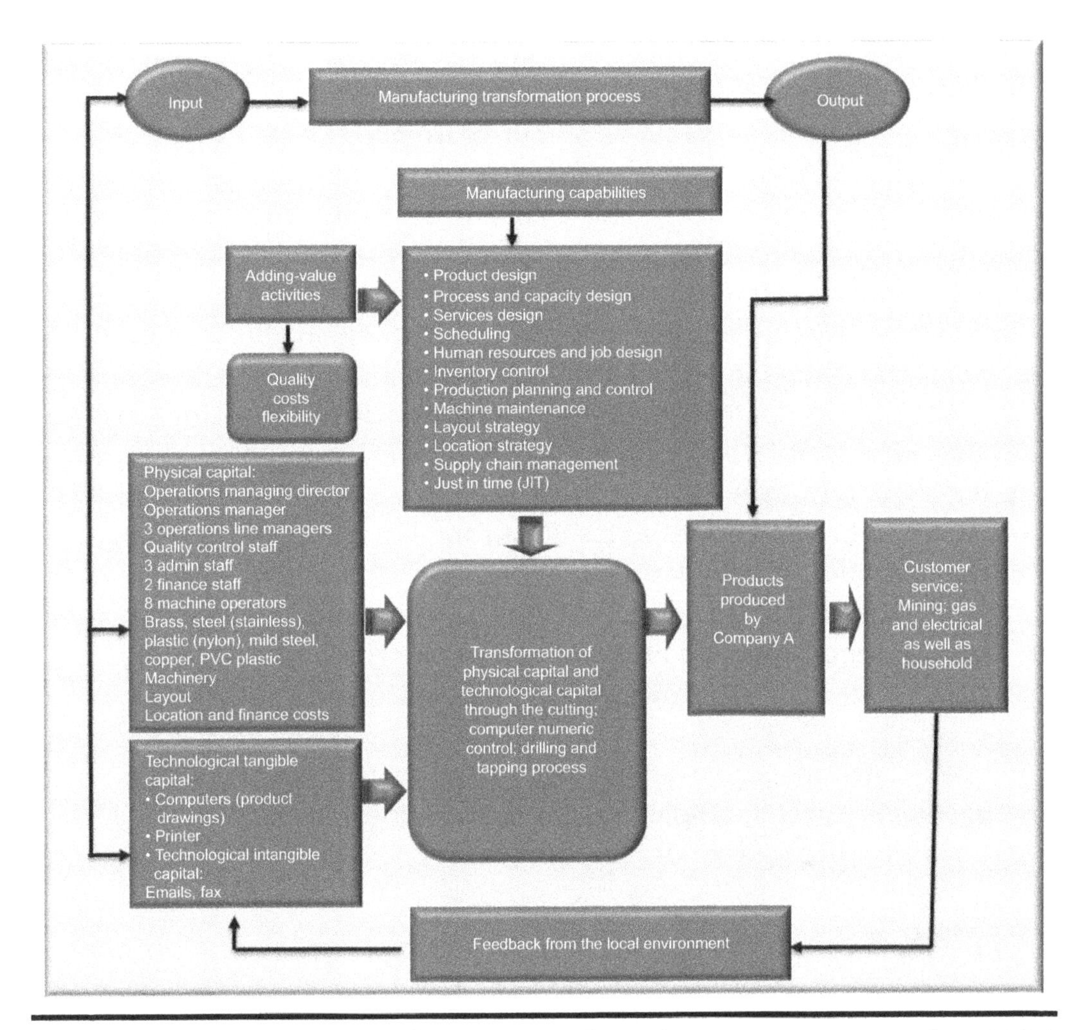

Figure 1.4 Company A: Manufacturing transformation process. (From Author, 2017.)

maintenance takes place when the section arranges service for the machinery on a monthly basis. Starting from the suppliers to the receiving section up to the work activities carried out in the manufacturing section for the final product being produced, this process is called the *supply chain process*. Three kinds of inspection take place at Company A.

Firstly, pre-inspection commences when the material is checked from the suppliers in the receiving section. Secondly, the on-line inspection takes place during the manufacturing process at various workstations. The last inspection is done when the final product is sent to stores before packaging for the customer to collect it or before it is delivered to the customer. Finally, the JIT tool is exercised from the time the material is delivered through the manufacturing process up to the customer delivery. The concept of JIT from the moment the supplier delivers material is not applicable in this case study. Cycle times to produce various components for Company A are used during the manufacturing process. The fastest time to deliver the components is between 40% and 20%–25% of the normal delivery time. Adding-value activities such as quality, cost, and flexibility needs to be considered by Company A to ensure efficient use of the latter capabilities in the manufacturing process for productivity improvement.

In addition, for these capabilities to be efficient in the manufacturing process, adding-value activities are also essential in order to boosting the productivity of Company A. The value activities involve quality; cost and flexibility. The manufacturing process in Figure 1.5 shows how Company A uses the ISO 9001 quality standard to ensure there is an efficient production process and achievement of the results for the business. The following procedure is followed, starting with the collection of the sample drawing of the part to be generated and ending with after sales in order to ensure appropriate manufacturing stages and the traceability of errors in the process.

The procedure of the flow of work being carried out is as follows:

Job: Manufacturing steel components
Chart begins: Collect sample drawing of the part to be generated
Chart ends: After sales of the part generated

1.3.1.6 Challenges Facing the Manufacturing Process of Company A

The challenges facing this company involved the supply side: management, the human element, inventory, technology, and transport.

1.3.1.6.1 The Suppliers

Company A is faced with a situation whereby suppliers deliver stock late and incorrect sizes are being delivered. Furthermore, when placing an order, the stock is sometimes not available for Company A.

1.3.1.6.2 Management

The challenge facing the operations managing director is shortcomings of technology, space, and inventory management. The company is still using aging machines and it has no plans to buy of machinery for contingency. When material is ordered, the company uses the same entrance as when the product is ready to be delivered to or collected by the customer. This is poor planning. There is also no control of inventory in terms of waste

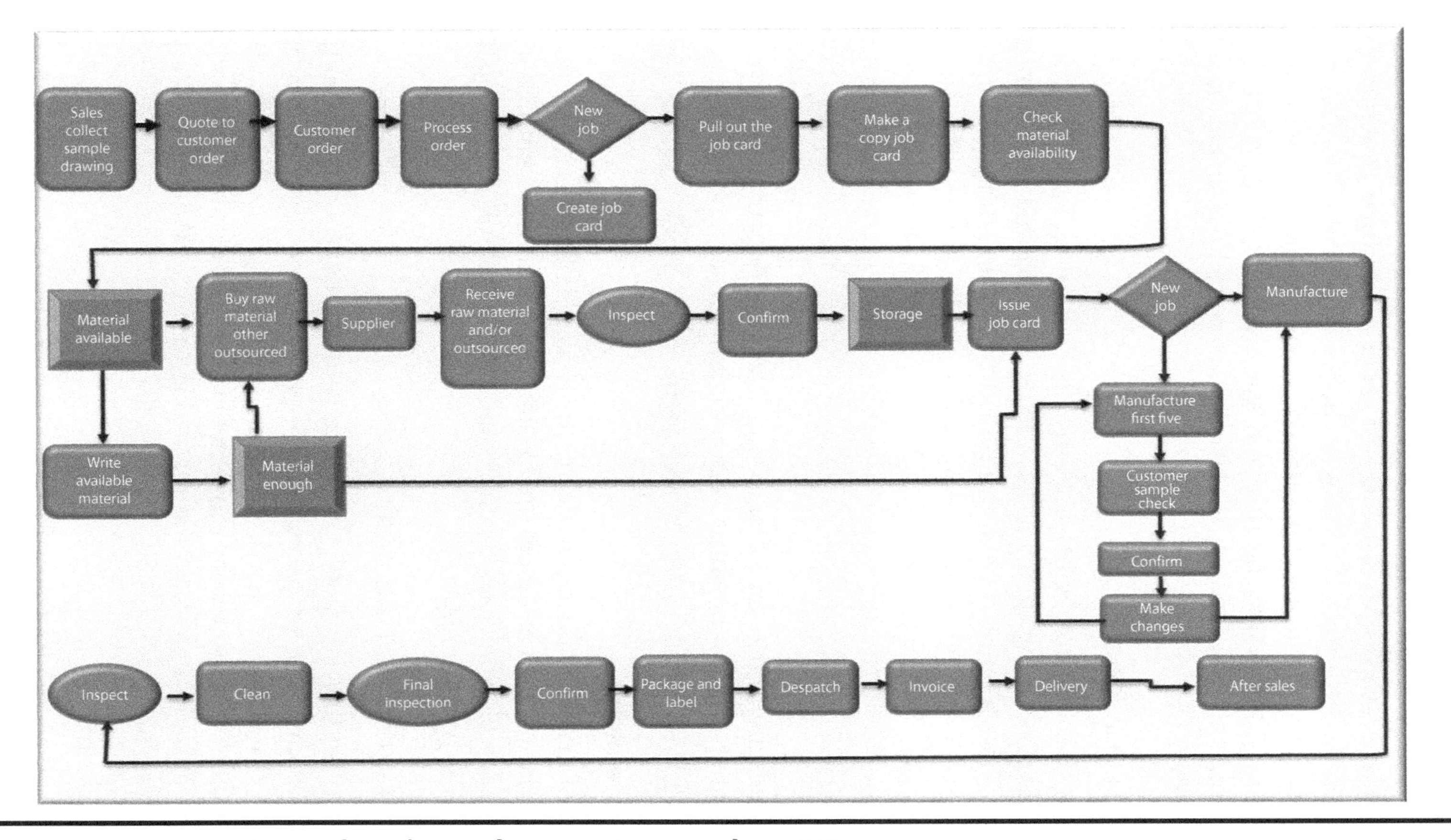

Figure 1.5 Company A manufacturing steel process. (From Author, 2017.)

incurred by the company. Tooling is also a serious matter facing the company whereby management makes a slow decision in ordering. Finally, managers are concerned with the resistance to change by employees on decisions made with regard to new technological programs introduced for productivity progress in the business. The shortcoming of management is that the operations managing director was slow to realize the importance of safety and the employees' working environment in terms of health in the company.

1.3.1.6.3 Human Element

The reason that employees resist change is due to aging, long-term experience, and fear of losing their job as well as a lack of incentives in the company. Even though employees are provided with test certificates for incoming material as well as SOP training, employees still fail to comply with SOPs in their workplace. Failure to comply with these procedures results in employees incurring waste in terms of the material used, which is costly for the company.

1.3.1.6.4 Machine

Old machines are in use, and as a result, delays are encountered by Company A as a manufacturing business. Competition is now becoming a serious dilemma for companies to survive and grow in the market.

1.3.1.6.5 Inventory

The types of material affected in Company A include bad material, cracks, and hardness. These problem areas are picked up late at the completion of the final product, which is costly for the company. As a result, the productivity of Company A is compromised.

1.3.1.6.6 Environment/Layout

Despite the fact that issues of safety, space, movement, housekeeping, working conditions, work-in-progress, and supply chain are normal, there is waste and outdated technology used. The environment is quite haphazard in the sense that there is unnecessary movement of employees and transportation of work-in-progress inventory taking place due to poor planning of the process layout in Company A.

1.3.1.6.7 Technology

Technology is not applicable in Company A in terms of physical capital except in the administrative department, where computers, printers, telephone, and fax are effectively used to communicate with suppliers, employees, and customers on information dissemination for the ordering of material as well as the delivery of products to customers, feedback from management to employees, and information sharing with all the parties involved. So a lack of technology in the manufacturing process delays the effective operational running of the company.

1.3.1.6.8 Transport

The type of transport used is a delivery truck. Of the deliveries, 40% are done early, before the time expected by the customer; and 20%–25% of deliveries are on time. The rest of the time is used inefficiently, whereby products are delivered late.

The type of material handling within the company involves the use of trolleys to transport an item from one operator to the other. Some of the material is then handled manually by the operator. The manual handling of material may lead to the operator be exposed to long-term injuries as these components are heavy to carry. In addition, manual material may result in employee fatigue, resulting in lateness and absenteeism.

The concern is not only how the operations managing director creates value to grow the business but also how the manager ensures continuous improvement of productivity in Company A. Company A has the opportunity to create an enabling environment through the use of work study for the continuous improvement of productivity in the business. Since work study is an exciting area of operations management that has a considerable effect on the productivity of Company A, the operations managing director can use this tool to determine how well the company is doing. As advised by Oeij et al. (2012:98), the improvement of productivity can be achieved through the application of work study in various ways.

1.3.2 Manufacturing at Medium-Sized Company B: Inefficiencies due to Failure of Management

Company B is a medium-sized company that was founded by five associates of shareholders with a combination of 110 years' experience, presided over by the operations managing director. This company began operating in early 2004.

The vision of Company B is to provide the best quality as determined by the authority of the African Bureau of Standards. This operation has been taking place for the past eight years.

The mission of Company B is to supply armed forces with the best possible variety of headwear for officers in the police and military as well as security in Southern Africa and other Asian countries, including islands in the east of South Africa. The types of material used by Company B to produce headwear include wool, fabric, and wire. The suppliers of Company B are located in Cape Town, Durban, China, Pakistan, and Taiwan.

This company is located in the industrial area of Northern Gauteng in South Africa. The company consists of 74 employees, of which 30 are highly qualified machine operators and 11 embroiderers, headed by the operations managing director and operations manager with the advice of one work study engineer. This company took a decision by liaising with Productivity SA in assisting on the use of toolkits to boost productivity. Some of these toolkits, such as employee empowerment, job descriptions, standards, time study, background of pictures for operational activities, and cause-and-effect analysis, have been applied and have contributed slightly to the productivity of Company B. Currently, the company has not yet benchmarked, but it is in the process of implementing the concept, including other work study tools such as brainstorming for competitiveness, preliminary surveys, use of process charts and diagrams, SOPs, ergonomics, value stream mapping, statistical process control, business process reengineering, and other work measurement tools as advised by the academic authors in work study research.

The operations managing director kept on advocating the use of integrated systems to ensure continuous improvement in the manufacturing process, but this was not enough to change the mind-set of employees. In order for Company B to ensure an improvement in productivity, the operations managing director kept on instilling the concept of various operational capabilities such as total quality improvement, supervisory and employees' training, supply chain and procurement, inventory control, three-months machine maintenance, design, process, and service, as well as adding-value activities such as quality, pricing, and reliability.

The assembly process of Company B consists of stores, the sample department, the cutting room, preparation, embroidery, quality check, and, finally, packaging. The average standard time for the whole process is 64.17 minutes. The company produces 1000 caps per day. Company B competes with other headwear companies in Malaysia as well as other East Asian countries. Company B is still battling in an attempt to improve the labor productivity of its businesses.

1.3.2.1 Challenges Experienced by Company B

The challenges facing this company are on the supply side: management, the human element, and manufacturing.

Lead times are very low for the delivery of material to Company B by suppliers due to imports done through shipping. The quality of service provided by Company B is beyond their control, since suppliers decide on the quantity of the material roll supplied. Sometime, Company B receives limited material, which is beyond their requirements. The challenge facing the operations managing director is deficiencies arising from the aging workforce and that fact that some are on maternity leave. Since most of these employees are aging, it is difficult for them to carry heavy boxes. Employees work from 07:00 to 16:30, including 30 minutes lunchtime and 20 minutes tea break. Incentives such as benefits and medical aids are not provided to employees. Communication between management and employees is a major problem in terms of involvement in decision making. Training is only provided internally and employees do not grow maturely.

Management lacks understanding of the manufacturing process due to the increased number of people on the assembly line, which is not easy controllable. The operations manager insists that there are as few employees on the line as possible, but during the operation, supervisors fail to follow the instructions and end up adding more workforce in the process. The more people there are on the line, the more expensive it is for the company to operate. The more confusing the line is, the less chance that Company B will reach the actual target as planned by the company. Due to the inefficient operation of the company, some employees leave and join other companies. The operations manager, on the instructions of upper management, only focuses on what is produced rather than also considering the costs incurred. There is also limited space for inventory and finished goods in this company. Material handling is done in the form of using trolleys.

The type of transport used is delivery truck. Of the deliveries, 80% are done on time. The rest of the time is used inefficiently, whereby products are delivered late. The concern is not only how the operations managing director creates value to grow the business but also how the manager ensures costs savings and improved employee career paths for continuous improvement of productivity in Company B. Company B has the opportunity to create its enabling environment through the use of other work study tools for the continuous improvement of productivity in the business. Since work study is a stimulating tool for business innovation, Company B can utilize these tools to determine the efficiency of its business manufacturing operation. Various case studies are provided next to facilitate an in-depth understanding of each case. In addition, essay questions and case study questions are addressed for the purposes of reader self-assessment.

1.3.3 Case of Packaging Operation at Company C

Company C was established in late January 2017 and has now been in existence for four months, focusing on the operational process of packaging for its customers in the south of Gauteng, South Africa. This company is situated in the central part of Gauteng and is administered by the operations director.

Company C produces packaging of various types of boxes for colored cocky pens, which are locally packed daily for export. Before the pens are packed in an envelope, a runner is responsible for carrying boxes full of pens, with different colors in each box, to be delivered to each workstation for each employee to insert two pens of different colors. These pens are then inserted in an envelope by each employee responsible for a certain color till the box is completed. The next employee, called the quality inspector, then does a quality check before sealing each envelope to avoid error and defects. These pens are then allocated and packed in an outbox and are placed on a conveyor belt for sealing. These boxes are packed on a pallet and wrapped with plastic, incorporating corners to protect finished products against damage. Company C packages 2250 cocky pens per hour, 12 pens per envelope, on three parallel lines.

1.3.3.1 Vision

The vision of Company C is to pack colored cocky pens through precise quality provision as well as management commitment to the customer.

1.3.3.2 Mission

The mission of the company is to become one of the most competitive companies to do a precise and efficient operation of packaging in the manufacturing industrial sector in the central part of the Gauteng region in South Africa.

1.3.3.3 Background of the Company

During the first four months of 2017, Company C managed to employ two purchasing supervisors, two packaging supervisors, 26 packers, and one bookkeeper, headed by an operations managing director. This company structure consists of five departments. The operations managing director assigns duties and makes certain that the supervisors in their section effectively supervise employees in their workplace to achieve the results of the company as requested. The operation structure of packaging for Company C is portrayed in Figure 1.6.

Company C needs to consider other avenues of the market to target other customers for business productivity growth. Its suppliers are Company E, patch pallet suppliers, Fiasco Suppliers, Strapper Suppliers, and Griffin Labels Suppliers.

The objectives and targets of Company C are indicated in Table 1.8 with the intention of measuring the actual results against the targeted results.

SOP procedures are not applicable in Company C, so it is a challenge for the operations managing director to consider how these procedures can be put in place for productivity measurement and improvement. Company C competes with other pen-packaging companies, such as Staedtler, in terms of pricing, flexibility, product quality, and delivery.

Even though there are competitors locally that are packaging pens, Company C is still going strong on packaging, identifying employees and finding the right machines to keep the business growing, due to the experience of the operations managing director. Employees in the company work on a job

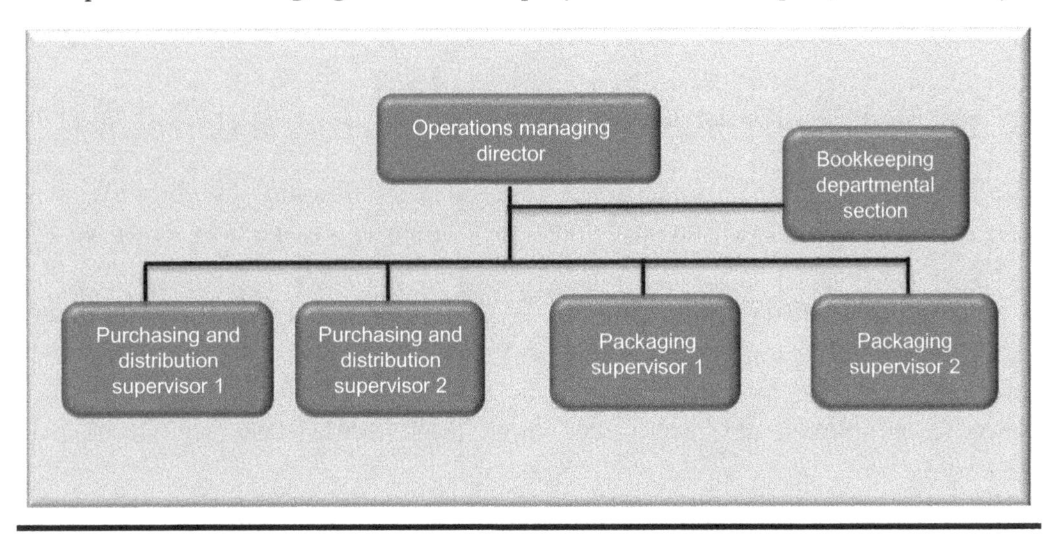

Figure 1.6 Organogram of Company C. (From Author, 2017.)

Table 1.8 Actual Results versus Planned Target

No.	Objective	Planned Targeted	Actual Results
1	Measure actual results against planned target	Only for months existing	Only for months existing

rotation basis, and the working hours are from 07:00 to 15:00, including one hour lunch time and 15 minutes tea time. Employees are provided with free lunch as an incentive.

The misconception of Company C is that the operations managing director confuses an increase in production with productivity improvement in the business. The concern is that the operations managing director focuses more on increasing output of products and profit without considering measuring the output of costs from period to period, employee training, safety, scheduling, inventory control, and capacity in order to see progress in productivity in the business. The operations managing director in Company C recruits candidates and provides on-the-job training in their workplace. The types of material that Company C purchases to make the packaging process a success include cellotape, wrapping plastics, palleting, and packaging box corners. The machinery used by Company C is pallet jacks, and forklifts are outsourced from another company. A manually operated conveyor belt is also used by employees to wrap boxes rather than an automated one. In terms of the location, Company C uses the same door for both receiving material and issuing packaged products. Company C also uses a truck to collect pens from Company E. Company C only utilizes technology to process orders through emails and to communicate with customers.

1.3.3.4 Background of the Manufacturing Process in Company C

The diagram of the manufacturing process for Company C, showing how resources are converted into products for company satisfaction, is presented in Figure 1.7.

The operations managing director in Company C uses operational strategies for packaging such as process and service design, assembly, employee on-the-job training and involvement, a small area for storage, a small work space for employee activity, and a small area for the flow of material on packaging.

The operations manager in Company C delegates employees to pack cocky pens of different colors using three parallel lines with the support of a

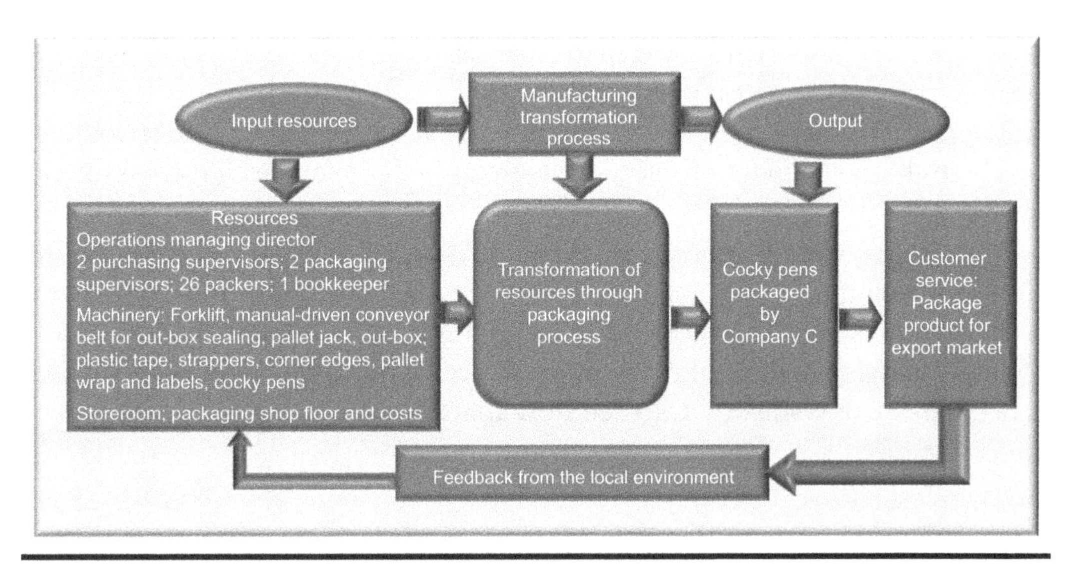

Figure 1.7 Company C: Manufacturing transformation process. (From Author, 2017.)

packaging line runner. This process is followed up until the last stage, involving the palletizing of wrapped boxes in preparation for a forklift to transport them into storage. The operations managing director advises Company E to collect the packaged product when the packaging is completed. The assembly process takes place when employees finish inserting all the colored cocky pens into the box, the quality check is done, and the outboxes are sealed.

For these capabilities to be cost-effective in the operational packaging process, adding-value activities such as quality, cost, and reliability are also vital in order to enhance the productivity of Company C.

There are no standards put in place for quality, safety, or environmental control in Company C.

The procedure of the work flow being carried out for the packaging process is indicated as follows:

Job: Packaging of colored cocky pens in Company C
Chart begins: Company C receives email from Company E (suppliers)
Chart ends: Company C delivers palletized product to Company E (customers)

1.3.3.5 The Manufacturing Process Followed in Company C

The type of process used in Company C is the computer numerical control (CNC) process. The standard of processing components for steel varies according to the different types of components ordered. This means that Company C must use flexibility in order to meet the requirements of any customers that have ordered components. The first stage of the CNC process starts from the turning point. At the completion of turning, operators prepare the item for drilling and timing. When the drilling process is completed, the next operator taps holes for adjustment of screws that are to be tightened in the components tapped.

The final product is then arranged for packaging, with different packaging processes being used depending on the customer procedure followed. Some of the products will be wrapped for each individual customer using the delivery note but also making the customer aware. There are still some rivals that benchmark with Company C to see how this company prepares the sketch of the prototype.

1.3.3.6 Challenges Facing the Manufacturing Process of Company C

The challenges facing this company are on the supply side: management, the human element, inventory, technology, and transport.

1.3.3.6.1 The Suppliers

Company C is faced with a situation whereby suppliers deliver stock late and incorrect sizes are being delivered. Furthermore, when placing an order, the stock is sometimes not available for Company C.

1.3.3.6.2 Management

The challenge facing the operations managing director is shortcomings in terms of technology, space, and inventory management. The company is still using aging machines, and there are no plans to buy machinery for contingency. When material is ordered, the company uses the same entrance as when the product is ready to be delivered to or collected by the customer. This is poor planning. There is also no control of inventory in terms of waste incurred by the company. The tooling is also a serious matter facing the company, whereby management is slow to make ordering decisions. Finally, managers are concerned with the resistance to change by employees in terms of decisions made with regard to new technological programs introduced for productivity progress in the business. The shortcoming of management is that the operations managing director was slow to realize the importance of safety and the employee working environment in terms of health in the company.

1.3.3.6.3 Human Element

The reason that employees resist change is due to aging, long-term experience, and fear of losing their job as well a lack of incentives in the company. Even though employees are provided with test certificate of incoming material as well as SOP training, employees still fail to comply with SOPs in their workplace. Failure to comply with these procedures results in employees incurring waste in terms of the material used, which is costly for the company.

1.3.3.6.4 Machine

Old machines are utilized, and as a result delays are encountered by Company C as a manufacturing business. Competition is now becoming a serious dilemma for companies if they are to survive and grow in the market.

1.3.3.6.5 Inventory

The types of material affected in Company C include bad quality material, cracks, and hardness. These problem areas are picked up late in the completion of the final product, which is costly for the company. As a result, the productivity of Company C is compromised.

1.3.3.6.6 Environment/Layout

Despite the fact that safety, space for materials, movement, housekeeping, working conditions, work in progress and supply chain are normal, there is waste, and outdated technology is used. The environment is haphazard in the sense that there is unnecessary movement of employees and transportation of work in progress and inventory taking place due to poor planning of the process layout in Company C.

1.3.3.6.7 Technology

Technology is not applicable in Company C in terms of physical capital except in the administrative department, where computers, printers, telephone, and fax are effectively utilized to communicate with suppliers, employees, and customers on information dissemination for ordering of material as well as delivery of products to customers, feedback from management to employees, and information sharing with all the parties involved. So a lack of technology in the manufacturing process delays the effective and operational running of the company.

1.3.3.6.8 Transport

The type of transport used is the delivery truck. Of the delivery, 40% is done early, before the time expected by the customer, and 20%–25% is delivered on time. The rest of the time is used inefficiently, whereby products are delivered late. The type of material handling within the company involves the use of trolleys to transport an item from one operator to another. In addition, some of the material is handled manually by the operator. The manual handling of material may lead to the operator be exposed to long-term injuries, as these components are heavy to carry. In addition, manual material may result in employee fatigue, resulting in lateness and absenteeism.

The concern is not only how the operations managing director creates value to grow the business but also how the manager ensures the continuous improvement of productivity in Company C. Company C has the opportunity to create an enabling environment through the use of work study for continuous improvement of productivity in the business. Since work study is an exciting area of operations management that has a considerable effect on the productivity of Company C, the operations managing director can use this tool to determine how well the company is doing.

1.3.4 Manufacturing at Medium-Sized Company D: Inefficiencies due to Failure of Management

Company D is a self-supporting, medium-sized company that has been in operation since the late 1990s, manufacturing wire and wireless security systems and developing innovative solutions for the local and international market. Company D is associated with certain professional bodies responsible for the updating of detection and monitoring systems.

With Company D, the information signals are converted into specific tones in order to diffuse signals over wire, wireless, and cable systems to a target point. This company produces components in the form of batches.

The vision of Company D is to be a leading champion in security solutions, innovation, quality product, and continuous technological improvement locally and worldwide. Company D supplies a wide range of radio communication systems for mobile communication based on sites and control centers. The strategy of Company D is based on the niche market. Parts generated are 2500 per hour. The mission of Company D is to provide the best quality product and to offer a high level of service to valued customers.

This company is located in the industrial area of central Gauteng in South Africa. The distribution points of this company are based in Cape Town, Durban, and East London, and these components are prepared for security companies. There are 102 employees, headed by the operations manager and supply chain managers. The culture of Company D is based on teamwork and a diverse work force in terms of age and gender. Exports are also done by Company D via courier to neighboring countries such as Botswana and Namibia.

The suppliers delivering to Company D are local companies such as Arrow Altech and Avnet, and the purchase order for Company D is done from month to month. In terms of SCM, tiers such as Arrow Altech and Avnet create a product sample and send it to Company D. Then, the purchasing department does a manual sample check. When Company D approves the tested sample A, sign off is done by this company and the company continues with the manufacturing process as guided by the approved sample. Any material sample that does not meet the requirements is put aside for retest or is returned to the suppliers. Companies competing with Company D locally are QD and RDC. The competition is based on price, good service, supplier–customer negotiation, and JIT or faster product and service delivery. Employees in Company D are motivated by receiving extra payment from overtime and earning bonuses by exceeding the company's expectations. All these incentives are paid according to what is determined by the manufacturing department at Company D.

Company D was using a single product. Due to the growth of the business, the company moved to using a multiple product, which forced the company to come up with changes. Machine maintenance is done and new technology is introduced to accelerate the manufacturing process in the company. Planning and scheduling is also taken seriously to ensure efficient manufacturing operation of Company D. The process starts with kitting components, followed by surface-mount device (SMD) manufacturing. The mechanical manufacturing process is then done through hole parts fitting. The board is placed in the form of plastic or metal. When plastic is used, a remote test is carried out, followed by assembly. The test is done through the mouse to check the radio frequency and cellular network in order to ensure that the system produced by Company D is working efficiently.

When assembly is completed, the packaging process is done. This process takes place for signals used on gates and garages and for other household requirements. The same process also applies to the use of metal. Before and during the manufacturing process as well as at the completion of the manufacturing process, some work study tools are applied by Company D, such as cause-and-effect diagrams and filming techniques with the intention of identifying problem areas such as poor planning. The delivery of items produced by Company D is done by the company's transport or through the use of couriers.

1.3.4.1 Challenges Experienced by Company D

Even though the challenges facing this company as indicated by electronic engineers were based on product quality, planning, logistics, and absenteeism due to chronic disease, research and development is encouraged to improve product design and remain competitive. There is a cycle time variation that makes it difficult to establish the efficiency of the manufacturing process in Company D. Waste and damages are experienced due to a poor quality system. Company D is also faced with a major change where international supply needs to be considered for a competitive edge. The company fails to control inventory due to the absence of a computerized system. The company has become more complex due to the increase in the range of products produced. Logistical problems are experienced where the supply chain is not attended in a timely manner from the first stage to the last stage. This problem may lead to late delivery and loss of customers. Safety issues in terms of employees' well-being and risk factors in terms of material damage are a major problem in Company D. This may arise due to poor planning or shortcomings of management.

Material handling is very poor; employees have to carry some of the material physically, leading these employees to severe ill-health. For Company D to manage inventory, senior officials foresee the use of a software package to track any component coming through the company to the completion of the product and until it reaches the customer. The aim of the company is to improve product quality within the business and ensure better service for the customer. The type of transport it uses is the delivery truck. The percentage of products that are delivered early for distribution is 20% and normally 50%. The next section to be discussed is the summary of case study results. The succeeding section is a summary of case study results for Companies A, B, C, and D.

1.4 Summary of Case Study Results

The following companies provide a comparison in terms of how they operate in achieving the results of the business and are indicated in Figure 1.8.

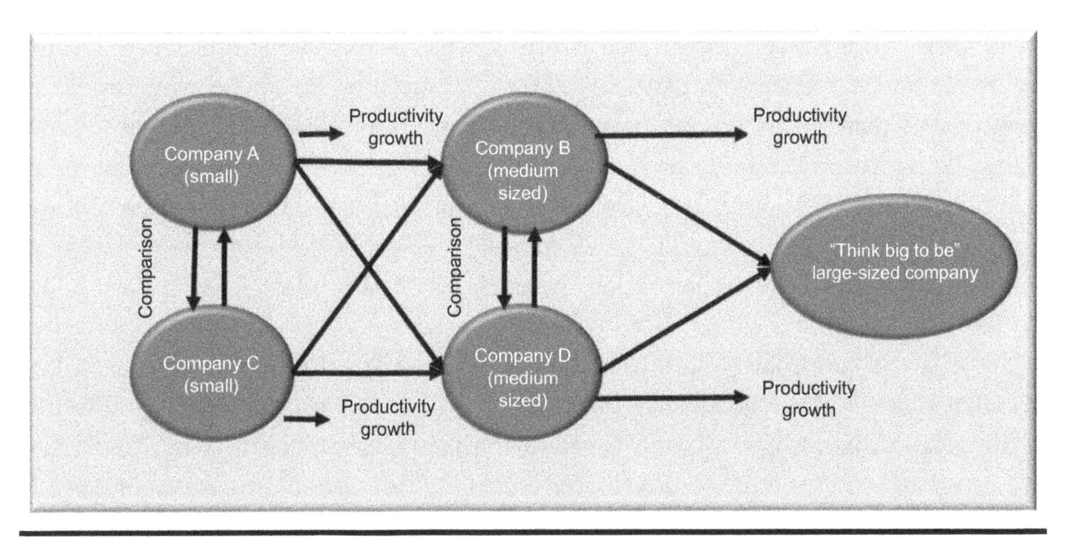

Figure 1.8 Company growth in terms of productivity improvement. (From Author, 2017.)

A comparison of areas of Companies A and C are presented as these companies are small-size manufacturing businesses as indicated in Figure 1.5. The similarities of the challenges these companies face include a lack of formal training, research and development, layout, space communication, incentives, technology, and planning. The difference between Company A and Company C is that Company A has an appropriate organization structure, which only needs some minor amendments as a small business, as well as capabilities such as production, process, and service, whereas Company C does not evidence any structure. Company C may have a serious problem of report relationship based on the chain of command and proper communication with employees, which may result in poor business performance. Company C focuses mainly on the packaging process of the product. The capability that is common to both companies is a focus on quality; however, Company C does not have certification of stock for ISO 9001 quality, which may lead to poor product packaging, waste, and poor service provided to the customer. Since Company C has only been in existence for four months, this company is still in a premature stage of growth to become a medium business, and work study could be considered in engaging this business to grow in terms of its productivity level. With Company A, certain capabilities need to be introduced first, such as technology, JIT, and standards such as ISO 140001 and ISO 18001 for safety measures and risk management on products generated, respectively. On the contrary, Company C needs to consider the size of the company in terms of employee recruitment, sales growth, profit, market share, and innovation if it is to continue to improve its productivity level.

On the other hand, Companies B and D may also have similarities in terms of areas that enable these businesses to exist in the market. These companies are regarded as medium-sized companies, as indicated in Figure 1.5. Both these companies focus on capabilities such as assembly of products from different items. In addition, they follow the SOP in terms of the material up to the completion of the product. The concept of quality in Company B is applicable but no standards or certification of ISO 9001 is in place, whereas Company D registered with SABS on ISO 9001 quality product. Both these companies need ISO 14001 for appropriate use of the workplace in terms of cleaning and good housekeeping and ISO 18001 for safety to avoid unnecessary injuries and risk to employee well-being. Company D has an appropriate formal organizational structure whereby duties are assigned to respective senior officials, such as the operations manager, supply chain manager, and electronic engineers, including supervisors, all headed by the CEO. Whereas Company B has a flat structure with the operations director and operations manager carrying out many of the managerial responsibilities, respectively, the challenges of decision-making may be difficult to reach the set objectives of the company. Even though Company B, has someone responsible for time studies, this time study specialists is still on a premature stage to do proper time studies. So, a qualified work study specialist is also needed to guide and mentor the intern work study specialist in gaining more experience to ensure a proper standard for the job being carried out. In order for Company B to grow in terms of productivity and to reach a stage of growth where it can be considered to be a large-sized business, more work has to be done in terms of standards, sufficient layout and space for material, proper chain of command in terms of work responsibilities, and involvement in research and development.

With Company D, however, the absence of time studies and absence of standards also may lead the company to a stagnant stage whereby no growth will happen in the business in terms of productivity and becoming a large-sized business. Even though work study tools such as standards can be in place to ensure productivity progress in all these companies, other work study tools need to be considered which will be mentioned in Chapter 2.

1.5 Book Delineation

This chapter has given an overview of research trends and challenges with regard to productivity problems facing manufacturing SMEs in South Africa (SA) and the rest of the world. Furthermore, it has looked at the research methodology with the support of the research findings. At the completion of the research findings, various areas of manufacturing were addressed to justify the in-depth consideration of manufacturing in manufacturing SMEs. These areas included the background of manufacturing SMEs, definitions of manufacturing SMEs, manufacturing systems and processes, and competitive priorities. Finally, this chapter presented a case study of Companies A, B, C, and D, addressing challenges facing the manufacturing process and the decisions that operations managing director, as well the operations manager, should take through the guidelines of work study in improving the productivity of the businesses. The purpose of using these areas of manufacturing indicated earlier was to express that there is a lack of understanding of the work study through the use of physical capital and technological capital to improve the productivity of manufacturing SMEs in SA.

References

Aagaard, A., Eskerod, P., Madsen, E.S. 2015. Key drivers for informal project coordination among sub-contractors. *International Journal of Managing Projects in Business*, 8(2): 222–240.

Ağan, Y., Kuzey, C., Acar, M.F., and Açıkgöz A. 2016. The relationships between corporate social responsibility, environmental supplier development, and firm performance. *Journal of Cleaner Production*, 112: 1872–1881.

Ali, M.H. and Suleiman, N. 2016. Sustainable food production: Insights of Malaysian Halal small and medium sized enterprises. *International Journal of Production Economics*, 181: 303–314.

Awheda, A., Rahman, M.N.A., Ramli, R., and Arshad, H. 2016. Factors related to supply chain network members in SMEs. *Journal of Manufacturing Technology Management*, 27(2): 312–335.

Ayeni, P., Ball, P., and Baines, T. 2016. Towards the strategic adoption of lean in aviation maintenance repair and overhaul (MRO) industry. *Journal of Manufacturing Technology Management*, 27(1): 38–61.

Bi, Z.M., Liu, Y., Baumgartner, B., Culver, E., Sorokin, J.N., Peters, A., Cox, B., Hunnicutt, J., Yurek, J., and O'Shaughnessey, S. 2015. Reusing industrial robots to achieve sustainability in small and medium-sized enterprises (SMEs). *Industrial Robot: An International Journal*, 42(3): 264–273.

Bhamu, J. and Sangwan, K.S. 2014. Lean manufacturing: Literature review and research issues. *International Journal of Operations and Production Management*, 34(7): 876940.

Bogue, R. 2014. Sustainable manufacturing: A critical discipline for the twenty-first century. *Assembly Automation*, 34(2): 117–122.

Brown, S. 2012. Operations master series: An interview with Terry Hill, Emeritus fellow at the University. *International Journal of Operations and Production Management*, 32(3): 375–384.

Chompu-inwai, R., Jaimjit, B., and Premsuriyanunt, P. 2015. A combination of material flow cost accounting and design of experiments techniques in an SME: The case of a wood products manufacturing company in Northern Thailand. *Journal of Cleaner Production*, 108: 1352–1364.

Choudhari, S.C., Adil, G.K., and Ananthakumar, U. 2012. Exploratory case studies on manufacturing decision areas in the job production system. *International Journal of Operations & Production Management*, 32(11): 1337–1361.

Coka, B. 2014. *Annual Report 2013–2014*. Midrand, Republic of South Africa: Productivity SA.

Debnath, R.M. and Sebastian, V.J. 2014. Efficiency in the Indian iron and steel industry: An application of data envelopment analysis. *Journal of Advances in Management Research*, 11(1): 4–19.

de Carvalho, M.M, Ho, L., and Pinto, S.H.B. 2014. The Six Sigma program: An empirical study of Brazilian companies. *Journal of Manufacturing Technology Management*, 25(5): 602–630.

De Snoo, C., Van Wezel, W., and Wortmann, J.C. 2011. Does location matter for a scheduling department? A longitudinal case study on the effects of relocating the schedulers. *International Journal of Operations & Production Management*, 31(12): 1332–1358.

Dora, M., Kumar, M., and Gellynck, X. 2016. Determinants and barriers to lean implementation in food-processing SMEs: A multiple case analysis. *Production Planning and Control*, 27(1): 1–23.

Drohomeretski, E., da Costa, S.E.G., de Lima, E.P., and da Rosa Garbuio, P.A. 2014. Lean, Six Sigma and Lean Six Sigma: An analysis based on operations strategy. *International Journal of Production Research*, 52(3): 804–824.

Duran, C., Cetindere, A., and Aksu, Y.E. 2015. Productivity improvement by work and time study technique for earth energy-glass manufacturing company. *Procedia Economics and Finance*, 26: 109–113.

Gashi, P., Hashi, I., and Pugh, G. 2014. Export behaviour of SMEs in transition countries. *Small Business Economics*, 42(1): 407–435 .

Gibb, M. and Luiz, J.M. 2011. Poverty, inequality and unemployment in South Africa: Context, issues and the way forward. *Economic Papers*, 30(3): 307–315.

Gobinath, S., Elangovan, D., and Dharmalingam, S. 2015. Lean manufacturing issues and challenges in manufacturing process: A review. *International Journal of ChemTech Research*, 8(1): 44–51.

Gröbler, A., Laugen, B.T., Arkader, B., and Fleury, A. 2013. Differences in outsourcing strategies between firms in emerging and in developed markets. *International Journal of Operations and Production Management*, 33(3): 296–321.

Grütter, A. 2010. *Introduction to Operations Management: A Strategic Approach*. Cape Town: Heinemann.

Gupta, S. and Jain, S.K. 2013. A literature review of lean manufacturing. *International Journal of Management Science and Engineering Management*, 8(4): 241–249.

Gupta, S. and Jain, K.S. 2015. An application of 5S concept to organize the workplace at a scientific instruments manufacturing company. *International Journal of Lean Six Sigma*, 6(1): 73–88.

Habidin, N.F. and Yusof, S.M. 2013. Critical success factors of Lean Six Sigma for the Malaysian automotive industry. *International Journal of Lean Six Sigma*, 4(1): 60–82.

Hamilton, B.H., Nickerson, J.A., and Owan, H. 2015. Diversity and productivity in production teams. In A. Bryson (ed.) *Advances in the Economic Analysis of Participatory and Labor-Managed Firms*, Volume 13: pp. 99–138. Bingley: Emerald Group.

Helkiö, P. and Tenhiäla, A. 2013. A contingency theoretical perspective to the product–process matrix. *International Journal of Operations & Production Management*, 33(2): 216–244.

Herrington, M., Kew, J., and Kew, P. 2014. Gem South African Report: the Crossroads: A Gold Mine or A Time Bomb? South Africa: Global Entrepreneurship Monitor. Available from: www.gemconsortium.org/country-profile/108.

Hilmola, O., Lorentz, H., Hilletofth, P., and Malmsten, J. 2015. Manufacturing strategy in SMEs and its performance implications. *Industrial Management and Data Systems*, 115(6): 1004–1021.

Hooi, L.W. and Leong, T.Y. 2017. Total productive maintenance and manufacturing performance improvement. *Journal of Quality in Maintenance Engineering*, 23(1): 2–21.

Hu, Q., Mason, M., Williams, S.J., and Found, P. 2015. Lean implementation within SMEs: A literature review. *Journal of Manufacturing Technology Management*, 26(7): 980–1012.

Huang, Y. and Handfield, R.B. 2015. Measuring the benefits of ERP on supply management maturity model: A "big data" method. *International Journal of Operations and Production Management*, 35(1): 2–25.

Huang, X., Tan, B.L., and Ding, X. 2015. An exploratory survey of green supply chain management in Chinese manufacturing small and medium-sized enterprises. *Journal of Manufacturing Technology Management*, 26(1): 80–103.

Hudnurkar, M., Rathod, U., and Jakhar, S.K. 2016. Multi-criteria decision framework for supplier classification in collaborative supply chains. *International Journal of Productivity and Performance Management*, 65(5): 622–640.

Huggins, R., Morgan, B., and Williams, N. 2015. Regional entrepreneurship and the evolution of public policy and governance. *Journal of Small Business and Enterprise Development*, 22(3): 473–511.

Hung, S., Hung, S., and Lin, M.J. 2015. Are alliances a panacea for SMEs? The achievement of competitive priorities and firm performance. *Total Quality Management and Business Excellence*, 26(2): 190–202.

ILO. 2015. World Employment and Social Outlook: Trends 2015. International Labour Organization. Available from: www.ilo.org/wcmsp5/groups/public/---dgreports/---dcomm/---publ/documents/publication/wcms_337069.pdf.

Jain, A., Bhatti, R., and Singh, H. 2014. Total productive maintenance (TPM) implementation practice. *International Journal of Lean Six Sigma*, 5(3): 293–323.

Lee, J., Lapira, E., Bagheri, B., and Kao, H. 2013. Recent advances and trends in predictive manufacturing systems in big data environment. *Manufacturing Letters*, 1: 38–41.

Lentes, J., Mandel, J., Schliessmann, U., Blach, R., Hertwig, M., and Kuhlmann, T. 2017. Competitive and sustainable manufacturing by means of ultra-efficient factories in urban surroundings. *International Journal of Production Research*, 55(2): 480–491.

Liu, Y. and Liang, L. 2015. Evaluating and developing resource-based operations strategy for competitive advantage: An exploratory study of Finnish high-tech manufacturing industries. *International Journal of Production Research*, 53(4): 1019–1037.

Małachowski, B. and Korytkowski, P. 2016. Competence-based performance model of multi-skilled workers. *Computers and Industrial Engineering*, 91: 165–177.

Manufacturing Indaba. 2014. Manufacturing innovation—growing South African manufacturers. Available from: http://manufacturingindaba.co.za/wp-content/uploads/2015/04/Report_Generic.pdf. [Accessed 2015-05-22.]

Mbohwa, C. 2017. Letter of permission to conduct an interview: Productivity improvement in manufacturing SMEs: Application of work study techniques. Faculty of Engineering and the Built Environment. University of Johannesburg. Johannesburg, South Africa. 12 April 2017.

Mellor, S., Hao, L., and Zhang, D. 2014. Additive manufacturing: A framework for implementation. *International Journal of Production Economics*, 149: 194–201.

Mishra, R., Pundir, A.K., and Ganapathy, L. 2016. Conceptualizing sources, key concerns and critical factors for manufacturing flexibility adoption. *Journal of Manufacturing Technology Management*, 27(3): 379–407.

National Credit Regulator (NCR). 2011. Literature Review on Small and Medium Enterprises' Access to Credit and Support in South Africa. Pretoria, South Africa. Available from: http://www.ncr.org.za/pdfs/Literature%20Review%20on%20SME%20Access%20to%20Credit%20in%20South%20Africa_Final%20Report_NCR_Dec%202011.pdf. [Accessed on 2015-05-20.]

Oakland, J.S. 2011. *Total Quality Management: Text with Cases*. London: Butterworth-Heinemann Publishers.

Oeij, P.R.A., De Looze, M.P., Have, K.T., Van Rhijn, J.W., and Kuijt-Evers, L.F.M. 2012. Developing the organisation's productivity strategy in various sectors of industry. *International Journal of Productivity and Performance Management*, 61(1): 93–109.

Ohu, I.P.N., Cho, S., Kim, D.H., and Lee, G.H. 2016. Ergonomic analysis of mobile cart-assisted stocking activities using electromyography. *Human Factors and Ergonomics in Manufacturing and Service Industries*, 26(1): 40–51.

Ojediran, T., Wintoki, A., and Odumade, O.E. 2016. Market & Liquidity Risk Management Department. Monthly Economic & Financial Markets Report: Review & Outlook. Available from: http://sme.firstbanknigeria.com/wp-content/uploads/2016/02/MONTHLY-ECONOMIC-AND-FINANCIAL-MARKETS-REPORT-JANUARY-2016.pdf.

O'Neill, P., Sohal, A., and Teng, C.W. 2016. Quality management approaches and their impact on firms' financial performance: An Australian study. *International Journal of Production Economics*, 171: 381–393.

Ozigbo, N.C. 2015. The dynamics of technological innovation capability on new product development and industry's performance: A study of Nigerian iron and steel industry. *Journal of Business Management and Economics*, 3: 1–8.

Ozturk, E., Karaboyacı, M., Yetis, U., Yigit, N.O., and Kitis, M. 2015. Evaluation of integrated pollution prevention control in a textile fiber production and dyeing mill. *Journal of Cleaner Production*, 88: 116–124.

Pandey, A., Singh, M., Soni, N., and Pachorkar, P. 2014. Process layout on advance CNG cylinder manufacturing. *International Journal of Application or Innovation in Engineering and Management (IJAIEM)*, 3(12): 113–116.

Panizzolo, R., Garengo, P., Sharma, M.K., and Gore, A. 2012. Lean manufacturing in developing countries: Evidence from Indian SMEs. *Production Planning and Control*, 23(10–11): 769–788.

Pedersen, P. and Slepniov, D. 2016. Management of the learning curve: A case of overseas production capacity expansion. *International Journal of Operations and Production Management*, 36(1): 42–60.

Piya, S., Khadem, M.M.R.K., and Shamsuzzoha, A. 2016. Negotiation based decision support system for order acceptance. *Journal of Manufacturing Technology Management*, 27(3): 443–468.

Prasad, K.D., Jha, S.K., and Prakash, A. 2015. Quality, productivity and business performance in home based brassware manufacturing units. *International Journal of Productivity and Performance Management*, 64(2): 270–287.

Prasad, S., Khanduja, D., and Sharma, S.K. 2016. An empirical study on applicability of lean and green practices in the foundry industry. *Journal of Manufacturing Technology Management*, 27(3): 408–426.

Republic of South Africa. 2003. *National Small Business Act 26 of 2003*. Pretoria: Government Printer.

Robinson, D.C., Sanders, D.A., and Mazharsolook, E. 2015. Ambient intelligence for optimal manufacturing and energy efficiency. *Assembly Automation*, 35(3): 234–248.

Sanjog, J., Patnaik, B., Patel, T., and Karmakar, S. 2016. Context-specific design interventions in blending workstation: An ergonomics perspective. *Journal of Industrial and Production Engineering*, 33(1): 32–50.

Saunila, M., Pekkola, S., and Ukko, J. 2014. The relationship between innovation capability and performance. *International Journal of Productivity and Performance Management*, 63(2): 234–249.

Schwab, K. 2015. The Global Competitiveness Report 2014–2015: Insight Report. World Economic Forum (WEF). Available from: www3.weforum.org/docs/WEF_GlobalCompetitivenessReport_2014-15.pdf.

SEDA. 2012. Analysis of the Needs, State and Performance of Small and Medium. Businesses in the Agriculture, Manufacturing, ICT and Tourism Sectors in South Africa. Pretoria, South Africa. Final Report 2012. Available from: http://www.seda.org.za/Publications/Publications/Success%20Story%20Booklet%202014.pdf. [Accessed 2015-05-21.]

SEDA. 2014. *Celebrating 10 Years of SEDA Success Stories.* Available at: http://www.seda.org.za/Publications/Publications/Success%20Story%20Booklet%202014.pdf. [Accessed on 2015-08-20.]

Sharma, S. and Shah, B. 2016. Towards lean warehouse: Transformation and assessment using RTD and ANP. *International Journal of Productivity and Performance Management*, 65(4): 571–599.

Shavarini, S.K., Salimian, H., Nazemi, J., and Alborzi, M. 2013. Operations strategy and business strategy alignment model (case of Iranian industries). *International Journal of Operations & Production Management*, 33(9): 1108–1130.

Silva, C., Stevenson, M., and Thurer, M. 2015. A case study of the successful implementation of workload control. *Journal of Manufacturing Technology Management*, 26(2): 280–296.

Singh, K. and Ahuja, I.S. 2015. An evaluation of transfusion of TQM-TPM implementation initiative in an Indian manufacturing industry. *Journal of Quality in Maintenance Engineering*, 21(2): 134–153.

Singh, B.J. and Bakshi, Y. 2014. Optimizing backup power systems through Six Sigma. *International Journal of Lean Six Sigma*, 5(2): 168–192.

Singh, J. and Singh, H. 2015. Continuous improvement philosophy: Literature review and directions. *Benchmarking: An International Journal*, 22(1): 75–119.

Soda, S., Sachdeva, A., and Garg, R.K. 2015. GSCM: Practices, trends and prospects in Indian context. *Journal of Manufacturing Technology Management*, 26(6): 889–910.

South African Reserve Bank. 2013. *Quarterly Bulletin.* March 2013. Pretoria. Available at: http://www.resbank.co.za/Lists/News%20and%20Publications/Attachments/5608/01Full%20Quarterly%20Bulletin.pdf. [Accessed on 2015-05-20.]

Srinivasan, S., Ikuma, L.H., Shakouri, M., Nahmens, I., and Harvey, C. 2016. 5S impact on safety climate of manufacturing workers. *Journal of Manufacturing Technology Management*, 27(3): 364–378.

Statistics SA. 2015. Quarterly Labour Force Survey: Quarter 2: 2015. Available at: http://www.statssa.gov.za/publications/P0211/P02112ndQuarter2015.pdf.

Stockes, D. and Wilson, N. 2010. *Small Business Management and Entrepreneurship.* 6th edn. Andover: South-Western Cencage Learning.

Sylla, N., Bonnet, V., Colledani, F., and Fraisse, P. 2014. Ergonomic contribution of ABLE exoskeleton in automotive industry. *International Journal of Industrial Ergonomics*, 44: 475–481.

Thomas, A., Byard, P., Francis, M., Fisher, R., and White, G.R.T. 2016. Profiling the resiliency and sustainability of UK manufacturing companies. *Journal of Manufacturing Technology Management*, 27(1): 82–99.

Urban, B. and Naidoo, R. 2012. Business sustainability: Empirical evidence on operational skills in SMEs in South Africa. *Journal of Small Business and Enterprise Development*, 19(1): 146–163.

Vinodh, S., Selvaraj, T., Chintha, S.K., and Vimal, K.E.K. 2015. Development of value stream map for an Indian automotive components manufacturing organization. *Journal of Engineering, Design and Technology*, 13(3): 380–399.

Wu, S.J., Melnyk, S.A., and Swink, M. 2012. An empirical investigation of the combinatorial nature of operational capabilities: Compensatory or additive? *International Journal of Operations and Production Management*, 32(2): 121–155.

Youn, S.H., Yang, M.G., Kim, J.H., and Hong, P. 2014. Supply chain information capabilities and performance outcomes: An empirical study of Korean steel suppliers. *International Journal of Information Management*, 34: 369–380.

Zhang, Y. and Gregory, M. 2011. Managing global network operations along the engineering value chain. *International Journal of Operations and Production Management*, 31(7): 736–764.

Zhou, Y. and Zhao, L. 2016. Analysis of the implementation of cleaner production for achieving the low-carbon transition for SMEs in the Inner Mongolian coal industry. *Journal of Cleaner Production*, 127: 418–424.

Chapter 2

Productivity Theory and Work Study: Groundwork Theories

2.1 Introduction

In this chapter, productivity theory and work study are explained and discussed from various literature sources based on their evolution and definition in order to provide a detailed examination of their differences.

2.2 Productivity and Work Study Theory: Groundwork Theories

2.2.1 Productivity Theory

Chapter 2 addresses the groundwork theories of productivity and work study. In terms of productivity, the evolution of productivity as well as the concept is discussed and its paradigm is presented from its origin to the current situation. Based on the literature studied and discussed, the groundwork of productivity theory focuses on its evolution, definition, and productivity applications in manufacturing small and medium enterprises (SMEs) around Gauteng, South Africa.

2.2.1.1 Evolution of Productivity Theory

In this section, the evolution of productivity theory is reviewed and examined on the basis of the work of different scientific authors to recognize existing in-depth knowledge and improve the schools of thought regarding the concept. As noted by Uddin (2015:241), productivity originates from the French mathematician Quesnay, who introduced the concept in a piece of writing in 1766. The reason was that Quesnay was emphasizing the value of productivity in manufacturing industries, one of which is SMEs. This mathematician saw productivity as the contributing factor to economic growth nationally and one that could obviously add value through cost-effectiveness, profitability, and an increase in competition at industry level.

In this section, information regarding the historical background of productivity in manufacturing SMEs from 1998 to 2015 is provided. Gunasekaran and Cecille (1998:311–190), as the first authors considered in this research book, proclaim the history of productivity, which has relevance in the manufacturing industry, in particular SMEs followed by other scientific authors in research. This consideration of these authors was based on physical capital such as employees' skills and knowledge, material, machine, layout, location, and finance; technologies such as automation and equipment as well as management abilities in ensuring productivity progress in manufacturing SMEs. Various authors (Heshmati, 2003:85; Subrahmanya, 2006:765; Tran, Grafton, & Kompas, 2009:275; Mathur, Mittal, & Dangayach, 2012:754–63; Klingner, Pravemann, & Becker, 2015:240) to name a few in this research book, were acknowledged to have contributed to the productivity improvement of manufacturing SMEs worldwide. So the definition of productivity is addressed next to reveal the productivity changes of SMEs.

2.2.1.2 Productivity Concept

In this section, the definition of productivity is explained and discussed on the basis of different literature sources, the aim being to provide a detailed background of the concept in manufacturing SMEs. For instance, Gunasekaran and Cecille (1998:312) describe productivity as "as the ratio of what is produced to what is required to produce it." What this definition implies is that productivity is based on the relationship between outputs, such as goods and services produced, and inputs, comprising employee resources, material, capital, and other resources in manufacturing SMEs.

Productivity is expressed as the relationship between the results of defects incurred during the generation of products and resources used under conditions such as accidents (in terms of employees) and breakdowns (in terms of machinery utilized) in manufacturing SMEs (Bamber, Sharp, & Hides, 1999:163–65). As discussed by Leseure (2000:1480–4); Martin, Jr., Horne, and Chan (2001:142–9);

and Rantanen (2001:85–8), productivity in manufacturing SMEs involves outputs indicating products produced, profitability, and customer satisfaction in relation to employees' knowledge; time spent; material; machinery; and reorganized structures. The outcome of the literature reviewed expresses the concept of productivity as the association between products generated and employee resources and capital utilized in manufacturing SMEs (Basu & Fernald, 2002:963–5; Heshmati, 2003:85; St-Pierre & Raymond, 2004:684–91). As stated by Li (2004:412–3), productivity focuses on products produced and profitability in association with employee resources in manufacturing SMEs.

Based on the literature read and discussed, productivity in manufacturing SMEs comprises the relationship between the impact of research and development and the size of the firm (Garengo, Biazzo, & Bititci, 2005:28–36; Tsai, 2005:796). Subrahmanya (2006a:765, 2006b:490–1) defines productivity in manufacturing SMEs as measuring the relationship of output to input in monetary terms. The manner in which productivity is expressed is referred to as output of processes versus input of energy. Halkos and Tzeremes (2007:715) describe productivity as the ratio of products to the size of the firm as well as that between manufacturing SMEs and similar SMEs in terms of their performances. According to Tran et al. (2009:275), productivity refers to the quantitative relationship between output of price earned and input of the number of employees hired and capital used in manufacturing SMEs. Productivity is a yardstick measuring the efficiency of a country's output, such as the economy, to input performance, such as employee security and the availability of skills needed for the job in manufacturing SMEs versus the efficiency of the other countries' output-to-input performance (Mahmood, 2008:52–3; O'Mahony & Timmer, 2009:374; Motohashia & Yuana, 2010:794).

Monreal-Pérez, Aragón-Sánchez, and Sánchez-Marín (2011:1) argue that productivity focuses on the measure of output in terms of innovation against inputs such as financial costs expended. Mathur et al. (2012:754–63); Grifell-Tatje and Lovell (2014:5–16); and Li and Zhao (2015:293–7) confirm that productivity is a yardstick through which the efficiency of the output-to-input performance of the country's economy or industry is measured against that of the other country's economy or industry; the previous period or set standards. As is emphasized by Andersen, Alston, and Pardey (2012:60–2); Gunasekaran and Spalanzani (2012:36–44); Oeij et al. (2012:94–102); and Karim and Arif-Uz-Zaman (2013:178–80), productivity is expressed as the ratio of output such as profit, products, sales, or market share into input resource factors such as capital, employee resources, machinery, material, technology, and energy used in manufacturing SMEs.

According to Chen and Zadrozny (2013:61–8), productivity involves the ratio of output profit compared to input such as capital, technology, or standards in manufacturing SMEs.

As is indicated by Hampf (2014:457) and Teng (2014:250), productivity involves the quantitative relationship between output in terms of profit and input such as costs in manufacturing SMEs. Ding, Guariglia, and Harris (2015:2–17) report that productivity in manufacturing SMEs is all about products generated in association with employees' abilities and skills, machinery, inventory, location, environment, financial costs, technology use, managerial abilities, government, and competition. Based on the literature sources discussed earlier in terms of the productivity concept, the following formula for productivity in manufacturing is presented using Klingner et al. (2015:240):

$$\text{Productivity}\left(y\right)=\frac{\text{Output}}{\text{Input}}$$

This formula is also used by Uddin (2015:241) and is addressed as follows:

$$\text{Productivity }\left(y\right)=\frac{\text{Actual results of the output}}{\text{Input resources consumed}}$$

Based on authors such as Klingner et al. and Udin, it can be said that one needs to measure productivity to ensure the progress (productivity improvement) of manufacturing SMEs. As indicated by other scholars from the literature studied in Section 2.2.1.2.1, productivity is measured looking at the output-to-input performance of the country, the current manufacturing industry, department, units, or sections comparing the same output-to-input performance with one of the other country, manufacturing industry, department, units or sections; with the previous period or set standards. Finally, the following two sections will provide an extensive background to productivity measurement and productivity improvement.

2.2.1.2.1 Productivity Measurement

Productivity measurement is presented differently depending on the author. Almström and Kinnander (2011:759–60) state that productivity measurement differs from one business to another reflecting the output in general and all input factors. In the case of motor industries, the products produced are measured in terms of productivity per employee working hours.

Phusavat et al. (2012:160–4) explain that productivity measurement in manufacturing SMEs is done based on the output attained, such as quality and profit, in relation to the nonfinancial and financial gain acquired.

Alolayyan, Ali, and Idris (2013:214) confirm that productivity among manufacturing SMEs measures products generated in relation to the environment. Hessels and Parker (2013:145) report that productivity is measured at the manufacturing SMEs' level in terms of both turnover and employment. Cagno et al. (2013:151) explain that productivity measurement is easily obtained from sales records, hours worked, or absenteeism records, which will exist in virtually all manufacturing SMEs. As reported by Kaur, Singh, and Ahuja (2013:69–71), manufacturing SMEs use productivity to measure machine downtime by scheduling through total preventative maintenance processes, inventory through economic order quantity, product specification to ensure quality in a product, and product time delivery to ensure customer satisfaction.

Belay et al. (2014:351–60) explain that productivity among manufacturing SMEs is measured focusing on profit, turnover, sales, and market share in relation to input resource factors such as employee ability, material, machine, location, layout, finance, technology, management, competition, and government. Similarly, Chauhan and Agrawal (2014:409) measure productivity relative to the quality of the product produced and the material used in manufacturing SMEs. Jourabchi et al. (2014:154) measure productivity in manufacturing SMEs as the value of the product in terms of cost and employee capital employed. Jain, Bhatti, and Singh (2014:294–8) state that productivity measures the availability of machine maintenance, production stoppage, downtime, reworks, environment, and delivery against capital costs incurred in terms of the number of damages, employee accidents, and late distribution incurred by customers. Sharma et al. (2015:385–7) contend that productivity is measured by visualizing the number of products yielded as compared to the material used to produce them.

Saunila, Pekkola, and Ukko (2014:238–49) use productivity to measure financial performance for return on investment and operations through networking and services rendered in manufacturing SMEs. As discovered by El Makrini (2015:130–39), productivity in manufacturing SMEs measures profitability and sales against costs. Furthermore, productivity also measures globalization such as export success. Bi et al. (2015:265) declare that productivity uses global demand (such as customers' better quality, personalized features, faster delivery time, and low cost) against technology that is modern and advanced. Whereas, on the other hand, productivity focuses on tools such as lean to measure quantifiable waste and maximize utilization of rates of resources. Finally, productivity measures profit made against costs incurred and products produced against material and machinery used.

Manufacturing SMEs use productivity to measure annual turnover (sales or income) through the utilization of employees and operational assets (Huang, Tan, & Ding, 2015:80). As explained by Al Serhan, Craig, and Ahmed (2015:295), productivity in manufacturing SMEs measures market share, growth, return on sales, and return on investment. Garza-Reyes, Ates, and Kumar (2015:1093) affirm that manufacturing SMEs use customer improvements in quality, satisfaction, and profitability to measure productivity. Prasad et al. (2015:227) uphold that financial measures in manufacturing SMEs are used in terms of delivery speed, cost, inventory turnover, and customer satisfaction. Shokri and Nabhani (2015:171–80) explain that manufacturing SMEs use productivity to measure skills; training; education; management; operation processes; and knowledge in their businesses.

Robinson, Sanders, and Mazharsolook (2015:234) report that productivity in manufacturing SMEs measures products and services compared to input resource factors such as employee resources, machinery, equipment, and the time used to carry out activities. The research will utilize multifactor productivity, sometimes referred as *total factor productivity* (TFP), for a better understanding of what really drives productivity.

O'Donnell (2012:255) explains that total productivity in manufacturing SMEs is a measure of the relationship between aggregate output and aggregate input. The formula used by O'Donnell (2012:255) and Bai and Sarkis (2014:279) for the total or multifactor productivity equation (MFP) is as follows:

$$= \frac{\text{Number of production output}}{\text{Human resources} + \text{material} + \text{overhead costs}}$$

From this equation, labor productivity can be developed as follows:

$$\text{Labor productivity} = \frac{\text{Output (manufactured products or units) or output per employee}}{\text{Employees spent hours}}$$

$$\text{or} \quad \text{Labor PI} = \frac{\text{Potential output per efficiency unit of workforce used at a time}^2}{\text{Output per efficiency unit of workforce used at a time}^1}$$

where PI is the productivity index, which is efficiency change, technology change, physical capital accumulation, and human capital accumulation (Gitto & Mancuso, 2015:3):

$$\text{Material productivity} = \frac{\text{Number of production output or output per material unit}}{\text{Material unit (input)}}$$

$$\text{or Material PI} = \frac{\text{Potential output per efficiency unit of material used at a time}^2}{\text{Potential output per efficiency unit of material used at a time}^1}$$

$$\text{Machinery productivity} = \frac{\text{Number of production output or output per material unit}}{\text{Machine hours (input)}}$$

or Machinery PI

$$= \frac{\text{Potential output per efficiency unit of machinery utilised at a time}^2}{\text{Potential output per efficiency unit of machinery utilised at a time}^1}$$

Capital / overhead productivity

$$= \frac{\text{Number of production output or output per capital costs}}{\text{Capital cost (input)}}$$

$$\text{or Capital PI} = \frac{\text{Potential output per efficiency unit of capital costs used at a time}^2}{\text{Potential output per efficiency unit of capital costs used at a time}^1}$$

As confirmed by Baporikar and Deshpande (2015:115–6), manufacturing SMEs depend on the effectiveness and efficiency of management in ensuring that employees' capabilities are engaged in the manufacturing process to achieve set goals for the businesses. Desai (2012:263) measures productivity in terms of the number of defects per million opportunities (DPMO), quality product, and competitiveness and market share. From the studied literature, manufacturing SMEs operate the manufacturing process efficiently, manufacturing quality products and exceeding customer expectations. Milana, Nascia, and Zeli (2013:103) emphasize manufacturing SMEs' productivity measurement in terms of output, turnover, volume, and return on investment. Medda and Piga (2014:419) assert that productivity in manufacturing SMEs is measured focusing on return on investment and research and development as compared to labor costs.

2.2.1.2.2 Productivity Improvement

In this section, the concept of productivity improvement is studied and discussed from various literatures to provide a detailed insight of what productivity improvement is all about. Chiang et al. (2012:181) and Sharma and Mishra (2012:351) refer to productivity improvement as a state whereby manufacturing

SMEs invest in employee training and research and development in order to contribute to improved manufacturing processes of the businesses.

Oeij et al. (2012:95–105) pose productivity improvement in manufacturing SMEs in different ways. Firstly, productivity improves when the output of tangible products and services increases at a constant input of employees, resources, material, and machinery. Secondly, this improvement occurs when the same outputs become constant with the decrease in the same input. Thirdly, improvement takes place when a similar output increases and inputs are reduced. Lastly, productivity improves when the same input decreases with the decrease in output being proportionally less. All these methods are exercised with the value of output and cost involved. These methods take place in manufacturing industries, one of which is manufacturing SMEs.

Mathur et al. (2012:754); Jagoda, Lonseth, and Lonseth (2013:389); and Hilmola et al. (2015:1008) describe productivity improvement as the efficient and effective utilization and management of input resource factors in manufacturing SMEs. Thus, when resources are efficiently utilized, the product cycle time reduces and the cost of manufacturing decreases, resulting in the optimization of productivity while maintaining a competitive edge. Abraham and Suganthi (2013:291) and Malik, Nasim, and Iqbal (2013:365) regard productivity improvement in manufacturing SMEs as the effective utilization of quality management systems to encourage responsibility through training to accomplish goals. Alolayyan et al. (2013:214) refers to productivity improvement as galvanizing safety and the elimination of injuries in the workplace and thus the improvement of productivity through safety standards.

Ali, Islam, and Howe (2013:409) assert that productivity in manufacturing can be improved; these activities primarily involve the simplification of production processes, chiefly through the elimination of waste. According to Krajewski, Ritzman, and Malhotra (2013), improving productivity means simplifying processes thorough the supply chain in order to accelerate material conversion.

According to Kumar, Singh, and Shankar (2013:90) and Lai and Chen (2013:2240), improved productivity entails creation of jobs, better salaries, improving standards of life, profitability, and reduction of costs in manufacturing SMEs. At the final stage, these achievements are realized through short delivery time, reduced lead time, just-in-time delivery, and shortened production cycle time in order to reach the agreed delivery date and to adhere to the cost limits. Based on the mathematical ratio of output and input defining productivity, improving productivity can only be achieved by improving the ratio, and productivity must be measured before any step is taken to improve it (Loosemore, 2014).

Belay et al. (2014:351–60) and Hong, Marvel, and Modi (2015:136–9) refer to productivity improvement as a means of improving employee knowledge through training, improved flow of material, machine maintenance, safety and easy access to tools and materials, and cost saving. Productivity improvement results from the application of business process reengineering, focusing on competitive priorities such as quality, cost, and delivery for customer satisfaction. Jain et al. (2014:293–306) substantiates that the effective utilization of skills is instrumental to productivity improvement in manufacturing SMEs. The formula for productivity improvement in manufacturing SMEs is presented by Gitto and Mancuso (2015:3) as follows:

$$\text{Productivity improvement} = \frac{\text{Present productivity results} - \text{previous productivity results}}{\text{Previous productivity results}} \times 100$$

According to Hamilton, Nickerson, and Owan (2015:101–4); Kafetzopoulos and Psomas (2015); and El-Khalil (2015:37–54), improving productivity in manufacturing is attained through new technology, employee training, process variation elimination, and elimination of waste. Isaga, Masurel, and Van Montfort (2015:6) and Parida et al. (2015:3) state that improved productivity in manufacturing SMEs is achieved through increased market share, better wages, and bonuses. Prasad et al. (2015:275) and Singh and Singh (2015:80) assert that productivity improvement entails the appropriate utilization of resources through skills development and focusing on competitive priorities in manufacturing SMEs.

Based on the literature studied, productivity improvement revolves around waste reduction and elimination, reducing defects, improving competitiveness, and increasing turnover, market share, and customer satisfaction. There is a need to formalize lean, Six Sigma, and lean Six Sigma techniques available to assist manufacturing SMEs in improving productivity. For the purpose of this book, examples of productivity will be based on specific resource productivity, focusing on manpower, material, machinery, and capital, whereas multifactor (total) factor productivity will cover both the total value of products produced and the total costs of all input resources. Productivity applications is the next section to be discussed, which is in the form of calculating the productivity of manufacturing SMEs. Case studies of Companies A, B, C, and D are used as practical examples. Productivity applications are illustrated in the next section in the form of companies in Gauteng, South Africa.

2.2.1.3 Productivity Applications

As indicated by Gitto and Mancuso (2015:3), an example of productivity with regard to output-to-input ratio is provided below, using formulas from Gitto and Mancuso.

Let's assume that in March 2016, Company A measured units in monetary value (Rands). Suppose the following:

- Turnover of the enterprise (output) was R1,200,000 and the capital (input) used to achieve this output was R700,000.

During the same month in 2016, Company C measured units in monetary value (Rands).

- The turnover is currently R1,400,000, and the capital (input) used to achieve this output is R700,000.

This increase in turnover is not due to output in value (price) increases.

The difference in the productivity of this company can be determined as follows:

One needs to remember that a factor for Company B is not done in isolation. Thus, both factors need to be compared to ensure a change in productivity.

Company C: March 2016	*Company A: March 2016*
$\text{Productivity} = \dfrac{\text{Output}}{\text{Input}}$	$\text{Productivity} = \dfrac{\text{Output}}{\text{Input}}$
$\text{Productivity} = \dfrac{\text{R1,200,000}}{\text{R700,000}}$	$\text{Productivity} = \dfrac{\text{R1,400,000}}{\text{R700,000}}$
=1.71 value of Rand per Rand costs	=2.00 value of Rand per Rand costs

When Company A measures the productivity of 2016 against that of Company C in 2016, the change in productivity is as follows:

$$\text{The productivity change} = \frac{2.00}{1.71} \times 100$$

$$= 116.96\%$$

If the percentage is greater than 100 in this instance, one should minus 100 from that greater number (in this case, 116.96%), since 100 is the standard percentage in South Africa.

For example:

116.9 – 100 = 16.96% increase in productivity. Therefore, the productivity of Company A has increased by 16% against that of Company C.

Company D would like to compare its productivity for 2016 with the productivity of Company B for 2016. The increase or decrease in productivity will determine whether this company will expand its existing business.

The following information is applicable to what Company B produces in a month for 2016:

OUTPUTS

Products manufactured	860 units manufactured @ R3,000 per unit
Inventory resources input	250 inventory units @ R200 per unit
Machine input	160 hours @ R880 per hour
Employee resources input	160 hours @ R100 per hour
Energy resources input	
Fixed costs	
Electricity, water and lights, and rent	Four weeks @ R3,750 whereby each month Company B spent R15,000 per month
Administrative Costs	
Equipment	10 components purchased @ R208.33 each whereby each component is purchased at R4,999.92 for two years
Stationery	10 items @ R600

The following information is applicable to what Company D produces in a month for 2016 compared with Company B.

OUTPUTS

Products manufactured	960 units manufactured @ R3,200 per unit
Inventory resources input	300 inventory units @ R200 per unit
Machine input	160 hours @ R940 per hour
Employee resources input	160 hours @ R120 per hour
Energy resources input	
Fixed costs	
Electricity, water and lights, and rent	Four weeks @ R4,000 whereby each month Company B spent R16,000 per month
Administrative Costs	
Equipment	10 components purchased @ R250 whereby each component is purchased at R6,000 for two years
Stationery	10 items @ R750

Company B versus Company D is required to determine the following:

1. The specific resource productivity (SRP) in Rands for each resource and the productivity index (PI) for the current year's results for Company B.
2. The total resource productivity (TRP) and PI for the current year's results for Company D.
3. Recommendations as to how work study experts will assist management in Company D to improve productivity after comparing the results of the TFP and PI with those of Company B.

The results for Company D are as follows:

1. SRP in Rands for each resource and the PI for Company D's current year's results.
2. The TRP and PI for the current year's results for Company D.

SRP	*Company B: 2016*	*Company D: 2016*	*PI*
Inventory	$=\frac{860\times R3{,}000}{250\times R200}$ $=\frac{R2{,}580{,}000}{R50{,}000}$ =51.60 Rand value per Rand costs	$=\frac{R960\times R3{,}200}{300\times R200}$ $=\frac{R3{,}072{,}000}{R60{,}000}$ =51.20 Rand value per Rand costs	$=\frac{51.20}{51.60}\times 100$ = 99.23% When the percentage results of the figure are less than 100%, it shows that productivity has decreased but profit has increased. Thus you subtract the figure obtained from 100, in this case 99.23%: = 100 − 99.23 = 0.77% decrease in productivity.
Machinery	$=\frac{860\times R3{,}000}{160\times R880}$ $=\frac{R2{,}580{,}000}{R140{,}800}$ =18.32 Rand value per Rand costs	$=\frac{R960\times R3{,}200}{160\times R940}$ $=\frac{R3{,}072{,}000}{R150{,}400}$ =17.15 Rand value per Rand costs	$=\frac{17.15}{18.32}\times 100$ = 93.61% When the percentage results of the figure are less than 100%, it shows that productivity is declining. Thus you subtract that figure, in this case 93.61%, from 100: =100−93.61=6.39% decrease in productivity.
Employee resources	$=\frac{860\times R3{,}000}{160\times R100}$ $=\frac{R2{,}580{,}000}{R16{,}000}$ = 161.25 Rand value per Rand costs	$=\frac{R960\times R3{,}200}{160\times R120}$ $=\frac{R3{,}072{,}000}{R25{,}600}$ =120.00 Rand value per Rand costs	$=\frac{120}{161}\times 100$ =74.53% =100 − 74.53 =25.47% decrease in productivity
Energy: Electricity, water and light, and rent	$=\frac{860\times R3{,}000}{160\times R93.75}$ $=\frac{R2{,}580{,}000}{R15{,}000}$ =172.00 Rand value per Rand costs	$=\frac{R960\times R3{,}200}{160\times R120}$ $=\frac{R3{,}072{,}000}{R16{,}000}$ =192.00 Rand value per Rand costs	$=\frac{192}{172}\times 100$ When the percentage results of the figure are greater than 100%, it shows that productivity is increasing. Thus you subtract that 100 from that figure, in this case: =111.63−100=11.63 increase in productivity.
Administrative costs	$=\frac{860\times R3{,}000}{24\text{ months}\times R208.33\times 10\text{ components}+10\times R600}$ $=\frac{R2{,}580{,}000}{R49{,}999.92+6{,}000}$ $=\frac{R2{,}580{,}000}{R55{,}999.92}$ =46.07 Rand value per Rand costs	$=\frac{R960\times R3{,}200}{24\text{ months}\times R250\times 10\text{ components}+10\times R750}$ $=\frac{R2{,}580{,}000}{R60{,}000+7{,}500}$ $=\frac{R3{,}072{,}000}{R67{,}500}$ =45.51 Rand value per Rand costs	$=\frac{45.51}{46.07}\times 100$ =98.79% =100 − 98.79 =1.21% decrease in productivity

The other way of calculating capital productivity is as follows:

TRP	*Company B: 2016*	*Company D: 2016*	*PI*
$TRP = \frac{\text{Total output}}{\text{Total input}}$ $TRP = \frac{\text{Total value}}{\text{Total costs}}$	$= \frac{860 \times R3{,}000}{(250 \times 200) + (120 \times 880) + (120 \times 100)}$ $= \frac{R2{,}580{,}000}{R167{,}600}$ =15.39 Rand value per Rand costs	$= \frac{R960 \times R3{,}200}{(250 \times 200) + (120 \times 880) + (120 \times 100)}$ $= \frac{R3{,}072{,}000}{R185{,}700}$ =16.54 Rand value per Rand costs	$= \frac{16.54}{15.39} \times 100$ =107.47% =107.47 – 100 =7.47% increase in productivity

Let's say manufacturing SMEs had a turnover of R2,580,000 and used capital of R167,600 in 2015. Under comparable situations, manufacturing SMEs increased turnover to R3,072,000 with capital of R185,700 in 2016. The PI of manufacturing SMEs can be calculated as follows:

Capital Productivity	*Company B: 2016*	*Company D: 2016*	*PI*
$= \frac{\text{Total output}}{\text{Total input}}$ $= \frac{\text{Total value}}{\text{Total costs}}$	$= \frac{R2{,}580{,}000}{R50{,}000}$ 140,800 16,000 15,000 55,999.92 $= \frac{R2{,}580{,}000}{R277{,}799.92}$ =9.29 Rand value per Rand costs	$= \frac{R2{,}580{,}000}{R60{,}000}$ 150,400 25,600 16,000 67,500 $= \frac{R3{,}072{,}000}{R319{,}500}$ =9.62 Rand value per Rand costs	$= \frac{9.62}{9.29} \times 100$ = 103.55% = 103.55 – 100 =3.55% increase in productivity of Company D

3. The manner in which work study experts will assist management in manufacturing SMEs to improve the productivity of their businesses after comparing the results of multifactor productivity and the PI of Company D in 2016 with those of Company B is addressed next.

Following the quality and quantity, as noted by Oeij et al. (2012:98), when making recommendations, be it SRP or multifactor productivity, productivity may improve in several ways:

- Output increase at constant input
- Input decrease at constant output
- Output increase and input decrease

- Output and input increase, with the increase in input proportionally less
- Input and output decrease, with the decrease in output proportionally less

The next section to be discussed is the work study theory used in manufacturing SMEs.

2.2.2 Work Study Theory

This section addresses the groundwork theories of productivity and work study. In terms of work study, the historical background of work study is discussed and its paradigm is presented from its origin to the current situation.

2.2.2.1 Historical Origin of Work Study

According to van Niekerk (1986:38), work study has been in use since as far back as 3500 B.C. through the ancient Egyptian civilization by building pyramids whereby religious order were given to people for control purposes based on the religious foundation of that time. Now, method study and work measurement started to prevail through the activity of business leaders with the intention of improving operations in their organizations. In the mid-1770s, a Frenchman, Jean R. Peronet, did measurement on time study to show how manufacturing is measured using cycle time. When work measurement was introduced in 1790, industry leaders in Britain such as Boulton and Watt checked how long it took for employees to do various work activities and how the completed activities were in line with the wages that had to be paid to the employees involved. Before pioneers such as Taylor, Gilbreth, and Bedeaux, Robert Owen conducted research on various operations with an emphasis on working conditions. The problem areas were resolved by means of method of study to ensure efficient methods of working that would reduce employee fatigue (van Niekerk, 1986:38).

According to Dos Santos, Powell, and Sarshar (2002:788–9), Frederick Winslow Taylor introduced work study with the help of various pioneers such as Gilbreth and Hawthorne. Due to acknowledgement by other pioneers of his popularity in the title "principles of scientific management," Taylor was regarded as the father of scientific management. Around 1911, Taylor's studies focused on the efficient use of time in the workplace, whereby he presented how various tasks should be accomplished. According to Vaszkun and Tsutsui (2012:364–8), the following tasks as required by Taylor need to be accomplished:

- Breaking the job into work activities to measure standard time
- Deciding on the appropriate work for the activity
- Encouraging a spirit of teamwork between management and employees
- Assigning responsibilities to management and employees, whereby management is responsible for planning and employees are responsible for carrying out work activities

As is indicated by van Niekerk (1986:38), Charles Bedeaux also contributed in performance ratings, whereby observed time could be converted into basic time. Bedeaux was then responsible for developing a time study method. An example of a rating scale is provided in Figure 2.1.

According to Hendry, Huang, and Stevenson (2013:80–1), studies were done by managers and university researchers at the Hawthorne works in the United States to confirm Taylor's principles that more lighting in the workplace would result in improved productivity among manufacturing industries, including SMEs. Gilbreth then contributed by analyzing the motions of operation, which was improved from 18 to five without using the effort to increase productivity (Dos Santos et al., 2002:789).

The literature studied in terms of Taylor's inputs with regard to efficiency, work measurement, delimitations, and management of tasks and piecework provided a constructive contribution to the productivity progress of manufacturing SMEs. However, Taylor's inputs were compared with those of Gilbreth and other theorists in management such as Cook, Gantt Fayol (Taneja, Pryor, & Leslie, 2011:2016), Abraham Maslow, and Douglas McGregor (Tanvir & Ahmed, 2013:52).

Based on the historical background in assisting management to improve ways of doing jobs in the workplace, theories such as Abraham Maslow's Hierarchy Theory; Douglas McGregor's X and Y Theory; Motivation Hygiene Theory; ERG (Existence, Recognition, Growth) Theory; Three Needs Theory; Goal Setting Theory; Expectancy Theory; Reinforcement Theory and Equity Theory assisted management to improve the productivity of industries in particular manufacturing (Tanvir & Ahmed, 2013:52). Thus this book will briefly describe groundwork theories to show how productivity adds value in enabling manufacturing SMEs to improve productivity in their businesses. Conyers (1977a:10–11) saw the importance of work study due to lack of communication between management and workers, which caused operational problems in the workplace. After the incidence in terms of operation in the workplace, work study courses were introduced to both management and workers so that they could acquire skills and knowledge and increase capability, which

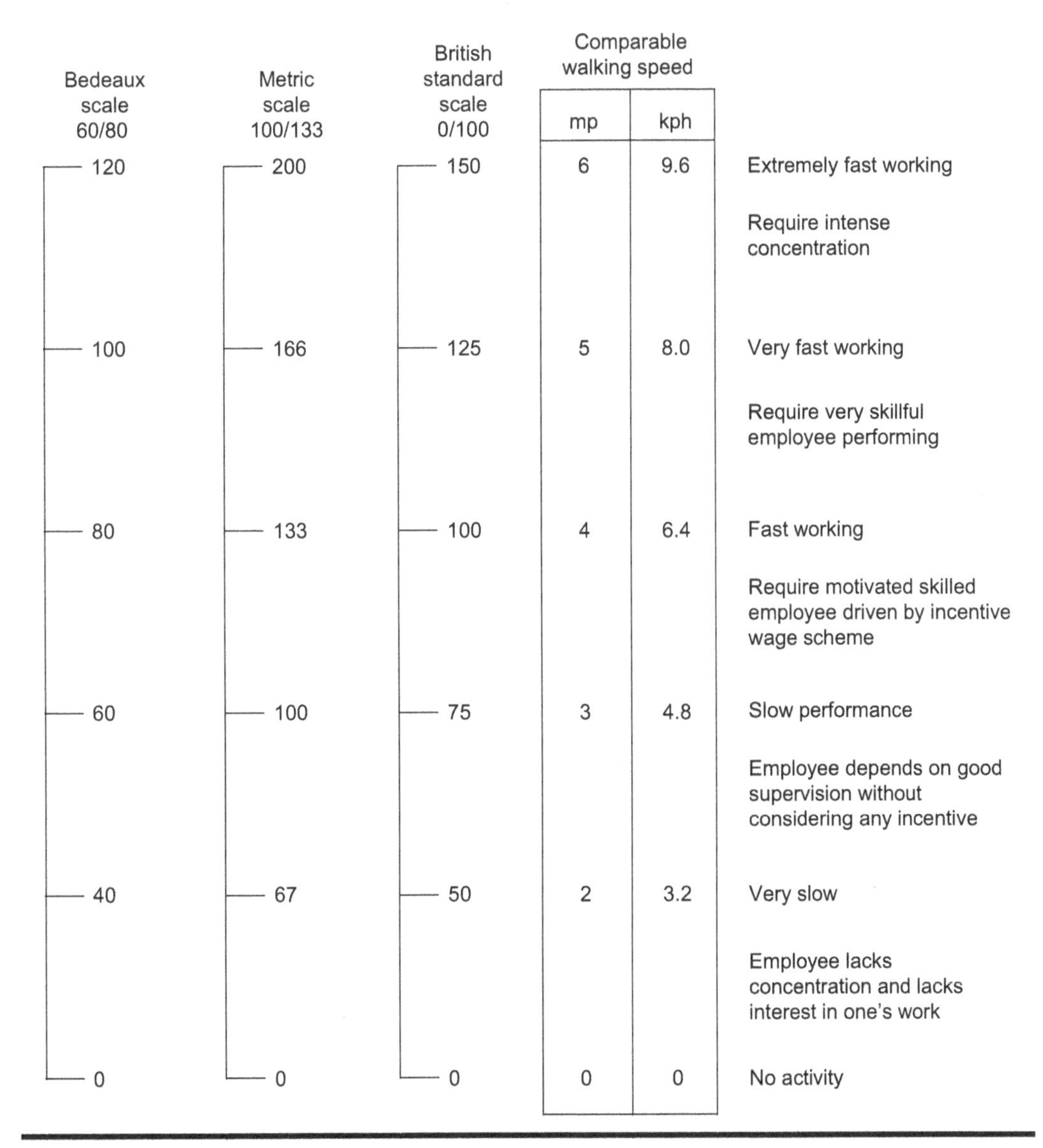

Figure 2.1 Rating at different levels of performance. (From van Niekerk, W.P., *Productivity and Work Study*. Durban: Butterworths Publishers, 1986.)

contributed to productivity improvement. The theory of work study is presented by various literature sources with the aim of providing a detailed background of what work study is.

Based on the literature studied, the original measure to be implemented in manufacturing was the introduction of work study principles in the apparel industries of Bangladesh. However, during the early nineteenth century, work study was regarded as the only methodology to stipulate a time frame to complete each task (also referred to as *work measurement*). Furthermore, work study also identified the precise method of doing the task (which is method study) (Tanvir & Ahmed, 2013:51). Work study was developed in 1911 by the well-known American scientist F.W. Taylor, who was generally known as the father of modem management science (Green, 1968:15; Conyers, 1977b:20; Barnes, 1980; Sandelands, 1994:12; Dos Santos et al., 2002:789; Towill, 2009:437; Vaszkun & Tsutsui, 2012:374; Tanvir & Ahmed, 2013:51). Therefore work study is described next by different authors to provide an understanding of work study in the workplace in manufacturing SMEs.

2.2.2.2 Work Study Concept

Work study is a powerful tool made up for management in manufacturing for the purpose of improving productivity. This means that in manufacturing industries, work study as a tool assists management and employees to operate efficiently and effectively with proper measure and experience (Garbett, 1958:13).

Green (1967:22–34; 1968:22–34) explain that work study in manufacturing focuses on methods of working such as identifying untidiness, bad lighting, dangerous areas, and measuring work to resolve problems in productivity improvement. Conyers (1977b:20–3) considers work study as an industrial tool to improve employment possibilities, encourage investment, add value, and remove bottlenecks with the purpose of improving productivity. Conyers (1987:8–44) states that work study in manufacturing involves method study (which involves the organised recording and diagnostic of factors and input quantity involved in an current or planned operation, in order to develop an easier and efficient method as well as minimising costs) and work measurement (which involves measuring and setting of standard time for the job) to improve productivity. Conyers (1988:43–46) considers work study as a method of planning inventory, managing safety, and preparing time for delivery and methods of payment to improve productivity in manufacturing.

Work study is "the systematic examination of the methods of carrying on activities so as to improve the effective use of resources and to set up standards of performance for the activities being carried out" (Kanawaty, 1992:92). Sandelands (1994:2–16) perceives work study as involving a better design of methods and measurements of work to improve productivity in manufacturing SMEs. According to Carter (1999:46–7), work study focuses on a bottom-up approach for management and workers to be part of the operational plan with the aim of measuring and improving productivity.

As explained by Grünberg (2003:89), work study in manufacturing includes method of operations improvement with the utilization of lean techniques such as total preventive maintenance (TPM); total quality management (TQM); Toyota production system (TPS); kaizen; 5S; theory of constrain (TOC); and benchmarking, as well as lean techniques involving work measurement tools such TQM for statistical control; business process reengineering with process mapping; and simulation methods to improve productivity in manufacturing to reduce waste and add value, which represents an improvement in productivity. Towill (2009:417–431) defines work study as encompassing a method of study involving the mapping of activities to determine the best way of carrying out a task; and work measurement as the standardizing of employee and machinery performance in enhancing productivity.

Finneran and O'Sullivan (2010:324–6) state that work study focuses on an ergonomic approach to identify and avoid risk factors such as impacts on physical health and injuries by using an economic and simple method for the task as well as investigating cycle times to determine an acceptable level strength of employees in repetitive work in order to improve productivity in manufacturing SMEs.

According to Ramanigopal, Hemalatha, and Murugan (2011:285), work study involves mapping and streamlining processes used in manufacturing to avoid bottlenecks, taking into consideration cycle time in order to improve productivity. Vaszkun and Tsutsui (2012:373–78) comment that work study is about systematic planning and critical examination in order to develop economic and easy methods of doing work as well as preparing standard times for jobs in order to increase productivity. Work study in manufacturing SMEs involves assisting management through the application of work measurements to set up standard times for employees to carry out jobs in the workplace at a particular standard of performance. Work measurement is also referred to as *quantitative technique* (Srinivasan et al., 2016:369; Trianni, Cagno, & Farné, 2016: 1539–51) to complete a task and establish the best method for carrying out the task in the simplest and most economic manner (which is method study) or qualitative technique (Srinivasan et al., 2016:369; Trianni et al., & Farné, 2016:1539–51) to improve productivity (Tanvir & Ahmed, 2013:51; Trianni et al., 2016:1539–51).

Muruganantham, Krishnan, and Arun (2014:455); Bechar and Eben-Chaime (2014:197–201); and Odesola, Okolie, and Nnametu (2015:2–3) assert that work study can be divided into method study and work measurement for improvement of productivity in manufacturing SMEs. With method study, the work systems are dealt with systematically with the aim of determining and using economic methods to simplify work activities for the employee.

In terms of work measurement, the standard time is determined through the application of time study using a stopwatch to ensure that a qualified worker is working at a normal pace, producing the required results for manufacturing SMEs. Based on the literature studied, in terms of theories and calculations, productivity is confused with utilization, effectiveness, and efficiency, whereas these concepts are not similar. In contrast, productivity measures output relative to a specific input. The other way of addressing productivity in simple terms is the sum of effectiveness and efficiency in manufacturing SMEs (O'Neill, Sohal, & Teng, 2016:386–9; Reverte et al., 2016:2871).

On the contrary, utilization is the extent to which resources are used in the manufacturing process in manufacturing SMEs. Effectiveness is how the output is reached against the target (Teng, 2014:251). Efficiency measures how well something is performing relative to existing standards. Efficiency and effectiveness are also distinct from each other in the sense that efficiency is measured in terms of maintaining a satisfactory relationship between cost and benefit, whereas effectiveness is measured in terms of doing the right things (Dora, Kumar, & Gellynck, 2016:13–6; Reverte et al., 2016:2871–80). So, for manufacturing SMEs to improve their productivity level, these SMEs need to be effective and efficient in their businesses (Schulze et al., 2016:3700); and, in order for manufacturing SMEs to be effective and efficient in their businesses, work study is the key for manufacturing SMEs to create their own work opportunities. In this case, manufacturing SMEs need to benchmark with other businesses, be they large or small, for them to grow their businesses. The key to improving the productivity of manufacturing SMEs is that if their performance cannot be measured, these SMEs cannot be managed. If they cannot be managed, the productivity of their businesses will never improve. This is referred to as thinking *convergently*, which is thinking within the bubble, and ultimately the business will fail. Papa, Cavaleiro de Ferreira, and Radulj (2015:878–9) emphasize the importance of measurement in manufacturing industries, be they big or small. These authors use the motto of Peter Drucker: "If you don't measure what you are doing, you can't control it, if you can't control it, you can't manage it, if you can't manage it, you can't improve." On the contrary, when manufacturing SMEs are learning from other successful SMEs measuring their businesses in all aspects of the manufacturing processes, they are able to realize the opportunity to use alternatives for improving these processes, resulting in productivity improvement. This process is a sign of *divergent* thinking, which is thinking outside the bubble.

Based on the literature reviewed and discussed, when businesses are in a stable environment, these businesses tend to relax without considering the turbulent environment that may affect them and put these businesses out of operation. When faced with such a situation, these businesses give up without looking at the opportunities and end up being inefficient and ineffective. These businesses have a convergent type of thinking, failing to think outside the box. On the contrary, businesses need to think divergently, whereby they generate ideas by providing different answers for any given problem in order for them to continue to improve in productivity; this is thinking outside the box (Daly, Mosyjowski, & Seifert, 2014:418–23). An example is provided, with various stages of the experiment, to show how the manufacturing SME that is innovative becomes more efficient and effective in improving productivity than the one that is not.

2.2.2.3 *Work Study Applications*

Various stages of a scenario of an experiment involving fish are used to provide a detailed understanding of a failing and a successful manufacturing SME. The diagram of Stages A, B, and C is depicted in Figure 2.2.

2.2.2.3.1 First Stage (A): First Experiment: Glass Cylinder with Food and Fish in It

An example of such a situation is shown in the first experiment, whereby a fish is placed in a transparent glass cylinder. This cylinder is filled with food for the fish to eat. This glass is depicted as *Cylinder A* in Figure 2.2.

2.2.2.3.2 Second Stage (Cylinder B): Second Experiment: Glass Cylinder with Fish in It Placed in Another Cylinder with Food Only in It

An example of such as situation is shown in the second experiment, whereby the fish in the cylinder has finished the food. A fish in an identical cylinder without food is inserted in another glass cylinder with food in it. This fish attempts to get food, but cannot reach it. The reason for this is that this food is placed inside another glass receptacle, which is indicated as *Cylinder B* in Figure 2.2. This fish keeps on striking the glass to reach the food and eventually gets to the stage of giving up without thinking of any possible way to reach the food and avoid death. Due to the fact that the fish was using

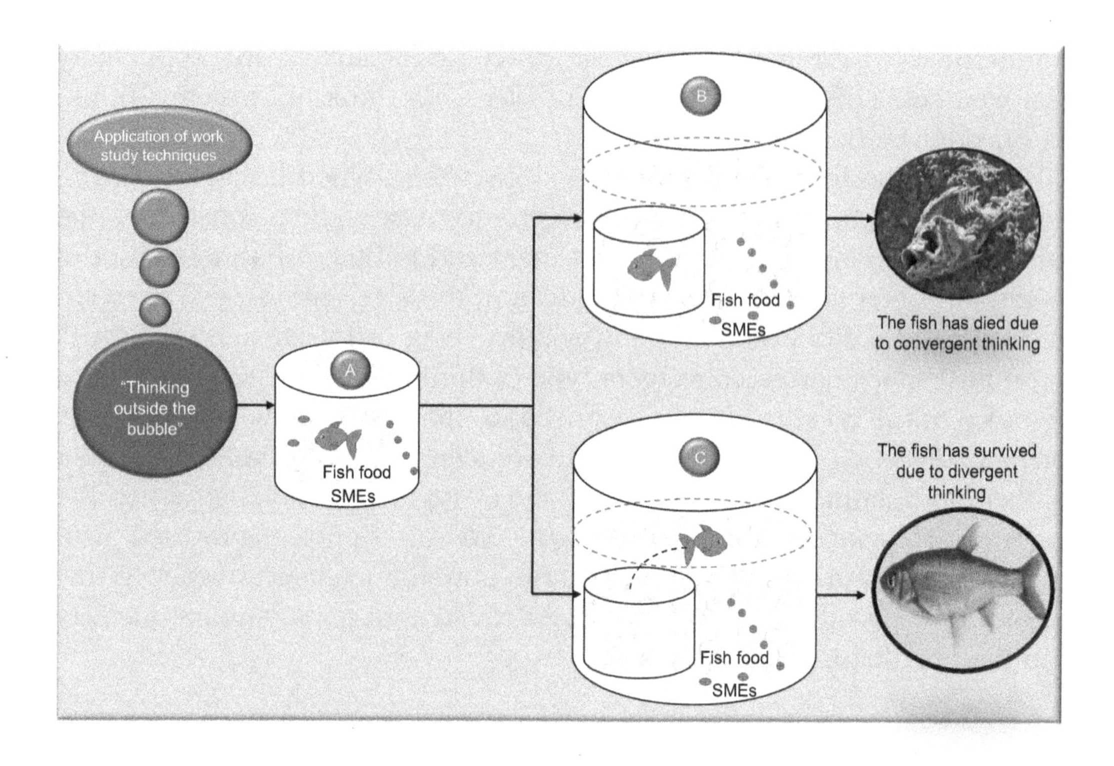

Figure 2.2 First experiment: Glass cylinder with food and fish in it.

convergent thinking, it kept giving up, thinking that the food was not attainable. The fish ends up dying.

2.2.2.3.3 Third Stage: Third Experiment: Glass Cylinder with Fish in It Placed in Another Cylinder with Food Only in It

In the third stage of the experiment, the fish uses an alternative way to reach the food, which is divergent thinking. This type of fish will survive in a turbulent environment due to its innovativeness in generating ideas to jump from the initial glass cylinder to the glass with the food in it in order to survive. This fish demonstrates efficiency and effectiveness in achieving its results. So, this type of innovativeness is feasible for manufacturing SMEs that are proactive. These SMEs tend to be flexible in adapting to the changing environment, aiming to use various alternative ideas in order to attain growth in terms of productivity. For manufacturing SMEs to be effective and efficient in their businesses, which involves focusing on productivity, these SMEs are ultimately attaining a high level of achievement and achieving a competitive edge in the market nationally and worldwide. This experiment is indicated as *Cylinder C* in Figure 2.2. So, given the low productivity in manufacturing SMEs as indicated earlier by research on the results provided by various government agencies in Chapter 1, the real challenge is to think "outside the bubble." In order to use this type of thinking, these manufacturing SMEs need to consider using various alternative tools exercised in work study into physical capital, technological capital, and management to ensure productivity progress in manufacturing SMEs worldwide. Therefore, a critical review of work study and productivity will be addressed in the subsequent section.

2.3 Critical Review of Work Study and Productivity Theory

This research book interprets the manner in which work study and productivity are defined and their historical background. The research done indicates that the definition of work study, as unpacked by the various authors, is similarly based on the improvement of work methods as well as the setting of an appropriate standard time for work activities being carried out by qualified workers with the aim of ensuring that the productivity of manufacturing SMEs improves. On the other hand, the concept of productivity involves the ratio of output to input as indicated by the literature sources from various academic scholars. The only difference is that the manner in which these scholars address productivity will depend on the type of company involved, since SMEs are categorized variously by the National Small Business Act, chapter 6 of 2003.

From the case studies conducted in March 2016, the results found in Company A and Company C show that the productivity of Company A has increased by 16% against Company C. The reason for the increase in productivity in Company A is that this company considered productivity improvement factors such the need for the human element which encourages compliance with standards in place such ISO 9001 even though ISO 14001 and 18001 are not yet in existence. Unlike Company C, which confuses an increase in production with an increase in productivity, this company increased its production but did not necessarily improve its productivity. Problem areas such as exposure of employees to the risk to harm from machinery in the form of dust and noise are not realized by Company C, and as a result, the productivity of the business is compromised. Such incidences result in a high rate of absenteeism and high costs for the company. This indicates that Company C does not have standards and appropriate work procedures in place for the improved well-being of its employees. As a result, the performance of employees drops, resulting in a decline in productivity.

Similar to Company B and Company D, the results also indicate that Company B has increased its productivity as compared to Company D. The reason for the progress of Company B in its productivity performance is that there are quality standards in place to ensure that there is sufficient control of inventory even through the company experiences quality, logistics, and supply chain challenges in the business. So, for companies to continue to improve in the productivity of their businesses, areas of work study such as standards need in-depth attention to ensure that employees follow the appropriate standard procedures and comply with standards. It is not only standards that will positively contribute to the productivity of manufacturing SMEs; other work study techniques are also import to focus on, and these are discussed in Chapters 6 and 7.

2.4 Summary

This chapter focused on the grounded theories of work study and productivity. Work study and productivity are the key concepts used in manufacturing SMEs in the twenty-first century. Work study presented in brief method study and work measurement followed by productivity theory consisted of productivity measurement and productivity improvement.

Finally, productivity applications are also provided. Since manufacturing SMEs are faced with low productivity, making it difficult for them to become part of the mainstream in developing the South African economy, more effort is needed on the part of manufacturing SMEs to improve their productivity levels. Physical capital, technological capital, and management are presently becoming global and economic problems, which lower the productivity of manufacturing SMEs in South Africa and the rest of the world.

As stated earlier, problems encountered by manufacturing SMEs are numerous and can be described among others as being operational, environmental, and competitive in nature.

A literature review focusing on the application of work study was undertaken among manufacturing SMEs abroad in improving productivity in their businesses, and the findings of these studies were favorable, which demonstrates the importance of work study in manufacturing SMEs operating in South Africa. The primary means of achieving this is to encourage the use of work study in physical capital, technological capital, and management to improve the productivity of manufacturing SMEs. The calculations based on productivity in manufacturing SMEs are also provided. The next issue to address is effectiveness versus efficiency in manufacturing SMEs, which is discussed in Chapter 3.

References

Abraham, P. and Suganthi, L. 2013. Intelligent quality management system using analytic hierarchy process and fuzzy association rules for manufacturing sector. *International Journal of Productivity and Quality Management*, 12(3): 287–311.

Ahmad, K.I., Shivastav, R.L., Pervez, S., and Khan, N.P. 2014. Analyzing quality and productivity improvement in steel rolling industry in central India. *IOSR Journal of Mechanical and Civil Engineering*, 6–11.

Ali, A.J., Islam, A., and Howe, L.P. 2013. A study of sustainability of continuous improvement in the manufacturing industries in Malaysia: Organizational self-assessment as a mediator. *Management of Environmental Quality: An International Journal*, 24(3): 408–426.

Almström, P. and Kinnander, A. 2011. The productivity potential assessment method. *International Journal of Productivity and Performance Management*, 60(7): 758–770.

Alolayyan, M.N., Ali, K.A.M., and Idris, F. 2013. Total quality management and operational flexibility impact on hospitals' performance: A structural modelling approach. *International Journal of Productivity and Quality Management*, 11(2): 212–227.

Al Serhan, Y.N. Craig, C., and Ahmed, J.Z. 2015. Time-based competence and performance: An empirical analysis. *Journal of Small Business and Enterprise Development*, 22(2): 288–301.

Andersen, M.A., Alston, J.M., and Pardey, P.G. 2012. Capital use intensity and productivity biases. *Journal of Productivity Analysis*, 37: 59–71.

Bai, C. and Sarkis, J. 2014. Determining and applying sustainable supplier key performance indicators. *Supply Chain Management: An International Journal*, 19(3): 275–291.

Bamber, C.J., Sharp, J.M., and Hides, M.T. 1999. Factors affecting successful implementation of total productive maintenance. *Journal of Quality in Maintenance Engineering*, 5(3): 162–181.

Baporikar, N. and Deshpande, M.V. 2015. Approaches and strategies of Pune auto component SMEs for excellence. *Journal of Science and Technology Policy Management*, 6(2): 114–126.

Barnes, R.M. 1980. *Motion and Time Study*. New York: Wiley.

Basu, S. and Fernald, J.G. 2002. Aggregate productivity and aggregate technology. *European Economic Review*, 46: 963–991.

Bechar, A. and Eben-Chaime, M. 2014. Hand-held computers to increase accuracy and productivity in agricultural work study. *International Journal of Productivity and Performance Management*, 63(2): 194–208.

Belay, A.M., Helo, P., Takala, J., and Welo, T. 2014. Considering BPR and CE for faster product delivery: A case study in manufacturing firms. *International Journal of Productivity and Quality Management*, 13(3): 349–375.

Bi, Z.M., Liu, Y., Baumgartner, B., Culver, E., Sorokin, J.N., Peters, A., Cox, B., Hunnicutt, J., Yurek, J., and O'Shaughnessey, S. 2015. Reusing industrial robots to achieve sustainability in small and medium-sized enterprises (SMEs). *Industrial Robot: An International Journal*, 42(3): 264–273.

Cagno, E., Michelli, G.J.L., Maisi, D., and Jacinto, C. 2013. Economic evaluation of OSH and its way to SMEs: A constructive review. *Safety Science*, 53: 134–152.

Carter, H. 1999. Strategic planning reborn. *Work Study*, 48(2): 46–48.

Chatzimichael, K. and Tzouvelekas, V. 2014. Human capital contributions to explain productivity differences. *Journal Productivity Analysis*, 41: 399–417.

Chauhan, P.S. and Agrawal, C.M. 2014. Identification of manufacturing defects leading to rejection during manufacturing of stabiliser bar. *International Journal of Productivity and Quality Management*, 14(4): 408–422.

Chen, B. and Zadrozny, P.A. 2013. Further model-based estimates of US total manufacturing production capital and technology, 1949–2005. *Journal of Productivity Analysis*, 39: 61–73.

Chiang, Y.H., Li, J., Choi, T.N.Y., and Man, K.F. 2012. Comparing China mainland and China Hong Kong contractors' productive efficiency. *Journal of Facilities Management*, 10(3): 179–197.

Coka, B. 2014. *Annual Report 2013–2014*. Midrand, Republic of South Africa: Productivity SA.

Conyers, F. 1977a. Work study: A journal which serves to promote productivity through time and motion study, job evaluation, process control and related subjects. *Work Study*, 26(4): 3–50.

Conyers, F. 1977b. Work study: A journal which serves to promote productivity through time and motion study, job evaluation, process control and related subjects. *Work Study*, 26(9): 3–46.

Conyers, F. 1987. Work study: A journal which serves to promote productivity through time and motion study, job evaluation, process control and related subjects. *Work Study*, 36(7): 3–53.

Conyers, F. 1988. Work study: A journal which serves to promote productivity through time and motion study, job evaluation, process control and related subjects. *Work Study*, 37(2): 3–55.

Daly, S.R., Mosyjowski, E.A., and Seifert, C.M. 2014. Teaching creativity in engineering courses. *Journal of Engineering Education*, 103(3): 417–449.

Desai, D.A. 2012. Quality and productivity improvement through Six Sigma in foundry industry. *International Journal of Productivity and Quality Management*, 9(2): 258–280.

Ding, S., Guariglia, A., and Harris, R. 2015. The determinants of productivity in Chinese large and medium-sized industrial firms, 1998–2007. *Journal of Productivity Analysis*, 45(2): 131–155.

Dora, M., Kumar, M., and Gellynck, X. 2016. Determinants and barriers to lean implementation in food-processing SMEs-a multiple case analysis. *Production Planning and Control*, 27(1): 1–23.

Dos Santos, A., Powell J.A., and Sarshar, M. 2002. Evolution of management theory: The case of production management in construction. *Management Decision*, 40(8): 788–796.

El-Khalil, R. 2015. Simulation analysis for managing and improving productivity. *Journal of Manufacturing Technology Management*, 26(1): 36–56.

El Makrini, H. 2015. How does management perceive export success? An empirical study of Moroccan SMEs. *Business Process Management Journal*, 21(1): 126–151.

Finneran, A. and O'Sullivan, L. 2010. The effects of force and exertion duration on duty cycle time: Implications for productivity. *Human Factors and Ergonomics in Manufacturing and Service Industries*, 20(4): 324–334.

Garbett, G.H.G. 1958. Time and motion study. *Work Study* 7(10): 13–60.

Garengo, P., Biazzo, S., and Bititci, U.S. 2005. Performance measurement systems in SMEs: A review for a research agenda. *International Journal of Management Reviews*, 7(1): 25–47.

Garza-Reyes, J.A., Ates, E.M., and Kumar, V. 2015. Measuring Lean readiness through the understanding of quality practices in the Turkish automotive suppliers industry. *International Journal of Productivity and Performance Management*, 64(8): 1092–1112.

Gitto, S. and Mancuso, P. 2015. The contribution of physical and human capital accumulation to Italian regional growth: A nonparametric perspective. *Journal of Productivity Analysis*, 43: 1–12.

Green, T.I.R. 1967. Work study, formerly time and motion study. *Work Study*, 16(9): 9–49.

Green, T.I.R. 1968. Work study. A journal devoted to work analysis, incentives, time study evaluation process control motion study, formerly time and motion study; process control, job evaluation, methods engineering and kindred subjects. *Work Study*, 17(2): 5–45.

Grifell-Tatje, E. and Lovell C.A.K. 2014. Productivity, price recovery, capacity constraints and their financial consequences. *Journal of Productivity Analysis*, 41: 3–17.

Grünberg, T. 2003. A review of improvement methods in manufacturing operations. *Work Study*, 52(2): 89–93.

Gunasekaran, A. and Cecille, P. 1998. Implementation of productivity improvement strategies in a small company. *Technovation*, 18(5): 311–320.

Gunasekaran, A. and Spalanzani, A. 2012. Sustainability of manufacturing and services: Investigations for research and applications. *International Journal of Production Economics*, 140: 35–47.

Halkos, G.E. and Tzeremes, N.G. 2007. Productivity efficiency and firm size: An empirical analysis of foreign owned companies. *International Business Review*, 16: 713–731.

Hamilton, B.H., Nickerson, J.A., and Owan, H. 2015. Diversity and productivity in production teams. In A. Bryson (ed.) *Advances in the Economic Analysis of Participatory and Labor-Managed Firms*, Volume 13: pp. 99–138. Bingley: Emerald Group.

Hampf, B. 2014. Separating environmental efficiency into production and abatement efficiency: A nonparametric model with application to US power plants. *Journal of Productivity Analysis*, 41: 457–473.

Hendry, L., Huang, Y., and Stevenson, M. 2013. Workload control: Successful implementation taking a contingency-based view of production planning and control. *International Journal of Operations and Production Management*, 33(1): 69–103.

Heshmati, A. 2003. Productivity growth, efficiency and outsourcing in manufacturing and service industries. *Journal of Economic Surveys*, 17(1): 79–112.

Hessels, J. and Parker, S. 2013. Constraints, internationalization and growth: A cross-country analysis of European SMEs. *Journal of World Business*, 48: 137–148.

Hilmola, O., Lorentz, H., Hilletofth, P., and Malmsten, J. 2015. Manufacturing strategy in SMEs and its performance implications. *Industrial Management and Data Systems*, 115(6): 1004–1021.

Hong, P., Marvel, J.H., and Modi, S. 2015. A survey of business network integration: Implications for quality and productivity performance. *International Journal of Productivity and Quality Management*, 15(2): 133–152.

Huang, X., Tan, B.L., and Ding, X. 2015. An exploratory survey of green supply chain management in Chinese manufacturing small and medium-sized enterprises. *Journal of Manufacturing Technology Management*, 26(1): 80–103.

Isaga, N., Masurel, E., and Montfort, K.V. 2015. Owner-manager motives and the growth of SMEs in developing countries. *Journal of Entrepreneurship in Emerging Economies*, 7(3): 190–211.

Jagoda, K., Lonseth, R., and Lonseth, A. 2013. A bottom-up approach for productivity measurement and improvement. *International Journal of Productivity and Performance Management*, 62(4): 387–406.

Jain, A., Bhatti, R., and Singh, H. 2014. Total productive maintenance (TPM) implementation practice. *International Journal of Lean Six Sigma*, 5(3): 293–323.

Jourabchi, S.M.M., Arabian, T., Leman, Z., and Ismail, Y.B. 2014. Contribution of Lean and Six Sigma to effective cost of quality management. *International Journal of Productivity and Quality Management*, 14(2): 149–165.

Kafetzopoulos, D. and Psomas, E. 2015. The impact of innovation capability on the performance of manufacturing companies. *Journal of Manufacturing Technology Management*, 26(1): 104–130.

Kanawaty, G. (ed.). 1992. *Introduction to Work Study*. 4th edn. Geneva: International Labour Office.

Karim, A. and Arif-Uz-Zaman, K. 2013. A methodology for effective implementation of lean strategies and its performance evaluation in manufacturing organizations. *Business Process Management Journal*, 19(1): 169–196.

Kaur, M., Singh, K., and Ahuja, I.S. 2013. An evaluation of the synergic implementation of TQM and TPM paradigms on business performance. *International Journal of Productivity and Performance Management*, 62(1): 66–84.

Klingner, S., Pravemann, S., and Becker, M. 2015. Service productivity in different industries–an empirical investigation. *Benchmarking: An International Journal*, 22(2): 238–253.

Krajewski, L.J., Ritzman, L.P., and Malhotra, M.K. 2013. *Operations Management: Processes and Supply Chains*. Abingdon: Pearson.

Kumar, R., Singh, R.K., and Shankar, R. 2013. Study on coordination of flexibility in supply chain of SMEs: A case study. *Global Journal of Flexible Systems Management*, 14(2): 81–92.

Lai, W. and Chen, H. 2013. Extreme internal–external industrial-service flexibilities and inter-firm cooperative networks in high-technology machine manufacturing. *Journal of Business Research*, 66: 2234–2244.

Legros, D. and Galia, F. 2012. Are innovation and R&D the only sources of firms' knowledge that increase productivity? An empirical investigation of French manufacturing firms. *Journal of Productivity Analysis*, 38: 167–181.

Leseure, M.J. 2000. Manufacturing strategies in the hand tool industry. *International Journal of Operations and Production Management*, 20(12): 1475–1487.

Li, M. 2004. Aggregate demand, productivity, and "disguised unemployment" in the Chinese industrial sector. *World Development*, 32(3): 409–425.

Li, S.K. and Zhao, L. 2015. The competitiveness and development strategies of provinces in China: A data envelopment analysis approach. *Journal of Productivity Analysis*, 44: 293–307.

Loosemore, M. 2014. Improving construction productivity: A subcontractor's perspective. *Engineering, Construction and Architectural Management*, 21(3): 245–260.

Mahmood, M. 2008. Labour productivity and employment in Australian manufacturing SMEs. *International Entrepreneurship and Management Journal*, 4: 51–62.

Malik, S.A., Nasim, K., and Iqbal, M.Z. 2013. TQM practices in electric fan manufacturing industry of Pakistan. *Journal of Productivity and Quality Management*, 12(4): 361–378.

Martin Jr., C.R., Horne, D.A., and Chan, W.S. 2001. A perspective on client productivity in business-to-business consulting services. *International Journal of Service Industry Management*, 12(2): 137–158.

Mathur, A., Mittal, M.L., and Dangayach, G.S. 2012. Improving productivity in Indian SMEs. *Production Planning and Control: The Management of Operations*, 23(10–11): 754–768.

Medda, G. and Piga, C.A. 2014. Technological spillovers and productivity in Italian manufacturing firms. *Journal of Productivity Analysis*, 41: 419–434.

Milana, C., Nascia, L., and Zeli, A. 2013. Decomposing multifactor productivity in Italy from 1998 to 2004: Evidence from large firms and SMEs using DEA. *Journal of Productivity Analysis*, 40: 99–109.

Monreal-Pérez, J., Aragón-Sánchez, A., and Sánchez-Marín, G. 2011. A longitudinal study of the relationship between export activity and innovation in the Spanish firm: The moderating role of productivity. *International Business Review*, 21(5): 862–877.

Mosheim, R. 2013. A shadow distance function decomposition of the environmental Kuznets curve: Comparing the South China Sea and the Caribbean. *Journal of Productivity Analysis*, 40: 457–472.

Motohashia, K. and Yuana, Y. 2010. Productivity impact of technology spillover from multinationals to local firms: Comparing China's automobile and electronics industries. *Research Policy*, 39: 790–798.

Muruganantham, V.R., Krishnan, P.N., and Arun, K.K. 2014. Integrated application of TRIZ with Lean in the manufacturing process in machine shop for productivity improvement. *International Journal of Productivity and Quality Management*, 13(4): 414–429.

O'Donnell, C.J. 2012. An aggregate quantity framework for measuring and decomposing productivity change. *Journal of Productivity Analysis*, 38: 255–272.

Odesola, I.A., Okolie, K.C., and Nnametu, J.N. 2015. A comparative evaluation of labour productivity of wall plastering activity using work study. *Project Management World Journal (PMWJ)*, IV(V): 1–10.

Oeij, P.R.A., De Looze, M.P., Have, K.T., Van Rhijn, J.W., and Kuijt-Evers, L.F.M. 2012. Developing the organisation's productivity strategy in various sectors of industry. *International Journal of Productivity and Performance Management*, 61(1): 93–109.

O'Mahony, M. and Timmer, M.P. 2009. Output, input and productivity measures at the industry level: The EU Klems Database. *The Economic Journal*, 119: 374–F403.

O'Neill, P., Sohal, A., and Teng, C.W. 2016. Quality management approaches and their impact on firms' financial performance: An Australian study. *International Journal of Production Economics*, 171: 381–393.

Papa, F., Cavaleiro de Ferreira, R., and Radulj, D. 2015. Pumps: Energy efficiency and performance indicators. *Water Practice and Technology*, 10(4): 872–885.

Parida, A., Kumar, U., Galar, D., and Stenström, C. 2015. Performance measurement and management for maintenance: A literature review. *Journal of Quality in Maintenance Engineering*, 21(1): 2–33.

Phusavat, K., Nilmaneenava, S., Kanchana, R., Wernz, C., and Helo, P. 2012. Identifying productivity indicators from business strategies' surveys. *International Journal of Productivity and Quality Management*, 9(2): 158–176.

Prasad, K.D., Jha, S.K., and Prakash, A. 2015. Quality, productivity and business performance in home based brassware manufacturing units. *International Journal of Productivity and Performance Management*, 64(2): 270–287.

Ramanigopal, C.S., Palaniappan, G., Hemalatha, N., and Murugan, T. 2011. Business process reengineering and its applications. *International Journal of Management Research and Review*, 1(5): 275–288.

Rantanen, H. 2001. Internal obstacles restraining productivity improvement in small Finnish industrial enterprises. *International Journal of Production Economics*, 69: 85–91.

Rehman, M.A.A., Shrivastava, R.R., and Shrivastava, R.L. 2014. Evaluating green manufacturing drivers: An interpretive structural modelling approach. *International Journal of Productivity and Quality Management*, 13(4): 471–494.

Reverte, C., Gomez-Melero, E., and Cegarra-Navarro, J.G. 2016. The influence of corporate social responsibility practices on organizational performance: Evidence from eco-responsible Spanish firms. *Journal of Cleaner Production*, 112: 2870–2884.

Robinson, D.C., Sanders, D.A., and Mazharsolook, E. 2015. Ambient intelligence for optimal manufacturing and energy efficiency. *Assembly Automation*, 35(3): 234–248.

Sandelands, E. 1994. Anbar abstracts issue. *Industrial Management and Data Systems*, 94(3): 1–32.

Saunila, M., Pekkola, S., and Ukko, J. 2014. The relationship between innovation capability and performance: The moderating effect of measurement. *International Journal of Productivity and Performance Management*, 63(2): 234–249.

Schulze, M., Nehler, H., Ottosson, M., and Thollander, P. 2016. Energy management in industry: A systematic review of previous findings and an integrative conceptual framework. *Journal of Cleaner Production*, 112: 3692–3708.

Sharma, V., Dixit, A.R., Qadri, M.A., and Kumar, S. 2015. An interpretive hierarchical model for lean implementation in machine tool sector. *International Journal of Productivity and Quality Management*, 15(3): 381–406.

Sharma, C. and Mishra, R.K. 2012. Export participation and productivity performance of firms in the Indian transport manufacturing. *Journal of Manufacturing Technology Management*, 23(3): 351–369.

Shokri, A. and Nabhani, F. 2015. LSS, a problem solving skill for graduates and SMEs. *International Journal of Lean Six Sigma*, 6(2): 176–202.

Singh, J. and Singh, H. 2015. Continuous improvement philosophy—literature review and directions. *Benchmarking: An International Journal*, 22(1): 75–119.

Srinivasan, S., Ikuma, L.H., Shakouri, M., Nahmens, I., and Harvey, C. 2016. 5S impact on safety climate of manufacturing workers. *Journal of Manufacturing Technology Management*, 27(3): 364–378.

St-Pierre, J. and Raymond, L. 2004. Short-term effects of benchmarking on the manufacturing practices and performance of SMEs. *International Journal of Productivity and Performance Management*, 53(8): 681–699.

Strobel, T. 2014. Directed technological change, skill complementarities and sectoral productivity growth: Evidence from industrialized countries during the new economy. *Journal of Productivity Analysis*, 42: 255–275.

Subrahmanya, M.H.B. 2006a. Labour productivity, energy intensity and economic performance in small enterprises: A study of brick enterprises cluster in India. *Energy Conversion and Management*, 47: 763–777.

Subrahmanya, M.H.B. 2006b. Energy intensity and economic performance in small scale bricks and foundry clusters in India: Does energy intensity matter? *Energy Policy*, 34: 489–497.

Taneja, S., Pryor, M.G., and Toombs, L.A. 2011. Frederick W. Taylor's scientific management principle: Relevance and validity. *The Journal of Applied Management and Entrepreneurship*, 16(3): 60–78.

Tanvir, S.I. and Ahmed, S. 2013. Work study might be the paramount methodology to improve productivity in the apparel industry of Bangladesh. *Industrial Engineering Letters*, 3(7): 51–60.

Teng, H.S.S. 2014. Qualitative productivity analysis: Does a non-financial measurement model exist? *International Journal of Productivity and Performance Management*, 63(2): 250–256.

Towill, D.R. 2009. Frank Gilbreth and health care delivery method study driven learning. *International Journal of Health Care Quality Assurance*, 22(4): 417–440.

Tran, B.T., Grafton, Q., and Kompas, T. 2009. Contribution of productivity and firm size to value-added: Evidence from Vietnam. *International Journal of Production Economics*, 121: 274–285.

Trianni, A., Cagno, E., and Farné, S. 2016. Barriers, drivers and decision-making process for industrial energy efficiency: A broad study among manufacturing small and medium-sized enterprises. *Applied Energy*, 162: 1537–1551.

Tsai, K. 2005. R&D productivity and firm size: A nonlinear examination. *Technovation*, 25: 795–803.

Uddin, N. 2015. Productivity growth, efficiency change, and technical progress of a corporate sector in Bangladesh: A malmquist output productivity index approach. *International Journal of Economics and Finance*, 7(8): 240–255.

van Niekerk, W.P. 1986. *Productivity and Work Study*. 2nd edn. Durban: Butterworths Publishers.

Vaszkun, B. and Tsutsui, W.M. 2012. A modern history of Japanese management thought. *Journal of Management History*, 18(4): 368–385.

Chapter 3

Effectiveness versus Efficiency in Manufacturing SMEs: A Productivity Perspective

3.1 Introduction

Work study can be used by manufacturing small and medium enterprises (SMEs) in South Africa to improve their productivity, as supported by other literature sources which were found to be successful worldwide. Thus, manufacturing SMEs need to focus on measuring effectiveness as well as efficiency through the use of work study techniques if they are to be competitive locally and internationally. This chapter addresses the detailed background of both effectiveness and efficiency in manufacturing SMEs.

3.2 Effectiveness versus Efficiency Concept

The background of both effectiveness and efficiency in manufacturing SMEs is debated in this section with the intention of providing a detailed look at the two concepts. The first concept to be addressed is effectiveness in manufacturing SMEs. The diagram in Figure 3.1 addresses both efficiency and effectiveness.

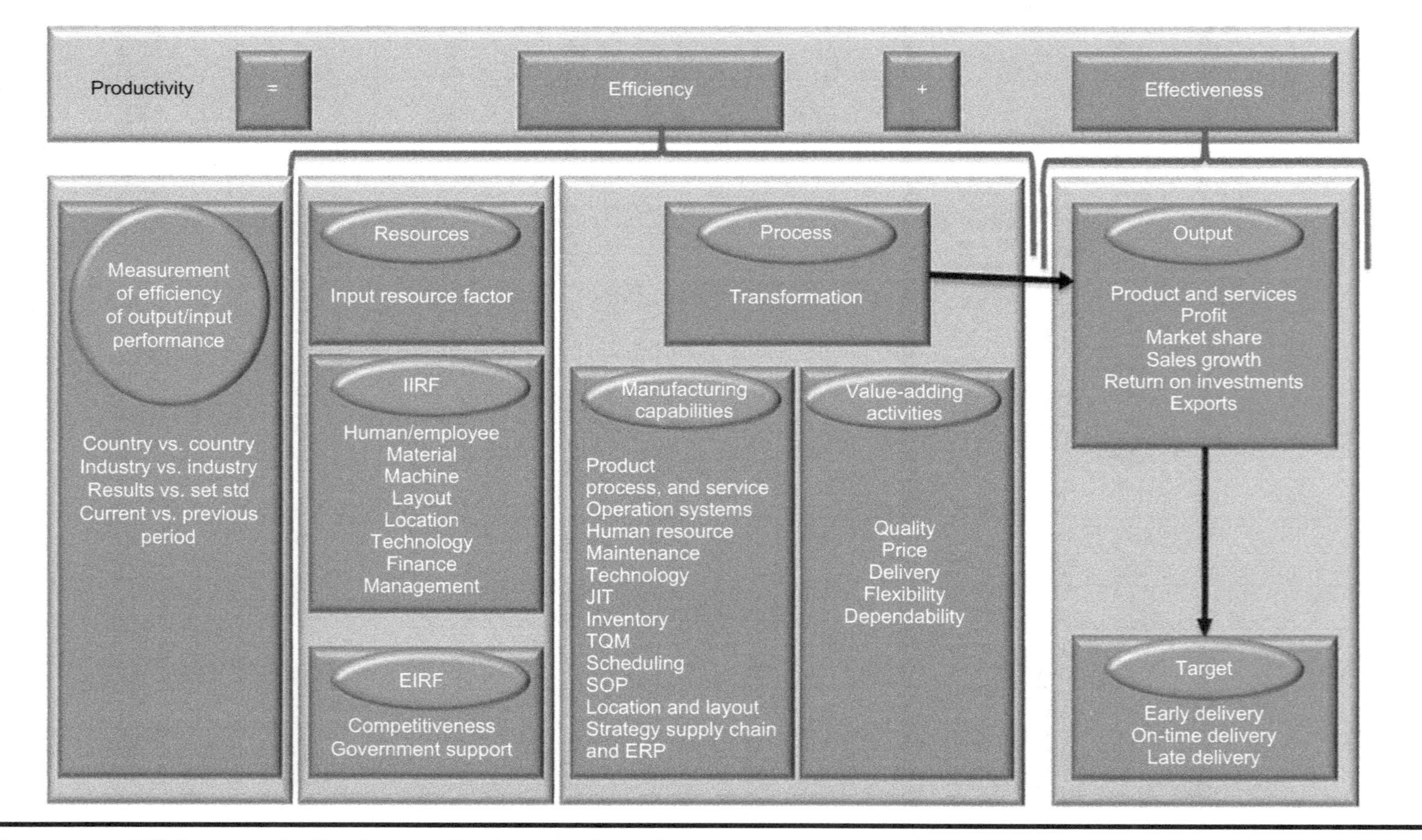

Figure 3.1 Efficiency and effectiveness in the manufacturing process. (From Author, 2017.)

3.2.1 Effectiveness in Manufacturing SMEs

Kristianto, Ajmal, and Sandhu (2012:30–36) emphasize that effectiveness entails attaining results for manufacturing SMEs through quality, delivery, and cost. According to Mathur et al. (2011:77–82) and Taylor and Taylor (2014:849–66), effectiveness means attaining results through quality, delivery, flexibility, dependability, and cost. Effectiveness in manufacturing SMEs is defined as the realization of product quality through information sharing, cost, time, delivery, innovation, manufacturer–supplier relationship, and customer satisfaction (Singh, 2011:622–31; McDermott & Prajogo, 2012:217–25; Brettel et al., 2014:37–9; del Río, Peñasco, & Romero-Jordán, 2016:2158–68).

As indicated by Valmohammadi (2011:498), effectiveness in manufacturing SMEs involves the achievement of customer requirements using employee teamwork, benchmarking, process and product design, and close ties with suppliers. Effectiveness in manufacturing SMEs involves reaching a goal that includes increasing market share, rate of return on investment, profit, sales, the good image of the company, and investors' confidence in the company through the manufacturing process (Gunasekaran & Spalanzani, 2012:44; Karim & Arif-Uz-Zaman, 2013:170–81; Leithold et al., 2016:263–74).

Habidin and Yusof (2013:63) state that the effectiveness of leadership characteristics in manufacturing SMEs depends on the improvement of quality through the communication and dissemination of information. Ruivo, Oliveira, and Neto (2012:1018) and Thirupathi and Vinodh (2016:6662) comment that effectiveness in manufacturing SMEs aims at achieving customer satisfaction through planning. Psomas, Pantouvakis, and Kafetzopoulos (2013:149); Sumaedi and Yarmen (2015:436); Dora and Gellynck (2015:274); Psomas and Antony (2015:2089); and Denton and Maatgi (2016:1) mention that effectiveness in manufacturing is considered as the achievement of the objectives of standard ISO 9001.

Garengo and Sharma (2014:221) note that effectiveness in manufacturing SMEs embraces the attainment of a competitive edge through management practices. Vinodh, Kumar, and Vimal (2014:291–300) state that effectiveness in manufacturing SMEs is perceived as achieving results for the business through the utilization of workers' abilities, material, and machinery. As explained by Shaharudin, Zailani, and Tan (2015:220–30) and Prajogo, Oke, and Olhager (2016:275), effectiveness in manufacturing SMEs aims at managing the supply chain through quality, delivery, flexibility, dependability, and cost. Chaplin, Heap, and O'Rourke (2016:130) affirm that effectiveness in manufacturing SMEs is perceived as achieving results of the business through the utilization of machinery. As argued by Dora, Kumar, and Gellynck (2016:7–18), effectiveness refers to the process of achieving results through the utilization of input resource factors such as human capital, material, machinery, location, layout, technology, management, infrastructure (i.e., water and energy), and competitiveness. Schulze et al. (2016:3693–702) report that effectiveness in manufacturing SMEs focuses on managing energy (e.g., utilization of electricity, gas, oil, and chemicals) through management practices (planning, organizing, leading, and control).

As found by Vivares-Vergara, Sarache-Castro, and Naranjo-Valencia (2016:116), effectiveness in manufacturing SMEs involves managing human resources (such as employees) through training. Effectiveness in manufacturing SMEs entails striving to deliver more value to the customer (Zelbst et al., 2012:330–4; Panizzolo et al., 2012:785). Li and Doolen (2013:19) and Jain, Adil, and Ananthakumar (2013:820) refer to effectiveness in manufacturing SMEs as achieving results for the business. Biloslavo, Bagnoli, and Figelj (2013:429) considered it to be indicated by the ratio of the result obtained to the one intended or expected. As noted by Jaca et al. (2013:223), effectiveness focuses on the process of achieving outcomes (results) through the use of organization, teams, and individuals. As pointed out by Digalwar, Tagalpallewar, and Sunnapwar (2013:70); Koyratty, Aumjaud, and Neeliah (2014:366); and Mahmood et al. (2014:668), effectiveness in manufacturing SMEs is about the achievement of quality products.

Singh and Ahuja (2014:415) report that effectiveness involves providing value to customers through identification of their expressed needs. According to Karim and Arif-Uz-Zaman (2013:180), the formula for effectiveness is as follows:

$$\text{Effectiveness} = \frac{\text{Actual output}}{\text{Planned output}} \times 100\%$$

This means, for example, that if the manufacturing company plans to produce 100 units and ends up producing 80, the effectiveness is at 80% rather than the expected percent:

$$\text{Effectiveness} = \frac{80}{100} \times 100\%$$

$$= 80\%$$

The importance of dividing actual output against standard/planned output, which is the formula for the definition of effectiveness, is to ensure that the units produced by manufacturing industries are produced as expected. Based on the foregoing calculation, there is a loss of 20% of units. According to Karim and Arif-Uz-Zaman (2013:174), the loss of units may be affected by production factors due to lack of employee training.

On the contrary, if the manufacturing company plans to produce 100 units and ends up producing 120, the effectiveness is at 80% rather than the expected percent:

$$\text{Effectiveness} = \frac{120}{100} \times 100\%$$
$$= 120\%$$

Therefore, the production units have increased by 20%.

Based on the foregoing calculation, there is an increase of 20% of units. Similarly, therefore, the increase of units may be positively affected by production factors due to improved efficiency. So, work study can be used by manufacturing SMEs in South Africa to improve their productivity, as supported by other literature sources worldwide. Thus, manufacturing SMEs need to focus on measuring efficiency through the use of work study techniques if they are to be competitive locally and internationally. Before effectiveness can be measured, manufacturing SMEs need to examine how workers and machinery are utilized. However, this process firstly focuses on machine utilization, or the total machine time required to manufacture the products (i.e., the machine running time), which is the numerator of the equation in this case, whereas the denominator is the total time allocated to make those outputs (i.e., machine available time) (Karim & Arif-Uz-Zaman, 2013:185–6).

Based on the literature studied, the formula of Karim and Arif-Uz-Zaman is as follows:

$$\text{Machine utilization} = \frac{\text{Machine running time}}{\text{Machine available time}} \times 100$$

Machine running time = machine available time minus any machine lost time (Karim & Arif-Uz-Zaman, 2013:185–6).

As is explained by Karim and Arif-Uz-Zaman (2013:185–6); Bevilacqua et al. (2015:22–7); and Vinodh et al. (2015:280–9), any machine lost time involves machine down time (breakdowns, maintenance, service, and repairs); machine idle time (shortage of work, material, workers, and machine wait time); and/or machine ancillary time (changeovers, set-up, cleaning, and meeting).

3.2.2 Effectiveness Applications

For example, an operator employed at Company A produces 100 units in an 8-hour-a-day shift, idle time has been recorded at 0.5 hours, and the standard manufacturing time for one unit is 4 minutes, then the machine utilization is calculate as follows:

$$\text{Machine utilization} = \frac{\text{Machine running time}}{\text{Machine available time}} \times 100$$

Analysis of the case:

Units manufactured = 100
Machine available time = 8 hours
Machine idle time = 0.5 hours
Standard time per unit = 4 minutes

where:

Machine utilization = MU
Machine running time = MRT
Machine available time = MAT
Firstly, MRT needs to be calculated, determining the machine utilization.

$$\text{MRT} = \text{MAT} - \text{any lost time (in this case machine idle time)}$$

Where machine idle time = MIT

$$\text{MRT} = 8 - 0.5 \text{ hours}$$

$$= 7.5 \text{ hours}$$

$$\text{Machine utilization} = \frac{7.5}{8} \times 100$$

$$= 93.75\%$$

Karim and Arif-Uz-Zaman (2013:178–80) report that in comparing actual output value generated by workers or machinery production, effectiveness is measured to improve productivity in manufacturing SMEs. A formula for production effectiveness in manufacturing SMEs to be measured follows:

$$\text{Effectiveness} = \frac{\text{Output value}}{\text{Input resource}} \times 100\%$$

$$\text{Effectiveness}_{\text{machinery @ standard}} = \frac{\text{No. of output} \times \text{average pitch}_{\text{standard time}}}{\text{No. of workers} \times \text{total allocated time}_{(\text{working time})}} \times 100\%$$

$$\text{Effectiveness}_{\text{machinery @ standard}} = \frac{\text{No. of output} \times \text{average pitch}_{\text{standard time}}}{\text{No. of machinery} \times \text{total allocated time}_{(\text{working time})}} \times 100\%$$

For example, if an operator employed at Company A produces 100 units in an 8-hour-a-day shift, idle time has been recorded at 0.5 hours, and the standard manufacturing time for one unit is 20 minutes, then the machine utilization is calculated as follows:

$$\text{Effectiveness}_{\text{machinery @ standard}} = \frac{\text{No. of output} \times \text{average pitch}_{\text{standard time}}}{\text{No. of machinery} \times \text{total allocated time}_{(\text{working time})}} \times 100\%$$

Analysis of the case:

Units manufactured = 100

Machine available time = 8 hours

Machine idle time = 0.5 hours

Standard time per unit = 4 minutes

where:

Machine utilization = MU

Machine running time = MRT

Machine available time = MAT

Firstly, MRT needs to be calculated, determining the machine utilization.

$$\text{MRT} = \text{MAT} - \text{any lost time (in this case machine idle time)}$$

where:

Machine idle time = MIT

$$\text{MRT} = 8 - 0.5 \text{ hours}$$

$$= 7.5 \text{ hours}$$

$$\text{Effectiveness}_{\text{machinery @ standard}} = \frac{\text{No. of output} \times \text{average pitch}_{\text{standard time}}}{\text{No. of machinery} \times \text{total allocated time}_{(\text{working time})}} \times 100\%$$

where:

No. of output = units manufactured = 100

Average pitch $_{\text{standard time}}$ = standard time per unit = 20 minutes

Total allocated time$_{(\text{working time})}$ = machine available time (MAT) = working time

$$\begin{aligned}\text{Effectiveness}_{\text{machinery @ standard}} &= \frac{100 \times 4}{1\text{ machine} \times 8 \times 60} \times 100 \\ &= \frac{400}{480} \times 100 \\ &= 83.33\%\end{aligned}$$

Therefore, this machine is effectively utilized for 83.33%, which is acceptable.

If the Effectiveness$_{\text{machinery @ standard}}$ is 75%, the standard time is calculated as follows:

$$75\% = \frac{100 \times (\text{std time})}{1\text{ machine} \times 8 \times 60} \times 100$$

$$= \frac{100 \times (\text{std time})}{480} \times 100$$

$$0.75 \times 480 = 100(\text{std time}), \quad \text{where std time} = X$$

$$360 = 100(\text{std time}), \quad \text{where std time} = X$$

$$\text{Therefore, } \frac{100X}{100} = \frac{360}{100}$$

$$X = 3.6\text{ minutes}$$

Manufacturing SMEs need not only be effective but also efficient in order to improve productivity in their businesses. So, the next issue to be debated is efficiency.

3.2.3 Efficiency in Manufacturing SMEs

In this section, the concept of efficiency is addressed, providing a comprehensive background of efficiency in manufacturing SMEs. This concept is also depicted in Figure 3.1.

Farooquie et al. (2012:2–3) maintain that efficiency in manufacturing SMEs entails improvements in cost, quality, and time in the manufacturing process by using safety and waste reduction in order to avoid waste inventory and injuries. Even though safety and waste reduction is presented in the form of signals for cost, quality, and time, efficiency is also considered when the growth of manufacturing SMEs is measured and monitored regularly. This is caused by the flexibility of the changing of resources through innovation in the product required by the customer.

As is argued by Tarí, Molina-Azorín, and Heras (2012:381–82), efficiency in manufacturing SMEs does not only introduce ISO 9001, which is a quality management system looking at how product quality and delivery adds value to customer requirements through continuous improvement, resulting in an efficient manufacturing process; it also encourages the use of ISO 14001, which is an environmental management system. With ISO 9001, the achievement of efficient manufacturing processes results in increased revenue through product reliability. This increase in revenue is the result of reduced costs due to efficiencies that positively contribute to the market share. With ISO 14001, however, the environmental management system is utilized in manufacturing SMEs for the prevention of pollution and emissions, which affect employees' health and cause safety hazards in the workplace. These preventive measures save costs and benefit manufacturing SMEs in terms of employee satisfaction, reduced product defects, customer satisfaction, market share, sales, return on sales, and return on assets.

Efficiency in manufacturing SMEs is "doing things right," which involves managing raw material, work in process, and finished goods inventories by reducing costs and eliminating waste and shortfall/oversupply via improved production planning and capacity utilization at both the manufacturing process level and the supply chain level (Ahmad et al., 2014:8). Efficiency refers to the optimal ability of firms to utilize resources in a cost-effective way to generate business output (Teng, 2014:251). According to Henriques and Catarino (2015:381–2), pollution in manufacturing SMEs includes contamination of air, water, and noise that negatively affect employees in the workplace. In addition, employees are also exposed to other pollution factors such as temperature, dust, and humidity. These pollution factors result in employee illness and absenteeism, errors and defects, late delivery, high costs, and customer complaints, which in turn lead to reduced market share, sales, return on sales, and return on assets (Tarí et al., 2012:381–82).

Gupta, Acharya, and Patwardhan (2013:635–38) report that efficiency in manufacturing SMEs focuses on waste elimination through the use of standard operating procedures such as proper planning and control; communication; skilled, knowledgeable and motivated employees; preventative maintenance of machinery and manufacturing processes; just-in-time delivery; and improved quality. Pazirandeh and Jafari (2013:891) refer to efficiency as the degree to which resources have been used economically. This simply means the product is manufactured without compromising the quality of the product and is delivered to the customer in an agreed time in a cost-effective manner.

As is conveyed by Georgiev and Georgiev (2015:1029–38), efficiency focuses on the application to international organizations of standard ISO 9001 to measure product quality in manufacturing SMEs in line with the quality material delivered by suppliers. This product quality is attributed to manufacturing process improvement in manufacturing SMEs. These manufacturing SMEs need to comply with ISO 9001 to ensure better customer satisfaction. In so doing, manufacturing SMEs are then in a better position to achieve a competitive edge. Johnson (2015:280) states that efficiency in manufacturing SMEs does not only focus on the application to an international organization of standard ISO 9001, but also to the implementation of standards such as ISO 18001. ISO 18001 refers to environmental and social standards.

As indicated by Narayanapillai (2014:606), efficiency in manufacturing SMEs centers on inventory management (IM) practices measuring technical factors (raw material ordering frequency, demand forecasting frequency, purchasing effectiveness, lead time, and inventory cost); managerial factors (IM practices pursued, safety stock planning of enterprise, employee training, management attitude); contingency factors (capacity utilization level of enterprise, product type and demand variability of the product) and organizational factors (interaction with suppliers, interaction with customers, supplier empowerment, and space limitation). This management practice results in competitiveness, effectiveness, and the minimization of inventory investment. The management effectiveness of inventory comprises cost, service level, and turnover ratio.

By integrating inventory with turnover ratio, inventory turnover ratio is formed. This ratio is regarded as a measure of how inventories are being efficiently managed through reduced material time spent in the manufacturing process, which ultimately results in inventory cost reduction. Based on the IM practices in manufacturing SMEs, raw material ordering frequencies play a pivotal role factor in inventory costs, whereby ordering cost is incurred when an order is placed with the suppliers. Furthermore, carrying or holding costs are avoided by ordering economic quantities of material required for managing inventory.

Els et al. (2012:39–40) introduce a formula for economic order quantity (EOQ) through the use of IM practices, which is presented as follows:

$$EOQ = \frac{\sqrt{2 \times C \times U}}{H}$$

where:

C = Cost of placing an order
U = Yearly usage
H = Inventory-holding cost per unit

EOQ involves the number of units that a firm should add to its inventory with each order to minimize the total costs of inventory, such as holding costs, order costs, and shortage costs (Teng, Min, & Pan, 2012:329–33). Ozturk et al. (2015:120–21) consider efficiency as adhering to the required standard working procedure in manufacturing outputs through the appropriate utilization of resources. These resources happen in terms of continuous and batch manufacturing processes in manufacturing SMEs. According to Martínez (2015:325–30), efficiency implies increasing output for a given set of input. However, increasing output with a given set of input involves products manufactured determining the number of workers used, material used, capital used, and amount of energy used, such as electricity.

Efficiency in manufacturing, which is also regarded as capacity utilization (Samarajeewa et al., 2012:314–18), involves the extent to which manufacturing industries are using their production potential (also referred to as *best use of production factors*) such as employees, material, machinery, capital (money), and energy (Gunasekaran & Spalanzani, 2012:35–45; Karim & Arif-Uz-Zaman, 2013:178–80; Wu et al., 2013:568–78; Bocken et al., 2014:43–54; Roh, Hong, & Min, 2014:198–204; Feiz et al., 2015:283–90; Schulze et al., 2016:3694–3700).

Based on the explanations given by the aforementioned authors, the formula for efficiency in manufacturing is presented in the next section.

3.2.4 Efficiency Applications

According to Uddin (2015:241), the formula for efficiency is

$$\text{Efficiency} = \frac{\text{Output value}}{\text{Input resource}} \times 100\%$$

However, it needs to be measured for continuous improvement. Thus, one needs to measure efficiency from time to time for improvement. In this case, efficiency is measured against the existing method.

This means, for example, if Company B plans to produce 400 units using four workers, each spending 8 hours a day, the efficiency is calculated as follows:

The manufacturing company's existing method for efficiency is

$$\text{Efficiency} = \frac{400}{4 \times 8 \text{ hours}}$$

$$= 12.5 \text{ units per hour}$$

If the company decides to produce 500 units using four workers for improvement, each spending 8 hours a day, the efficiency is calculated as follows:

The manufacturing company's actual efficiency is

$$\text{Efficiency} = \frac{500}{4 \times 8 \text{ hours}}$$

$$=15.63 \text{ units per hour}$$

Therefore

$$\text{Efficiency} = \frac{15.63}{12.5} \times 100\%$$

$$= 125\%$$

The importance of dividing actual output against input resources, where the actual results are measured against the set standard, is that it shows how the manufacturing company has improved. Based on the previous calculation, there is an efficiency gain of 25% in terms of units. On the contrary, according to Karim and Arif-Uz-Zaman (2013:181), the failure of a manufacturing company to improve its business may be affected by production factors due to inefficiency.

On the contrary, if Company B plans to produce 400 units using four workers, each spending 8 hours a day, the efficiency is calculated as follows:

The manufacturing company's set standard for efficiency is

$$\text{Efficiency} = \frac{400}{4 \times 8 \text{ hours}}$$

$$= 12.5 \text{ units per hour}$$

If the company decides to produce 295 units using three workers for improvement, each spending 8 hours a day, the efficiency is calculated as follows:

The manufacturing company's actual efficiency is

$$\text{Efficiency} = \frac{295}{3 \times 8 \text{ hours}}$$

$$= 12.30 \text{ units per hour}$$

Therefore

$$\text{Efficiency} = \frac{12.3}{12.5} \times 100\%$$

$$= 98.4\%$$

Based on this calculation, there is an efficiency loss of 1.6% in terms of units. The manufacturing company needs to resolve the problem to ensure that there is stability or improvement in the business. Sometimes, manufacturing companies may reach stability, which means that these companies are still progressing but need to go further in order to improve. An example of stability of results for manufacturing companies follows. If the company in this case decides to produce 300 units using three workers for improvement, each spending 8 hours a day, the efficiency is calculated as follows:

The manufacturing company's actual efficiency is

$$\text{Efficiency} = \frac{300}{3 \times 8 \text{ hours}}$$

$$= 12.50 \text{ units per hour}$$

Therefore

$$\text{Efficiency} = \frac{12.5}{12.5} \times 100\%$$

$$= 100\%$$

Based on this calculation, the results are constant, which is 0%. The manufacturing company needs to resolve the problem to ensure that there is stability or improvement in the business. So, if manufacturing companies stay at the stability boundary, this indicates that these companies may be in the process of dropping in terms of efficiency.

So, work study can be used by manufacturing SMEs in South Africa to improve their productivity, as supported by other literature sources which were found to be successful worldwide. Thus, manufacturing SMEs need to focus on measuring efficiency through the use of work study techniques if they are to be competitive locally and internationally. Before efficiency can be measured, manufacturing SMEs need to examine how workers and machinery are utilized. However, this process firstly focuses on machine utilization, regarded as the total machine time required to manufacture the products (i.e., machine running time), which is the numerator of the equation in this case, whereas the denominator is the total time allocated to make those outputs (i.e., machine available time) (Karim & Arif-Uz-Zaman, 2013:185–6).

3.3 Application of Machine and Employee Utilization Using Effectiveness and Efficiency

Based on the literature studied, the formula of Karim and Arif-Uz-Zaman is addressed as follows:

$$\text{Machine utilization} = \frac{\text{Machine running time}}{\text{Machine available time}} \times 100$$

Machine running time = machine available time minus any machine lost time (Karim & Arif-Uz-Zaman, 2013:185–6).

As explained by Karim and Arif-Uz-Zaman (2013:185–6); Bevilacqua et al. (2015:22–7); and Vinodh et al. (2015:280–9), any machine lost time involves machine downtime (breakdowns, maintenance, service, and repairs);

machine idle time (shortage of work, material, workers, and machine waiting time); and/or machine ancillary time (changeovers, set-up, cleaning, and meeting).

For example, if an operator employed at Company C produces 100 units in an 8-hour-a-day shift, idle time has been recorded at 0.5 hours, and the standard manufacturing time for one unit is 20 minutes, then the machine utilization is calculated as follows:

$$\text{Machine utilization} = \frac{\text{Machine running time}}{\text{Machine available time}} \times 100$$

Analysis of the case:

Units manufactured = 100

Machine available time = 8 hours

Machine idle time = 0.5 hours

Standard time per unit = 20 minutes

where:

Machine utilization = MU

Machine running time = MRT

Machine available time = MAT

Firstly, MRT needs to be calculated, determining the machine utilization.

$$\text{MRT} = \text{MAT} - \text{any lost time (in this case machine idle time)}$$

where:

Machine idle time = MIT

$$\text{MRT} = 8 - 0.5 \text{ hours}$$
$$= 7.5 \text{ hours}$$

$$\text{Machine utilization} = \frac{7.5}{8} \times 100$$
$$= 93.75\%$$

Karim and Arif-Uz-Zaman (2013:178–80) report that in comparing actual output value generated from workers or machinery production, efficiency is measured to improve productivity in manufacturing SMEs. The formulae to measure production efficiency in manufacturing SMEs are

$$\text{Efficiency} = \frac{\text{Output value}}{\text{Input resource}} \times 100\%$$

$$\text{Efficiency}_{\text{worker/ load factor}} = \frac{\text{No. of output} \times \text{average pitch}_{\text{standard time}}}{\text{No. of workers} \times \text{total allocated time}_{(\text{working time})}} \times 100\%$$

$$\text{Efficiency}_{\text{machinery}} = \frac{\text{No. of output} \times \text{average pitch}_{\text{standard time}}}{\text{No. of machinery} \times \text{total allocated time}_{(\text{working time})}} \times 100\%$$

For example, if an operator employed at a certain manufacturing SME produces 100 units in an 8-hour-a-day shift, idle time has been recorded at 0.5 hours, and the standard manufacturing time for one unit is 20 minutes, then the machine utilization is calculated as follows:

$$\text{Efficiency}_{\text{machinery}} = \frac{\text{No. of output} \times \text{average pitch}_{\text{standard time}}}{\text{No. of machinery} \times \text{total allocated time}_{(\text{working time})}} \times 100\%$$

Analysis of the case:

Units manufactured = 100
Machine available time = 8 hours
Machine idle time = 0.5 hours
Standard time per unit = 4 minutes

where:

Machine utilization = MU
Machine running time = MRT
Machine available time = MAT
Firstly, MRT needs to be calculated, determining the machine utilization.

$$\text{MRT} = \text{MAT} - \text{any lost time (in this case machine idle time)}$$

where:

Machine idle time = MIT

$$\text{MRT} = 8 - 0.5 \text{ hours}$$

$$= 7.5 \text{ hours}$$

$$\text{Efficiency}_{\text{machinery}} = \frac{\text{No. of output} \times \text{average pitch}_{\text{standard time}}}{\text{No. of machinery} \times \text{total allocated time}_{(\text{working time})}} \times 100\%$$

where:

No. of output = units manufactured = 100

Average $\text{pitch}_{\text{standard time}}$ = standard time per unit = 20 minutes

Total allocated $\text{time}_{\text{(working time)}}$ = machine running time (MRT) = working time

$$\text{Efficiency}_{\text{machinery}} = \frac{100 \times 4}{1 \text{ machine} \times 7.5 \times 60} \times 100$$

$$= \frac{400}{450} \times 100$$

$$= 88.89\%$$

Therefore, this machine is efficiently utilized for 88.89%, which is acceptable.

$$\text{Load factor}_{\text{worker}} = \frac{\text{Total work content}}{\text{Total cycle time}}$$

As stated by Chen, Li, and Shady (2012:1071–2); Karim and Arif-Uz-Zaman (2013:179–85); Hasle (2014:44); and Chueprasert and Ongkunaruk (2015:2315), load factor involves a portion of the total cycle time of a machine or process measured operation that the skilled worker is expected to do at normal operation point.

The formula for the calculation of the load factor is as follows:

$$\text{Load factor}_{\text{worker}} = \frac{\text{Total work content}}{\text{Total cycle time}}$$

or

$$= \frac{\text{Normal time}}{\text{Standard time}}$$

For example, an operator operates a machine at the normal time of 1.6 minutes per cycle, and the total cycle time (which is standard time) is 2 minutes. The load factor can be calculated as follows:

$$\text{Load factor}_{\text{worker}} = \frac{\text{Normal time}}{\text{Standard time}}$$

where:

Normal time = 1.6 minutes
Total cycle time = 2 minutes

$$\text{Load factor}_{\text{worker}} = \frac{\text{Normal time}}{\text{Standard time}}$$

$$\text{Load factor}_{\text{worker}} = \frac{1.6 \text{ minutes}}{2 \text{ minutes}} \times 100$$

$$= 80\% \text{ (acceptable)}$$

Machine available time focuses on the time a machine could work based on attendance time, that is, a working day or a week plus overtime. Machine running time is the time during which the machine is actually operating, which is machine available time less any lost time. Machine running time at standard involves the optimum speed at which the machine must run to be able to achieve certain predetermined production outputs. Lastly, a load factor entails that part of the total cycle time of a machine (or process) controlled time that the worker is required to carry out at a standard performance level.

Work study and productivity are the key concepts used in manufacturing SMEs that can compete in the twenty-first century. Since manufacturing SMEs are faced with low productivity, making it difficult for them to become part of the mainstream of developing the South African economy, more effort is needed on the part of manufacturing SMEs if they are to improve in their productivity levels. Physical capital and technological capital are presently becoming global and economic problems which results to lower productivity of manufacturing SMEs in South Africa and the rest of the world. As stated earlier, the problems encountered by manufacturing SMEs are numerous and can be described among others as being operational, environmental, and competitive in nature.

For example, if an operator employed at Company D produces 120 units in an 8-hour-a-day shift, idle time has been recorded at 0.75 hours, and the standard manufacturing time for one unit is 25 minutes, then the machine utilization is calculated as follows:

$$\text{Machine utilization} = \frac{\text{Machine running time}}{\text{Machine available time}} \times 100$$

Analysis of the case:

Units manufactured = 120

Machine available time = 8 hours

Machine idle time = 0.75 hours

Standard time per unit = 25 minutes

where:

Machine utilization = MU

Machine running time = MRT

Machine available time = MAT

The following indices need to be calculated such as machine utilization index, load factor worker, and Machine Efficiency Index (which is $\text{Efficiency}_{\text{machine@standards}}$), to check whether the machine is efficiently utilized in Company B to improve the productivity of the business.

3.4 Critical Review of the Nexus between Efficiency and Effectiveness in Productivity

This research book explains the association between efficiency and effectiveness in productivity. These concepts are interrelated, but effectiveness, as indicated earlier in the research discussed, focuses more on achieving results irrespective of the method, whereas efficiency goes deeper by centering on the best method to achieve results for manufacturing SMEs. This simply means that manufacturing SMEs need to be efficient by doing the job well in the manufacturing process within various workstations using the minimum application of resources; this results in improved cost, improved product and service quality, saved cycle time, waste and defect reduction, safe working conditions, appropriate standards, and minimum inventory based on the operation planning and requirements. Furthermore, these applications include innovation skills and recent technology for the benefit of the customers and the improvement of the economy in the country as well as worldwide. Effectiveness means attaining the results of manufacturing SMEs through quality, delivery, cost, dependability, flexibility, innovation, employee teamwork, benchmarking, close ties with suppliers, communication and dissemination of information, planning, utilization of machinery, management practices (planning, organizing, leading and control), and training. For manufacturing SMEs, as indicated earlier in this chapter, being effective as a business is not enough; they also need to operate efficiently in order to boost productivity.

For example, with effectiveness, Companies A, B, C, and D may accomplish the aim of producing the expected results, but with efficiency, these companies function in the best possible manner to improve their productivity level with less waste of time and effort. The concern about Companies B, C, and D was that these companies focused mainly on effectiveness, that is, assembling units and preparing the product for the customer, which is not enough to ensure productivity progress in the business. On the contrary, Company A tried its best to use appropriate standards considering the human element in terms of innovation, which is mainly based on efficiency in the business. Even though the ISO 9001 quality standard is the main standard used by Companies A, B, C, and D, these companies need to improve their productivity level by maintaining both effectiveness and efficiency. This collaboration may be done through the application of resources such as safety, cleanliness, and waste reduction; these need to be considered through the establishment of ISO 14001 and ISO 18001, which were not exercised.

3.5 Summary

This chapter primarily presented concepts such as effectiveness and efficiency that need to be comprehended before addressing factors influencing the productivity of SMEs. The purpose was to clarify these concepts, as efficiency can be confused with effectiveness. Efficiency and effectiveness also are found to be distinct from each other in the sense that efficiency is measured in terms of maintaining a satisfactory relationship between cost and benefit. Since manufacturing SMEs need to focus on measuring effectiveness as well as efficiency through the use of work study techniques if they are to be competitive locally and internationally, these concepts to be defined to provide a detailed background. Furthermore, calculations based on effectiveness and efficiency in manufacturing SMEs are presented.

References

Ahmad, K.I., Shivastav, R.L., Pervez, S., and Khan, N.P. 2014. Analyzing quality and productivity improvement in steel rolling industry in central India. *IOSR Journal of Mechanical and Civil Engineering*, 6–11.

Bevilacqua, M., Ciarapica, F.E., Sanctis, I.D., and Paciarotti, G.M.C. 2015. A changeover time reduction through an integration of lean practices: A case study from pharmaceutical sector. *Assembly Automation*, 35(1): 22–34.

Biloslavo, R., Bagnoli, C., and Figelj, R.R. 2013. Managing dualities for efficiency and effectiveness of organisations. *Industrial Management and Data Systems*, 113(3): 423–442.

Bocken, N.M.P., Short, S.W., Rana, P., and Evans, S. 2014. A literature and practice review to develop sustainable business model archetypes. *Journal of Cleaner Production*, 65: 42–56.

Brettel, M., Friederichsen, N., Keller, M., and Rosenberg, M. 2014. How virtualization, decentralization and network building change the manufacturing landscape: An industry 4.0 perspective. *International Journal of Mechanical, Aerospace, Industrial, Mechatronic and Manufacturing Engineering*, 8(1): 37–44.

Chaplin, L., Heap, J., and O'Rourke, M.T.J. 2016. Could "Lean Lite" be the cost effective solution to applying lean manufacturing in developing economies? *International Journal of Productivity and Performance Management*, 65(1): 126–136.

Chen, Y.C., Li, Y., and Shady, B.D. 2010. From value stream mapping toward a lean/sigma continuous improvement process: An industrial case study. *International Journal of Production Research*, 48(4): 1069–1086.

Chueprasert, M. and Ongkunaruk, P. 2015. Productivity improvement based line balancing: A case study of pasteurized milk manufacturer. *International Food Research Journal*, 22(6): 2313–2317.

del Río, P., Peñasco, C., and Romero-Jordán, D. 2016. What drives eco-innovators? A critical review of the empirical literature based on econometric methods. *Journal of Cleaner Production*, 112: 2158–2170.

Denton, P.D. and Maatgi, M.K. 2016. The development of a work environment framework for ISO 9000 standard success. *International Journal of Quality and Reliability Management*, 33(2): 1–13.

Digalwar, A.K., Tagalpallewar, A.R., and Sunnapwar, V.K. 2013. Green manufacturing performance measures: An empirical investigation from Indian manufacturing industries. *Measuring Business Excellence*, 17(4): 59–75.

Dora, M. and Gellynck, X. 2015. House of lean for food processing SMEs. *Trends in Food Science and Technology*, 44: 272–281.

Dora, M., Kumar, M., and Gellynck, X. 2016. Determinants and barriers to lean implementation in food-processing SMEs: A multiple case analysis. *Production Planning and Control*, 27(1): 1–23.

Els, G., van der Walt, R., de Wet, S.R., and Meyer, L. 2012. *Fundamentals of Cost and Management Accounting*. 6th edn. Durban: LexisNexis.

Farooquie, P., Gani, A., Zuberi, A.K., and Hashmi, I. 2012. An empirical study of innovation–performance linkage in the paper industry. *Journal of Industrial Engineering International*, 8: 23:1–6.

Feiz, R., Ammenberg, J., Baas, L., Eklund, M., Helgstrand, A., and Marshall, R. 2015. Improving the CO2 performance of cement, part II: Framework for assessing CO2 improvement measures in the cement industry. *Journal of Cleaner Production*, 98: 282–291.

Garengo, P. and Sharma, M.K. 2014. Performance measurement system contingency factors: A cross analysis of Italian and Indian SMEs. *Production Planning and Control*, 25(3): 220–240.

Georgiev, S. and Georgiev, E. 2015. Motivational factors for the adoption of ISO 9001 standards in Eastern Europe: The case of Bulgaria. *Journal of Industrial Engineering and Management (JIEM)*, 8(3): 1020–1050.

Gunasekaran, A. and Spalanzani, A. 2012. Sustainability of manufacturing and services: Investigations for research and applications. *International Journal of Production Economics*, 140: 35–47.

Gupta, S., Acharya, P., and Patwardhan, M. 2013. A strategic and operational approach to assess the lean performance in radial tyre manufacturing in India: A case based study. *International Journal of Productivity and Performance Management*, 62(6): 634–651.

Habidin, N.F. and Yusof, S.M. 2013. Critical success factors of Lean Six Sigma for the Malaysian automotive industry. *International Journal of Lean Six Sigma*, 4(1): 60–82.

Hasle, P. 2014. Lean production: An evaluation of the possibilities for an employee supportive lean practice. *Human Factors and Ergonomics in Manufacturing and Service Industries*, 24(1): 40–53.

Henriques, J. and Justina Catarino, J. 2015. Sustainable value and cleaner production e research and application in 19 Portuguese SME. *Journal of Cleaner Production*, 96: 379–386.

Jaca, C., Viles, E., Paipa-Galeano, L., Santos, J., and Mateo, R. 2014. Learning 5S principles from Japanese best practitioners: Case studies of five manufacturing companies. *International Journal of Production Research*, 52(15): 4574–4586.

Jain, B., Adil, G.K., and Ananthakumar, U. 2013. An instrument to measure factors of strategic manufacturing effectiveness based on Hayes and Wheelwright's model. *Journal of Manufacturing Technology Management*, 24(6): 812–829.

Johnson, M.P. 2015. Sustainability management and small and medium-sized enterprises: Managers' awareness and implementation of innovative tools. *Corporate Social Responsibility and Environmental Management*, 22: 271–285.

Karim, A. and Arif-Uz-Zaman, K. 2013. A methodology for effective implementation of lean strategies and its performance evaluation in manufacturing organizations. *Business Process Management Journal*, 19(1): 169–196.

Koyratty, B.N.S., Aumjaud, B., and Neeliah, S.A. 2014. Food additive control: A survey among selected consumers and manufacturers. *British Food Journal*, 116(2): 353–372.

Kristianto, Y., Ajmal, M.M., and Sandhu, M. 2012. Adopting TQM approach to achieve customer satisfaction. *The TQM Journal*, 24(1): 29–46.

Leithold, N., Woschke, T., Haase, H., and Kratzer, J. 2016. Optimising NPD in SMEs: A best practice approach. *Benchmarking: An International Journal*, 23(1): 262–284.

Li, J. and Doolen, T.L. 2013. A study of Chinese quality circle effectiveness. *International Journal of Quality and Reliability Management*, 31(1): 14–31.

Mahmood, K., Mahmood, I., Qureshi, A., and Nisar, A. 2014. An empirical study on measurement of performance through TQM in Pakistani aviation manufacturing industry. *International Journal of Quality and Reliability Management*, 31(6): 665–680.

Martínez, C.L.P. 2015. Estimating and analyzing energy efficiency in German and Colombian manufacturing industries using DEA and data panel analysis. Part I: Energy-intensive sectors. *Economics, Planning, and Policy*, 10(3): 322–331.

Mathur, A., Dangayach, G.S., Mittal, M.L., and Sharma, M.K. 2011. Performance measurement in automated manufacturing. *Measuring Business Excellence*, 15(1): 77–91.

McDermott, C.M. and Prajogo, D.I. 2012. Service innovation and performance in SMEs. *International Journal of Operations and Production Management*, 32(2): 216–237.

Narayanapillai, R. 2014. Factors discriminating inventory management performance: An exploratory study of Indian machine tool SMEs. *Journal of Industrial Engineering and Management (JIEM)*, 7(3): 605–621.

Ozturk, E., Karaboyacı, M., Yetis, U., Yigit, N.O., and Kitis, M. 2015. Evaluation of integrated pollution prevention control in a textile fiber production and dyeing mill. *Journal of Cleaner Production*, 88: 116–124.

Panizzolo, R., Garengo, P., Sharma, M.K., and Gore, A. 2012. Lean manufacturing in developing countries: Evidence from Indian SMEs. *Production Planning and Control*, 23(10–11): 769–788.

Pazirandeh, A. and Jafari, H. 2013. Making sense of green logistics. *International Journal of Productivity and Performance Management*, 62(8): 889–904.

Prajogo, D., Oke, A., and Olhager, J. 2016. Supply chain processes. *International Journal of Operations and Production Management*, 36(2): 220–238.

Psomas, E. and Antony, J. 2015. The effectiveness of the ISO 9001 quality management system and its influential critical factors in Greek manufacturing companies. *International Journal of Production Research*, 53(7): 2089–2099.

Psomas, E.L., Pantouvakis, A., and Kafetzopoulos, D.P. 2013. The impact of ISO 9001 effectiveness on the performance of service companies. *Managing Service Quality: An International Journal*, 23(2): 149–164.

Roh, J., Hong, P., and Min, H. 2014. Implementation of a responsive supply chain strategy in global complexity: The case of manufacturing firms. *International Journal of Production Economics*, 147: 198–210.

Ruivo, P., Oliveira, T., and Neto, M. 2012. ERP use and value: Portuguese and Spanish SMEs. *Industrial Management and Data Systems*, 112(7): 1008–1025.

Samarajeewa, S., Hailu, G., Jeffrey, S.R., and Bredahl, M. 2012. Analysis of production efficiency of beef cow/calf farms in Alberta. *Applied Economics*, 44(3): 313–322.

Schulze, M., Nehler, H., Ottosson, M., and Thollander, P. 2016. Energy management in industry: A systematic review of previous findings and an integrative conceptual framework. *Journal of Cleaner Production*, 112: 3692–3708.

Shaharudin, M.R., Zailani, S., and Tan, K.C. 2015. Barriers to product returns and recovery management in a developing country: Investigation using multiple methods. *Journal of Cleaner Production*, 96: 220–232

Singh, R.K. 2011. Developing the framework for coordination in supply chain of SMEs. *Business Process Management Journal*, 17(4): 619–638.

Singh, K. and Ahuja, I.S. 2014. Effectiveness of TPM implementation with and without integration with TQM in Indian manufacturing industries. *Journal of Quality in Maintenance Engineering*, 20(4): 415–435.

Sumaedi, S. and Yarmen, M. 2015. The effectiveness of ISO 9001 implementation in food manufacturing companies: A proposed measurement instrument. *Procedia Food Science*, 3: 436–444.

Tarí, J.J., Molina-Azorín, J.F., and Heras, I. 2012. Benefits of the ISO 9001 and ISO 14001 standards: A literature review. *Journal of Industrial Engineering and Management (JIEM)*, 5(2): 297–322.

Taylor, A. and Taylor, M. 2014. Factors influencing effective implementation of performance measurement systems in small and medium-sized enterprises and large firms: A perspective from contingency theory. *International Journal of Production Research*, 52(3): 847–866.

Teng, H.S.S. 2014. Qualitative productivity analysis: Does a non-financial measurement model exist? *International Journal of Productivity and Performance Management*, 63(2): 250–256.

Teng, J., Min, J. and Pan, Q. 2012. Economic order quantity model with trade credit financing for non-decreasing demand. *Omega*, 40: 328–335

Thirupathi, R.M. and Vinodh, S. 2016. Application of interpretive structural modelling and structural equation modelling for analysis of sustainable manufacturing factors in Indian automotive component sector. *International Journal of Production Research*, 54(22): 6661–6682.

Uddin, N. 2015. Productivity growth, efficiency change, and technical progress of a corporate sector in Bangladesh: A malmquist output productivity index approach. *International Journal of Economics and Finance*, 7(8): 240–255.

Valmohammadi, C. 2011. The impact of TQM implementation on the organizational performance of Iranian manufacturing SMEs. *The TQM Journal*, 23(5): 496–509.

Vinodh, S., Kumar, S.V., and Vimal, K.E.K. 2014. Implementing lean sigma in an Indian rotary switches manufacturing organisation. *Production Planning and Control*, 25(4): 288–302.

Vinodh, S., Selvaraj, T., Chintha, S.K., and Vimal, K.E.K. 2015. Development of value stream map for an Indian automotive components manufacturing organization. *Journal of Engineering, Design and Technology*, 13(3): 380–399.

Vivares-Vergara, J.A., Sarache-Castro, W.A., and Naranjo-Valencia, J.C. 2016. Impact of human resource management on performance in competitive priorities. *International Journal of Operations and Production Management*, 36(2): 114–134.

Wu, D., Greer, M.J., Rosen, D.W., and Schaefer, D. 2013. Cloud manufacturing: Strategic vision and state-of-the-art. *Journal of Manufacturing Systems*, 32: 564–579.

Zelbst, P.J., Green, K.W., Sower, V.E., and Reyes, P.M. 2012. Impact of RFID on manufacturing effectiveness and efficiency. *International Journal of Operations and Production Management*, 32(3): 329–350.

Chapter 4

Factors Influencing Productivity in Manufacturing SMEs

4.1 Introduction

Factors influencing growth in manufacturing small and medium enterprises (SMEs) are studied and debated to provide an insight into how manufacturing SMEs face challenges in growing their businesses. These factors comprise physical capital components such as human resources, material, machinery, building (environment/ergonomics), location, and finance; and technological capital components such as tangible assets and intangible assets as well as management.

A diagram of factors influencing productivity in manufacturing SMEs is depicted in Figure 4.1.

4.2 Factors Influencing Productivity in Manufacturing SMEs

The purpose of this study is to identify factors influencing productivity of manufacturing SMEs.

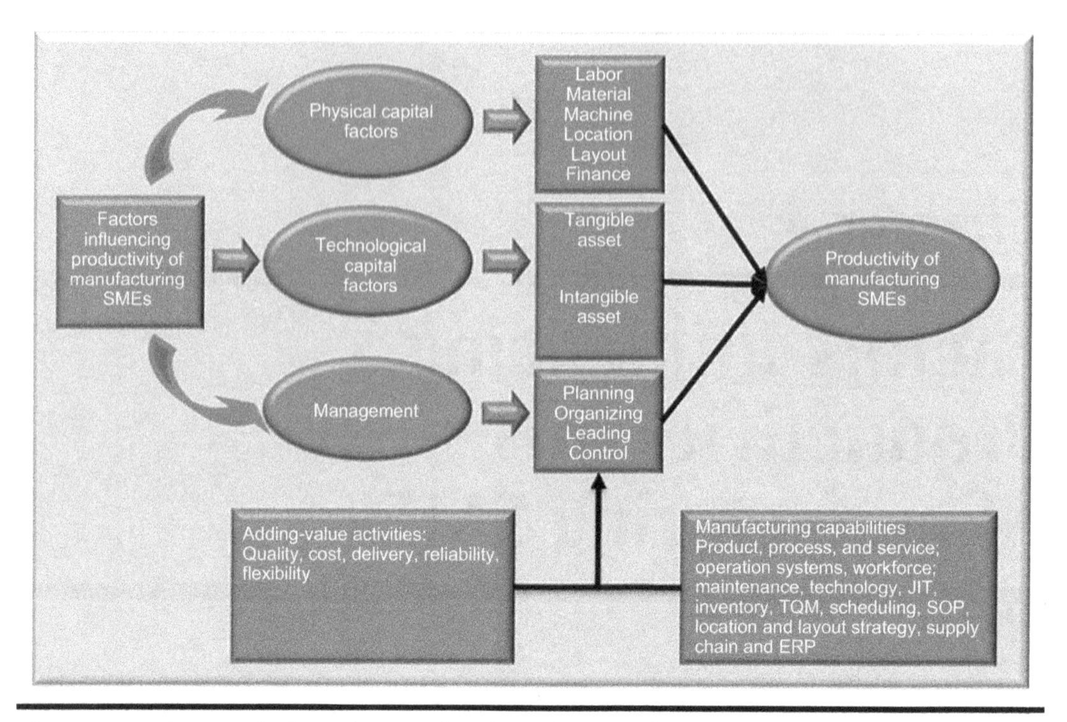

Figure 4.1 Factors influencing productivity in manufacturing SMEs. (From Author, 2017.)

4.2.1 Physical Capital

This section has a breakdown structure of physical capital factors influencing productivity of manufacturing SMEs, such as human resources, material, machinery, building, location, and finance.

4.2.1.1 Human Resources

Jun, Seo, and Son (2013:1005) explain that knowledge, skills, innovation, funding, and time play a pivotal role in contributing toward the productivity of manufacturing SMEs. Based on the literature studied, a shortage of sufficient knowledge, skills, human resources, funding, and time delays the productivity of manufacturing SMEs in their businesses. Mitchelmore, Rowley, and Shiu (2014:590–1) comment that knowledge, motives, traits, self-image, social roles, and skills contributed to improved productivity of manufacturing SMEs.

As is explained by Tan, Smyrnios, and Xiong (2014:324–9), manufacturing SME growth is influenced by various human resources elements, such as employee knowledge, skills, training (learning), education, and motivation, as well as other factors such as information sharing, environmental sustainability, and competition (foreign direct investment). Furthermore, the literature studied recommends that manufacturing SMEs need to engage creatively toward their employees in order to improve productivity in their businesses. This is a challenge to most manufacturing SMEs—to encourage employees to use their innovative skills in order to achieve the expected results. By using innovation, the productivity of manufacturing SMEs improves. Finally, manufacturing SMEs need to create a strong learning culture in order to provide opportunities for personal development and the communication of knowledge toward improving productivity.

Akinboade (2015:185–6) states that background and access to resources, employee motivation, and education influence the productivity of manufacturing SMEs in their businesses.

4.2.1.2 Material

According to Granly and Welo (2014:199–201), appropriate use of material is critical to the productivity of manufacturing SMEs. Based on the literature studied, the research reports that reduced waste material cuts costs in manufacturing SMEs, thereby contributing to their productivity. As advised by Ren et al. (2015:190), material resources play a vital role in the productivity of manufacturing SMEs. These resources need to be managed and made available to facilitate the manufacturing process. Findings from the literature report that the availability of material in the right classification influences the productivity of manufacturing SMEs.

Tarí, Molina-Azorín, and Heras (2012:300) determine factors driving the productivity of manufacturing SMEs. For manufacturing SMEs to sustain their businesses, ISO 9001 standard needs to be exercised. In terms of ISO 9001, manufacturing SMEs inspect incoming material to be used in the manufacturing process. The purpose is to guarantee that this process meets the product specification that meets the customers' requirements.

4.2.1.3 Machinery

As is pointed out by Panizzolo et al. (2012:770), the use of machinery in the manufacturing process manipulates the productivity of manufacturing SMEs. Based on the literature studied, the research provides information that the utilization of machinery is governed by the bottlenecks taking place in the workplace. Radziwon et al. (2014:1188) explain the extent to which the use of machinery contributes to the manufacturing process in manufacturing SMEs, which in turn influences the productivity of these SMEs in their businesses.

4.2.1.4 Building

Granly and Welo (2014:201) discovered that ISO 14001 plays a pivotal role in enabling manufacturing SMEs to avoid being exposed to chemicals and hazardous substances. However, manufacturing SMEs, by complying with safety measures, enable material to be saved from damage and ensure that employees are not exposed to injuries and accidents. Thus, compliance with safety measures in the environment where material is stored and where employees are working plays an important role in the productivity of manufacturing SMEs. Government intervention in manufacturing SMEs contributes to the growth of manufacturing SMEs. In cases where safety precautions are needed, government needs to communicate with environmental agencies of consultants to ensure that every employee member complies with safety rules and regulations in the manufacturing SMEs (Granly & Welo, 2014:201).

4.2.1.5 Location

Granly and Welo (2014:201) advise that energy such as oil and gas as well as the utilization of water is crucial in areas where these chemicals are used in manufacturing SMEs. Findings from the literature studied indicate that areas where oil, gas, and water are stored contribute to the productivity of manufacturing SMEs.

Akinboade (2015:185–6) argues that not only does the background and access to resources and employees' motivation and educational background influence the productivity of manufacturing SMEs; identification of the location is also essential to these SMEs, in particular where there is appropriate infrastructure such as access to reliable power, communication, water, and transport services. José Rodríguez-Gutiérrez, Moreno, and Tejada (2015:208) emphasize that the growth of manufacturing SMEs depends on the territory of the area in which the business is located and where these SMEs are operating, including the internal areas where work activities are carried out.

José Rodríguez-Gutiérrez et al. (2015:196) state that not only is location is looked at but also competitiveness in terms of the change in size of manufacturing SMEs between two time periods. Findings from the literature studied indicate that competitiveness based on the size of the manufacturing SMEs between time periods adds value to the productivity of manufacturing SMEs.

4.2.1.6 Finance

Barnes and Hinton (2012:503) argue that financial investment in innovative activities is critical to the growth of manufacturing SMEs. Findings from the literature studied indicate that manufacturing SMEs, by investing in finance for innovative activities, influence the productivity of their businesses more than those that do not.

As discovered by Teeratansirikool et al. (2013:171–5), finance plays a pivotal role in the operations of manufacturing SMEs. The outcome of the findings in the literature studied indicates that measuring finance contribute to the productivity of manufacturing SMEs in their businesses. As emphasized by Yazdanfar and Öhman (2015:2), a common challenge facing manufacturing SMEs is access to financial resource. Based on the literature studied, financial constraints impact on the productivity of manufacturing SMEs in their businesses.

4.2.2 Technological Capital

The next section discusses the technological capital factors influencing the productivity of manufacturing SMEs, such as tangible assets and intangible assets.

4.2.2.1 Tangible Assets

As is pointed out by Panizzolo et al. (2012:776), electronic equipment such as barcodes play an important role before, during, and after the manufacturing process for the purpose of monitoring and control. Monitoring and controlling the manufacturing process through barcodes results in productivity progress in manufacturing SMEs.

Despite extensive consumption of energy among manufacturing SMEs, the literature studied reports that the recording of energy bills using energy-saving equipment, investment in new energy equipment, energy tracking through meters (equipment), energy tracking for plant, and energy tracking for equipment results in efficiency improvement. By improving efficiency through these energy-saving technologies, productivity progress prevails in the business of manufacturing SMEs (Kostka, Moslener, & Andreas, 2013:59–61).

Poba-Nzaou, Raymond, and Fabi (2014:495) advise that the use of automated administrative processes and compliance with standards for these processes eliminate errors in manufacturing SMEs. Elimination of errors using automated administrative processes and adhering to standards can only be carried out by employing barcodes for tracking. As a result, tracking drives the productivity of manufacturing SMEs. The literature reviewed further indicates that tangible elements such as hardware equipment also play an important role in the productivity of manufacturing SMEs. When hardware is available to ensure information flow from one section to the other for the collective achievement of results in manufacturing SMEs, the productivity of manufacturing SMEs increases (Guo & Wang, 2014:258; José Rodríguez-Gutiérrez et al., 2015:204).

Al-Somali, Gholami, and Clegg (2015:4–19) advise that one tangible asset contributing to the productivity of manufacturing SMEs, allowing easy telecommunication with other businesses, is the use of telephones by employees in terms of issues relating to the procurement of materials and payments.

Chaplin, Heap, and O'Rourke (2016:130) emphasize the importance of energy-saving equipment in manufacturing SMEs. Findings from the literature indicate that energy-saving equipment helps manufacturing SMEs to function efficiently, which contributes to the productivity of these SMEs.

4.2.2.2 Intangible Assets

4.2.2.2.1 Software

As it is described by Wu, Yu, and Wu (2012:164), technological capital in manufacturing SMEs such as hardware (equipment) and software (information) play a significant role in the productivity of manufacturing SMEs. As is stated by Guo and Wang (2014:258) and José Rodríguez-Gutiérrez et al. (2015:204), material such as software also contributes to the productivity of manufacturing SMEs.

As is stated by Heidrich and Tiwary (2013:5884–6) and Cuerva, Triguero-Cano, and Córcoles (2014:105–8), technologies such as material recycling play an important role in influencing the productivity of manufacturing SMEs. Based on the literature studied, recycling of material as a tangible element reduces environmental impact and maximizes resource efficiency for cost and waste reduction in manufacturing SMEs.

4.2.2.2.2 Technological Tools

Cantamessa, Montagna, and Neirotti (2012:209–10) report that a tool such as information and communication technology (ICT) ensures that the product lifecycle management system supports the management of information for new product design and development, which influence the productivity of manufacturing SMEs. May, Stahl, and Taisch (2016:633) advise that the use of ICT is to ensure monitoring and energy consumption patterns that support operational management to make informed decisions on the use per product. This monitoring and identification of energy consumption pattern through the use of sufficient information reduces costs and contributes to the productivity of manufacturing.

4.2.2.2.3 Technological Systems

As is perceived by Panizzolo et al. (2012:769–76), technological capital, involving tools such as just-in-time (JIT) and total quality control (TQC) in manufacturing SMEs' processes, influences material flow and product quality, which in turn impact on the productivity of manufacturing SMEs. Lourens and Jonker (2013:162–75) assert that technological capital such as creating product design, process design, manufacturing, materials handling and management, control, and planning through the use of office system and costs identified in generating a product using financial systems are among the influences on the productivity of manufacturing SMEs.

As is detected by Poba-Nzaou et al. (2014:495–7) and Huang and Handfield (2015:3), technological capital, involving a tool such as the enterprise resource planning (ERP) to be aligned with manufacturing SMEs' processes, influences operational costs and, in turn, the productivity of manufacturing SMEs. Kumar, Heustis, and Graham (2015:131–3) advise that technological capital systems such as radio frequency identification (RFID) contribute to the productivity of manufacturing SMEs. When RFID is effectively used to track the inventory in the manufacturing process, the productivity of manufacturing SMEs accelerates.

The link between lean production and environmental and sustainable manufacturing actions is generally a positive one, with exceptions related to end-of-pipe abatement requiring additional resources to minimize emissions. The examples generally given for energy reduction support the principle of lean waste reduction and, in turn, flow.

Ball (2015:414–9) declares that, by adding value to technology, manufacturing SMEs follow the waste management hierarchy for categorizing waste management strategies containing a number of steps in sequence, such as prevention, reduction, reuse, recycling, recovery (energy), and disposal. This categorization is referred to as material, product, and packing in preparation for the customer. On the other hand, waste management hierarchy systems in terms of energy include prevention, reduction, reuse, and disposal. This hierarchy can only function by means of the JIT tool to reduce cost, improve quality product, and ensure quick delivery. As a result, JIT used in this hierarchy contributes to the productivity of manufacturing SMEs.

4.2.2.2.4 Innovation

Gao et al. (2011:436–8) investigate how technological capital such as innovation influences productivity on the assembly line among manufacturing SMEs. The result of the literature studied reports that when lack of innovation, such as the use of an unknowledgeable and unskilled cheap workforce, as well as non-availability of network arise, high costs are incurred and ultimately the productivity of manufacturing SMEs is affected.

Furthermore, Guo and Wang (2014:262–3) state that competitiveness is critical to the productivity of manufacturing SMEs. Manufacturing SMEs, by participating in international markets via export, are able to have access to potential sources of innovation. As a result, this type of innovation helps these SMEs to acquire more external knowledge, which enables their businesses to sustain their productivity. These manufacturing SMEs become innovative through benchmarking with other innovative manufacturing SMEs; hiring employees with experience and training, coming from innovative firms; collaborating with universities on research and development and/or with other research agents; and holding business or university exhibitions.

In terms of intangible assets such as searching knowledge and exposure to innovation, costs are reduced by manufacturing SMEs and contribute to the productivity of their businesses (Guo & Wang, 2014:263; José Rodríguez-Gutiérrez et al., 2015:196). According to Yazdanfar and Öhman (2015:2), one of the factors influencing the productivity of manufacturing SMEs is competiveness, whereby one SME improves its manufacturing operational process through innovation, which is then regarded as the prerequisite for another SME's survival.

4.2.2.2.5 Networking

As is pointed out by Panizzolo et al. (2012:776), networks such as electronic data interchange, internet, extranet, intranet, websites, barcoding, and fax are exercised to ensure productivity enhancement in manufacturing SMEs. Panizzolo et al. (2012:776) emphasize the importance of team work through networking for decision making in manufacturing SMEs. The literature shows that working as a team from top to bottom improves the degree of integration within the business as well as between the buyer and the supplier in manufacturing SMEs, encourages innovation through brainstorming, and cuts down manufacturing costs by examining manufacturing operations, which influences the productivity of manufacturing SMEs.

One of the components of innovation influencing growth in manufacturing SMEs is networking. When knowledge is shared among the members of these SMEs as well as stakeholders in the growth of manufacturing SMEs, these SMEs manage to sustain the productivity of their businesses (Brink, 2015:264). May et al. (2016:633–6) suggest that knowledge transfer plays a pivotal role in the manufacturing process of manufacturing SMEs. Transferring knowledge through networking among the members of the manufacturing SMEs becomes critical to the productivity of their SMEs' businesses.

4.2.2.2.6 Technological Skills

Cantamessa et al. (2012:209–10) identify technological capital factors such as education in IT. Findings from the literature studied report that employees exposed to IT skills improve the productivity of manufacturing SMEs. According to Wu et al. (2012:164), technological skills in manufacturing SMEs influence the productivity of their businesses. When there is availability of technical knowledge or managerial skills, manufacturing SMEs add value to their productivity.

Kostka, Moslener, and Andreas (2013:59–61) endorse that training in technical know-how and the skill of energy efficiency investments is essential to employee members in manufacturing SMEs. However, manufacturing SMEs that have acquired specialized knowledge of energy efficiency saving contribute to the productivity of their businesses more than those that do not. May et al. (2016:633–6) set a guideline that technological skills play an important role in the manufacturing processes of manufacturing SMEs. These skills are embedded by training employees to prepare them to be engaged in achieving the results of manufacturing SMEs by contributing to the productivity of their SMEs' businesses.

4.2.3 Management Challenges

Management challenges are studied and addressed below in order to show how these challenges impact on the productivity of manufacturing SMEs.

Management manufacturing capabilities such as technology for communication, improvement, employee involvement, and inventory contribute to the productivity of manufacturing SMEs. Not only these capabilities but also value activities such as quality, delivery, cost, and reliability add value to these capabilities and are also a challenge to the productivity of manufacturing SMEs (Bayo-Moriones, Billón, & Lera-López, 2013:119–25).

As is debated by Panizzolo et al. (2012:769–81); Bhamu, Singh, and Sangwan (2014: 877–918); and Dora and Gellynck (2015:274), management is a serious challenge to the productivity of manufacturing SMEs. Manufacturing capabilities exercised by management, such as JIT; total quality management (TQM); product, process, and service design; maintenance; manufacturing planning and control; employees' abilities and involvement; supply chain; scheduling; technology and innovation; and inventory management stimulate the productivity of manufacturing SMEs. In addition, the adding-value activities such as quality, cost, delivery, reliability, and flexibility also are the driving force of manufacturing capability, influencing the productivity of manufacturing SMEs.

Awheda et al. (2016:313–23) state that management challenges, such as manufacturing capabilities and adding-value activities, reinforce the productivity of manufacturing SMEs. Manufacturing capabilities used are inventory; total quality; technology; JIT delivery; production planning and control; standard operating procedures; and innovation within the manufacturing operations of manufacturing SMEs. These capabilities are integrated with adding-value activities such as costs, delivery, quality, reliability, and flexibility in order to allow manufacturing SMEs to compete locally and globally.

According to Oguntoye and Evans (2017:76–7), management plays an important role in the productivity of manufacturing SMEs. Management challenges such as the availability of manufacturing capabilities that include the environment for employees' health and safety, technology, maintenance, supply chain, innovation, and scheduling along with adding-value activities such as cost, speed, quality, and reliability impacts on the productivity of manufacturing SMEs. The background of physical capital in manufacturing SMEs is the next section to be discussed.

4.3 Understanding of Physical Capital in Manufacturing SMEs

In this section, the meaning of physical capital in manufacturing SMEs is studied and described with the purpose of providing an understanding of what physical capital is in manufacturing. Physical capital in manufacturing SMEs involves physical infrastructures of supply chain, such as road, rail, and air (Sharma & Sehgal, 2010:103).

Johnson and Templar (2011:95–100) state that physical capital in manufacturing SMEs encompasses physical assets such as material inventory; access to finance and costs; infrastructure (such as supply chain); machinery; location; environment; and management, as well as technological capital such as equipment, knowledge, and information.

Mandal and Madheswaran (2011:59–72) consider physical capital in manufacturing SMEs as capital involving skill; knowledge; material; environment; location; machinery and energy such as power, fuel, and coal; and technological capital, which entails information. As explained by Naude and Matthee (2011:63–83), physical capital in manufacturing SMEs focuses on location in terms of transport costs and infrastructure in terms of determining economic activity and trade, which involves ports, airports, roads, and rail as well as information and communication technologies used in trade transactions and in customs. Plant is meant to be a well-maintained physical facility, which provides a safe and comfortable working environment for its workers in the manufacturing SMEs (Sethi et al., 2011:500).

Akroush (2012:344) describes physical capital as assets controlled by the manufacturing SMEs and used as inputs in the manufacturing process. Physical capital, also referred to as *tangible resources*, includes the financial resources and physical assets identified and valued in the financial statements of manufacturing SMEs (Jardon & Martos, 2012:464). According to Manderson and Kneller (2012:319), physical capital focuses on the location where the production process is taking place in the manufacturing SMEs. Physical capital refers to "the plant; location; knowledge; skills; training, material, management, information and design in manufacturing SMEs" (Mensah, 2012:44). As is indicated by Saxer, de Beer, and Dimitrov (2012:145–9), the focal points of physical capital in manufacturing SMEs are training, machinery, material, finance, and information.

Benkraiem and Gurau (2013:159) consider physical capital to be physical assets used in manufacturing SMEs. Physical capital is all manufacturing SMEs' tangible assets, also referred to as *physical assets* (Singh, Oberoi, & Ahuja, 2013:1446). Physical capital in manufacturing SMEs includes scarce physical assets such as land, housing, and mines (Vuong & Napier, 2014:644). Damoa (2013:273) refers to *location* as the choice of place where a firm is located.

4.4 Understanding of Technological Capital in Manufacturing SMEs

In this section, the theoretical background of technological capital is studied and presented to provide an insight of what it entails. Technological capital is considered to be an application of an innovative tool such as benchmarking in measuring the manufacturing process and software to enable manufacturing SMEs to be creative in incorporating processes in order to improve their operational performances (Asrofah, Zailanis, & Fernando, 2010:119). As defined by Asrofah et al. (2010:116), benchmarking is "a technique that identifies captures and implements best practices such as adapting outstanding practices from within the manufacturing SMEs or from other manufacturing SMEs." According to Grigoriev et al. (2014:57), innovation means realizing an idea and putting it into practice.

Ramdass and Pretorius (2011:171) point out that technology capital involves technology such as automation, which helps reduce the time of the machining process in the manufacturing SMEs. Invention differs from innovation in the sense that it is a process of discovering something or suggesting something new. The literature studied also refers to invention as the term *novelty* or *creativity*.

Technological capital in manufacturing SMEs comprises innovation adoption, such as information and communication technology (Barnes & Hinton, 2012:504). Cantamessa et al. (2012:191–201) refer to technological capital in manufacturing SMEs as tools such as ICT that are used to ensure that the product lifecycle management system supports the management of information for new product design and development. Jain and Ahuja (2012:793–803) consider technological capital as a standardized system, such as ISO 9001, that benefits the manufacturing operation of manufacturing SMEs. These benefits include improved process flow, standardization of the production process, improved method, and effective use of existing technology.

McDermott and Prajogo (2012:219) explain that technological capital in manufacturing SMEs focuses on the effective use of existing knowledge to satisfy existing customers in the market. Pedersen and Sudzina (2012:4–5) point out that technological capital in manufacturing SMEs entails performance measurement (PM) systems in order to gain an understanding of how internal (organizational capabilities) such as knowledge and skills, managerial systems, and technical systems, as well as external factors such as delivery and quality of customers, shape the productivity rate of manufacturing SMEs. Skills and knowledge base involve scientific understanding of the organization's members; management systems focus on creating and controlling knowledge; and technical systems involve knowledge embedded in a database.

As described by Wu et al. (2012:164), technological capital in manufacturing SMEs comprises a process of acquiring technical knowledge; the managerial or organizational skills that firms need to efficiently utilize the hardware (equipment) to automate the manufacturing process; and software (information) to disseminate it to other members of the business and other manufacturing SMEs.

Findings from Lourens and Jonker (2013:57) divide technological capital into technology relating to product design; process design; manufacturing; materials handling and management; and the control and planning of office systems and financial systems among manufacturing SMEs. Grigoriev et al. (2014:58) refer to technology as a set of means, processes, operations, and methods by means of which the input elements, constituting the manufacturing process, are transformed into those of the output. In addition, technology encompasses machinery, mechanisms, tools, skills, and knowledge used in manufacturing SMEs.

Technological capital in manufacturing SMEs is described as a total of two components: the tangible component, including the active part of the firm's tangible fixed assets (equipment such as tool; hardware and machinery manufacturing processes and manufacturing systems), and the intangible component, comprising intangible assets related to products manufacturing and production management (such as knowledge, skills, experience, abilities, new [innovative] ideas; software) (Grigoriev et al., 2014:56–61).

According to Oliveira, Thomas, and Espadanal (2014:498–502), technological capital involves hardware (such as tools or machinery) that embodies the technology which is used to allow the innovation and information in the software to be used effectively. Manufacturing SMEs use software to share information or knowledge among the members of the business and with other manufacturing SMEs for competitiveness. Technological capital focuses on the introduction of new technologies such as robotics and networking, whereby equipment communicates and coordinates the operations of manufacturing SMEs (Brennan et al., 2015:2). Robotics involves computerized machinery that is designed to facilitate the manufacturing processes in manufacturing SMEs (Brennan et al., 2015:13).

As discovered by Huang and Handfield (2015:3–4), technological capital involves systems such as ERP developed from other systems such as material requirements planning (MRP) and manufacturing resource planning to ensure that the decisions of manufacturing SMEs will add value to the business of these SMEs. These systems bring forth the collaboration of operational activities and the corresponding accounting transactions within manufacturing SMEs. According to Kumar et al. (2015:131–3), technological capital involves radio frequency identification technology (RFID) systems to reduce stock-outs by tracking; managing inventory; avoiding shrinkage; correcting in-stock; identifying stock; and authentication by identifying the person responsible for the stock level in the warehouse and the process for the final product. Furthermore, this system also deals with traceability on processes, packaging, and stock kept from stock safety through the use of barcodes and by making use of a product that has a short life span to avoid waste and damages, which are costly for manufacturing SMEs.

May et al. (2016:632–3) say that technological capital in manufacturing SMEs comprises tools such as ICT, which monitor energy and identify consumption patterns to support operational management processes and allow informed decisions on the use of energy per product.

4.5 Understanding of Management in Manufacturing SMEs

In this section, management is explained and defined on the basis of various literature sources in order to present a detailed background of the concept in manufacturing SMEs. Management focuses on planning (such as strategic decisions on establishing policies, maximizing the strength of the team, and continuous improvement in quality and standards setting); organizing (through work teams); leading (through employee involvement and customer satisfaction); monitoring (through audit of costs, inventory, and safety and health); exercising capabilities such as product, process, and service design; TQM; and standards to achieve the results of manufacturing SMEs (Psomas, Kafetzopoulos, & Fotopoulos, 2012:54–66).

As explained by Govindan, Khodaverdi, and Jafarian (2013:347), management entails the planning of environmental objectives for health and safety regulations for employees; the establishment of environmental policies; and the monitoring and continuous analysis of environmental and supplier selection in manufacturing SMEs. Bhamu et al. (2014: 876–902) state that management in manufacturing SMEs refers to production planning through scheduling of quality and supply chain coordination, over and above the control of systems, in order to make the mass production of goods competitive through the business. According to Chatha and Butt (2015:604), management involves the process of achieving the objectives of manufacturing SMEs through planning, organizing, leading, and controlling.

Management involves the planning and implementation of work activities by a management team involving employees to ensure minimum costs, the reduction of waste, the elimination of pollution, and saving energy using organization policies and procedures in order to make a profit for manufacturing SMEs (Schulze et al., 2016:3694–700). As explained by Mageswari, Sivasubramanian, and Dath (2017:506–21), management brings forth the planning of manufacturing processes through policy implementation and motivation of employees in the form of formal training and by enabling these employees to use their skills in engaging their effort in their individual performance. Case study findings and results on factors affecting productivity in manufacturing SMEs in South Africa is the subsequent section to address.

4.6 Case Study Findings and Results on Factors Affecting Productivity in Manufacturing SMEs in South Africa

In this section, the findings of factors affecting productivity in manufacturing SMEs in the literature studied are addressed to find out which factors affect companies visited in Gauteng, South Africa.

Based on the research book, it appears that factors influencing the productivity of manufacturing SMEs are found to be physical, technological, and managerial in nature. A breakdown structure of physical capital factors such as human resources, material, machinery, building, location, and finance impact on the productivity of manufacturing SMEs, whereas technological capital factors influencing the productivity of manufacturing SMEs in their businesses are tangible and intangible assets. Tangible assets involve equipment such as hardware, scanners, and trackers, whereas intangible assets comprise software, tools, systems, innovation, networking, and technological employees' knowledge and skills. Finally, management, through planning, organizing, leading, and control, also contributes to the productivity of manufacturing SMEs.

Companies A, B, C, and D provided the information regarding factors affecting their business performances in Gauteng, South Africa. Human elements, such as resistance to change, are being seen as a serious challenge facing Company A and Company B. Furthermore, knowledge and skills provided to employees were not contributing enough to the efficient running of the business in Companies C and D. Product and service quality is a serious challenge facing all these companies, whereby there is waste, defects, and late delivery of products to customers. Transports and movement of material is a disaster for all these companies, since material is carried manually, which may cause life-long injuries to employees. Even though quality is a serious challenge to these companies, maintenance scheduling is done effectively, but only in Companies A, B, and D. Company C is in an early start-up stage whereby it has been only four months in business. Storage space is lacking for all companies, and the receiving and issuing of inventory is done at the door, which may cause trafficking and hazards for employees. Safety is not yet in the process of being exercised since no signage and demarcations are available to guide the employees as to where to walk. Costs may be incurred if factors such as employee exposure to hazards and lack of motivation, poor inventory control, inappropriate storage, and lack of safety measures are not addressed by management at the top in order to ensure productivity progress in the business in Gauteng, South Africa.

4.7 Summary

The purpose of this study is to identify factors influencing productivity of manufacturing SMEs. These factors comprise physical capital components such human resources, material, machinery, building, location, and finance, as well as technological capital components such as tangible assets and intangible assets. Factors influencing growth in manufacturing SMEs were studied and debated to provide an insight into how manufacturing SMEs face challenges in growing their businesses. Finally, physical capital and technological capital were also addressed to provide their detailed background in manufacturing SMEs. Tangible assets used by manufacturing SMEs comprised of equipment such as trackers, hardware, and telephone, whereas intangible assets used in manufacturing SMEs include computer software; information and communication technology systems; employees' innovative skills (through research and development); networking; and employees' knowledge.

References

Akinboade, O.A. 2015. Determinants of SMEs growth and performance in Cameroon's central and littoral provinces' manufacturing and retail sectors. *African Journal of Economic and Management Studies*, 6(2): 183–196.

Akroush, M.N. 2012. Organisational capabilities and new product performance: The role of new product competitive advantage. *Competitive Review: An International Business Journal*, 22(4): 343–365.

Al-Somali, S.A., Gholami, R., and Clegg, B. 2015. A stage-oriented model (SOM) for ecommerce adoption: A study of Saudi Arabian organisations. *Journal of Manufacturing Technology Management*, 26(1): 2–35.

Asrofah, T., Zailani, S., and Fernando, Y. 2010. Best practices for the effectiveness of benchmarking in the Indonesian manufacturing companies. *Benchmarking: An International Journal*, 17(1): 115–143.

Awheda, A., Rahman, M.N.A., Ramli, R., and Arshad, H. 2016. Factors related to supply chain network members in SMEs. *Journal of Manufacturing Technology Management*, 27(2): 312–335.

Ball, P. 2015. Low energy production impact on lean flow. *Journal of Manufacturing Technology Management*, 26(3): 412–428.

Barnes, D. and Hinton, C.M. 2012. Reconceptualising e-business performance measurement using an innovation adoption framework. *International Journal of Productivity and Performance Management*, 61(5): 502–517.

Bayo-Moriones, A., Billón, M., and Lera-López, F. 2013. Perceived performance effects of ICT in manufacturing SMEs. *Industrial Management and Data Systems*, 113(1): 117–135.

Benkraiem, R. and Gurau, C. 2013. How do corporate characteristics affect capital structure decisions of French SMEs? *International Journal of Entrepreneurial Behaviour and Research*, 19(2): 149–164.

Bhamu, J. and Sangwan, K.S. 2014. Lean manufacturing: Literature review and research issues. *International Journal of Operations and Production Management*, 34(7): 876–940.

Brennan, L., Ferdows, K., Godsell, J., Ruggero, G., Keegan, R., Kinkel, S., Srai, S., and Taylor, M. 2015. Manufacturing in the world: Where next? *International Journal of Operations and Production Management* 35(9): 1–35.

Brink, T. 2015. Passion and compassion represent dualities for growth. *International Journal of Organizational Analysis*, 23(1): 41–60.

Cantamessa, M., Montagna, F., and Neirotti, P. 2012. Understanding the organizational impact of PLM systems: Evidence from an aerospace company. *International Journal of Operations and Production Management*, 32(2): 191–215.

Chaplin, L., Heap, J., and O'Rourke, M.T.J. 2016. Could "Lean Lite" be the cost effective solution to applying lean manufacturing in developing economies? *International Journal of Productivity and Performance Management*, 65(1): 126–136.

Chatha, K.A. and Butt, I. 2015. Themes of study in manufacturing strategy literature. *International Journal of Operations and Production Management*, 35(4): 604–698.

Cuerva, M.C., Triguero-Cano, A., and Córcoles, D. 2014. Drivers of green and non-green innovation: Empirical evidence in Low-Tech SMEs. *Journal of Cleaner Production*, 68: 104–113.

Damoa, O.B.O. 2013. Strategic factors and firm performance in an emerging economy. *African Journal of Economic and Management Studies*, 4 (2): 267–287.

Dora, M. and Gellynck, X. 2015. House of lean for food processing SMEs. *Trends in Food Science and Technology*, 44: 272–281.

Gao, J., Yao, Y., Zhu, V.C.Y., Sun, L., and Lin, L. 2011. Service-oriented manufacturing: A new product pattern and manufacturing paradigm. *Journal of Intelligent Manufacturing*, 22: 435–446.

Govindan, K., Khodaverdi, R., and Jafarian, A. 2013. A fuzzy multi criteria approach for measuring sustainability performance of a supplier based on triple bottom line approach. *Journal of Cleaner Production*, 47: 345–354.

Granly, B.M. and Welo, T. 2014. EMS and sustainability: Experiences with ISO 14001 and Eco-Lighthouse in Norwegian metal processing SMEs. *Journal of Cleaner Production*, 64: 194–204.

Grigoriev, S.N., Yeleneva, J.Y., Golovenchenko, A.A., and Andreev, V.N. 2014. Technological capital: A criterion of innovative development and an object of transfer in the modern economy. *Procedia CIRP*, 20: 56–61.

Guo, B. and Wang, Y. 2014. Environmental turbulence, absorptive capacity and external knowledge search among Chinese SMEs. *Chinese Management Studies*, 8(2): 258–272.

Heidrich, O. and Tiwary, A. 2013. Environmental appraisal of green production systems: Challenges faced by small companies using life cycle assessment. *International Journal of Production Research*, 51(19): 5884–5896.

Huang, Y. and Handfield, R.B. 2015. Measuring the benefits of ERP on supply management maturity model: A "big data" method. *International Journal of Operations and Production Management*, 35(1): 2–25.

Jain, S.K and Ahuja, I.S. 2012. An evaluation of ISO 9000 initiatives in Indian industry for enhanced manufacturing performance. *International Journal of Productivity and Performance Management*, 61(7): 778–804.

Jardon, C.M. and Martos, M.S. 2012. Intellectual capital as competitive advantage in emerging clusters in Latin America. *Journal of Intellectual Capital*, 13(4): 462–481.

Johnson, M. and Templar, S. 2011. The relationships between supply chain and firm performance: The development and testing of unified proxy. *International Journal of Physical Distribution and Logistics Management*, 41(2): 88–103.

José Rodríguez-Gutiérrez, M.J., Moreno, P., and Tejada, P. 2015. Entrepreneurial orientation and performance of SMEs in the services SME. *Journal of Organizational Change Management*, 28(2): 194–212.

Jun, S., Seo, J.H., and Son, J. 2013. A study of the SME technology road mapping program to strengthen the R&D planning capability of Korean SMEs. *Technological Forecasting and Social Change*, 80: 1002–1014.

Kostka, G., Moslener, U., and Andreas, J. 2013. Barriers to increasing energy efficiency: Evidence from small- and medium-sized enterprises in China. *Journal of Cleaner Production*, 57: 59–68

Kumar, S., Heustis, D., and Graham, J.M. 2015. The future of traceability within the U.S. food industry supply chain: A business case. *International Journal of Productivity and Performance Management*, 64(1): 129–146.

Lourens, A.S. and Jonker, J.A. 2013. An integrated approach for developing a technology strategy framework for small-to-medium sized furniture manufacturers to improve competitiveness. *South African Journal of Industrial Engineering*, 23(2): 144–153.

Mageswari, S.D.U., Sivasubramanian, R.C., and Dath, T.N.S. 2017. A comprehensive analysis of knowledge management in Indian manufacturing companies. *Journal of Manufacturing Technology Management*, 28(4): 506–530.

Mandal, S.K. and Madheswaran, S. 2011. Energy use efficiency of Indian cement companies: A data envelopment analysis. *Energy Efficiency*, 4: 57–73.

Manderson, E. and Kneller, R. 2012. Environmental regulations, outward FDI and heterogeneous firms: Are countries used as pollution havens. *Environmental Resource Economics*, 51: 17–352.

May, G., Stahl, B., and Taisch, M. 2016. Energy management in manufacturing: Toward eco-factories of the future: A focus group study. *Applied Energy*, 164: 628–638.

McDermott, C.M. and Prajogo, D.I. 2012. Service innovation and performance in SMEs. *International Journal of Operations and Production Management*, 32(2): 216–237.

Mensah, M.S.B. 2012. Access to market of a manufacturing small business sector in Ghana. *International Journal of Business and Management*, 7(12): 36–46.

Mitchelmore, S., Rowley, J., and Shiu, E. 2014. Competencies associated with growth of women-led SMEs. *Journal of Small Business and Enterprise Development*, 21(4): 588–601.

Naude, W. and Matthee, M. 2011. The impact of transport costs on new venture internalisation. *Journal of International Entrepreneurship*, 9: 62–89.

Oguntoye, O. and Evans, S. 2017. Framing manufacturing development in Africa and the influence of industrial sustainability. *Procedia Manufacturing*, 8: 75–80.

Oliveira, T., Thomas, M., and Espadanal, M. 2014. Assessing the determinants of cloud computing adoption: An analysis of the manufacturing and services sectors. *Information and Management*, 51: 497–510.

Panizzolo, R., Garengo, P., Sharma, M.K., and Gore, A. 2012. Lean manufacturing in developing countries: Evidence from Indian SMEs. *Production Planning and Control*, 23(10–11): 769–788.

Pedersen, E.R.G and Sudzina, F. 2012. Which firms use measures? Internal and external factors shaping the adoption of performance measurement systems in Danish firms. *International Journal of Operations and Production Management*, 32(1): 4–27.

Poba-Nzaou, P., Raymond, L., and Fabi, B. 2014. Risk of adopting mission-critical OSS applications: An interpretive case study. *International Journal of Operations and Production Management*, 34(4): 477–512.

Psomas, E.L., Kafetzopoulos, D.P., and Fotopoulos, C.V. 2012. Developing and validating a measurement instrument of ISO 9001 effectiveness in food manufacturing SMEs. *Journal of Manufacturing Technology Management*, 24(1): 52–77.

Radziwon, A., Bilberg, A., Bogers, M., and Madsen, E.S. 2014. The smart factory: Exploring adaptive and flexible manufacturing solutions. *Procedia Engineering*, 69: 1184–1190.

Ramdass, K. and Pretorius, L. 2011. Implementation of modular manufacturing in the clothing industry in Kwazulu Natal: A case study. *South African Journal of Industrial Engineering*, 22(1): 167–181.

Ren, L., Zhang, L., Tao, F., Zhao, C., Chai, X., and Zhao, X. 2015. Cloud manufacturing: From concept to practice. *Enterprise Information Systems*, 9(2): 186–209.

Saxer, M., de Beer, N., and Dimitrov, D.M. 2012. High speed 5-axis machining for tools applications. *South African Journal of Industrial Engineering*, 23(2): 144–153.

Schulze, M. Nehler, H., Ottosson, M., and Thollander, P. 2016. Energy management in industry: A systematic review of previous findings and an integrative conceptual framework. *Journal of Cleaner Production*, 112: 3692–3708.

Sethi, S.P., Veral, E.A., Shapiro, H.J., and Emelainova, O. 2011. Mattel, Inc.: Global manufacturing principles (GMP): A life-cycle analysis of a company-based code of conduct in the toy industry. *Journal of Business Ethics*, 99: 483–517.

Sharma, C. and Sehgal, S. 2010. Impact of infrastructure on output, productivity and efficiency: Evidence from the Indian manufacturing industry. *Indian Growth and Development Review*, 3(2): 100–121.

Singh, D., Oberoi, J.S., and Ahuja, I.S. 2013. An empirical investigation of dynamic capabilities in managing strategic flexibility in manufacturing organizations. *Management Decisions*, 51(7): 1442–1461.

Tan, C.S.L., Smyrnios, K.X., and Xiong, L. 2014. What drives learning orientation in fast growth SMEs? *International Journal of Entrepreneurial Behavior and Research*, 20(4): 324–350.

Tarí, J.J., Molina-Azorín, J.F., and Heras, I. 2012. Benefits of the ISO 9001 and ISO 14001 standards: A literature review. *Journal of Industrial Engineering and Management*, 5(2): 297–322.

Teeratansirikool, L., Siengthai, S., Badir, Y., and Charoenngam, C. 2013. Competitive strategies and firm performance: The mediating role of performance measurement. *International Journal of Productivity and Performance Management*, 62(2): 168–184.

Vuong, Q.H. and Napier, N.K. 2014. Resource cure or destructive creation in transition: Evidence from Vietnam's corporate sector. *Management Research Review*, 37(7): 642–657.

Wu, W., Yu, B., and Wu, C. 2012. How China's equipment manufacturing firms achieve successful independent innovation. *Chinese Management Studies*, 6(1): 160–183.

Yazdanfar, D. and Öhman, P. 2015. The growth–profitability nexus among Swedish SMEs. *International Journal of Managerial Finance*, 11(4): 1–26.

Chapter 5

Identifying the Environment for Manufacturing SMEs

5.1 Introduction

Manufacturing small and medium enterprises (SMEs) are challenged with variable factors in their working environment, which delay the manufacturing process in their business in South Africa and the rest of the world. Globally, these factors involve noise (Aurich et al., 2012:337; Attarchi et al., 2013:243–4; Sinay & Balážikova, 2014:301–6; Yang & Bagheri, 2015:3–6; Dehaghi et al., 2016:1); pollution (Driessen et al., 2012:144–9; Muduli et al., 2013:336; Jiang, Lin, & Lin, 2014:119; Ocampo, 2015:40–8; Židonienė, 2016:9–11); vibration (Spillane, Oyedele, & von Meding, 2012:412–14; Barve & Muduli, 2013:1105; Sinay & Balážikova, 2014:298–302; Moatari-Kazerouni, Chinniah, & Agard, 2015:4459–75; Dejanovi & Heleta, 2016:440–46); hazardous working conditions (Spillane et al., 2012:412–14; Attarchi et al., 2013:244; Pagell et al., 2014:1161–72; Moatari-Kazerouni et al., 2015:4461; Dejanovi & Heleta, 2016:441–42); excessive temperatures (Kolb, Gockel, & Werth, 2012:621–27; Amponsah-Tawiah et al., 2013:76–8; Sharma & Sidhu, 2014:620–1; Lee et al., 2015:20–4; Ozturkoglu, Saygılı, & Ozturkoglu, 2016:74); dust (Gupta, Acharya, & Patwardhan, 2012:195–201; Amponsah-Tawiah et al., 2013:75–82; Taiwo, 2014:71; Reinhold, Järvis, & Tint, 2015:282–9; Echt & Mead, 2016:1–2); unavailability of personal protective clothing (Potter & Hamilton, 2014:11; Ceballos, Mead, & Ramsey, 2015:217; Top, Adanur, & Öz, 2016:377–80); poor lighting (Mak et al., 2014:1155–6; Reinhold et al., 2015:287; He et al., 2016:1–12); poor employee posture (Wang et al., 2014:161–3; Basahel, 2015:4643–45;

Gonen, Oral, & Yosunlukaya, 2016:616); and poor air quality (Pietroiusti & Magrini, 2014:321; Akram, 2015:41–9; Antuña-Rozado et al., 2016:1). Since these factors affect manufacturing SMEs in the rest of the world, it is evident that South Africa will also be affected (Fore & Mbohwa, 2010:316–24; Chavalala, 2013:3–41; Ledwaba, 2015:13–51). So where a suitable manufacturing environment is lacking, these factors negatively affect the productivity of manufacturing SMEs.

Standard working procedures needed to ensure a safe and healthy environment in manufacturing SMEs are studied and addressed below. These procedures can be followed by focusing on techniques such as ergonomics (Andani et al., 2014:274–8; Longoni & Cagliano, 2015:1335–50) and good housekeeping in manufacturing SMEs (Amin & Karim, 2013:1156; Sangwan, Bhamu, & Mehta, 2014:583; Gupta & Jain, 2015:73–80; O'Neill, Sohal, & Teng, 2016:383).

5.2 Factors to Consider When Selecting the Environment for Manufacturing SMEs

In this section, various factors influencing the environment of manufacturing SMEs are addressed. These factors include noise, pollution, vibration, hazardous working conditions, excessive temperatures, dust, unavailability of personal protective clothing, poor lighting, poor employee posture, and poor air quality. An example of problem areas facing manufacturing SMEs with regard to the working environment is depicted in Figure 5.1.

Factors addressed and learned from different literature sources, which are indicated in Figure 5.1, to name a few, include poor material handling by employees without gloves; the vibration and noise of machinery, also used with unprotected hands; poor safety measures protecting the employee when transporting goods manually; unsafe working conditions such as inappropriate use of a step ladder or tools and materials being scattered around the workplace. Poor material handling and manual transport may result in employees being exposed to musculoskeletal injury when the employee is bending frequently. Secondly, the vibration of machinery may result in long-term conditions involving trembling. Employees may also be at risk of losing limbs and fingers, which cannot be replaced. Scattered tools result in loss, and accessibility becomes difficult when the employee is supposed to use those tools again. Since there are costs involved, manufacturing SMEs may be affected in terms of profitability, thereby reducing productivity in their businesses. Manufacturing SMEs end up not competing well in the market,

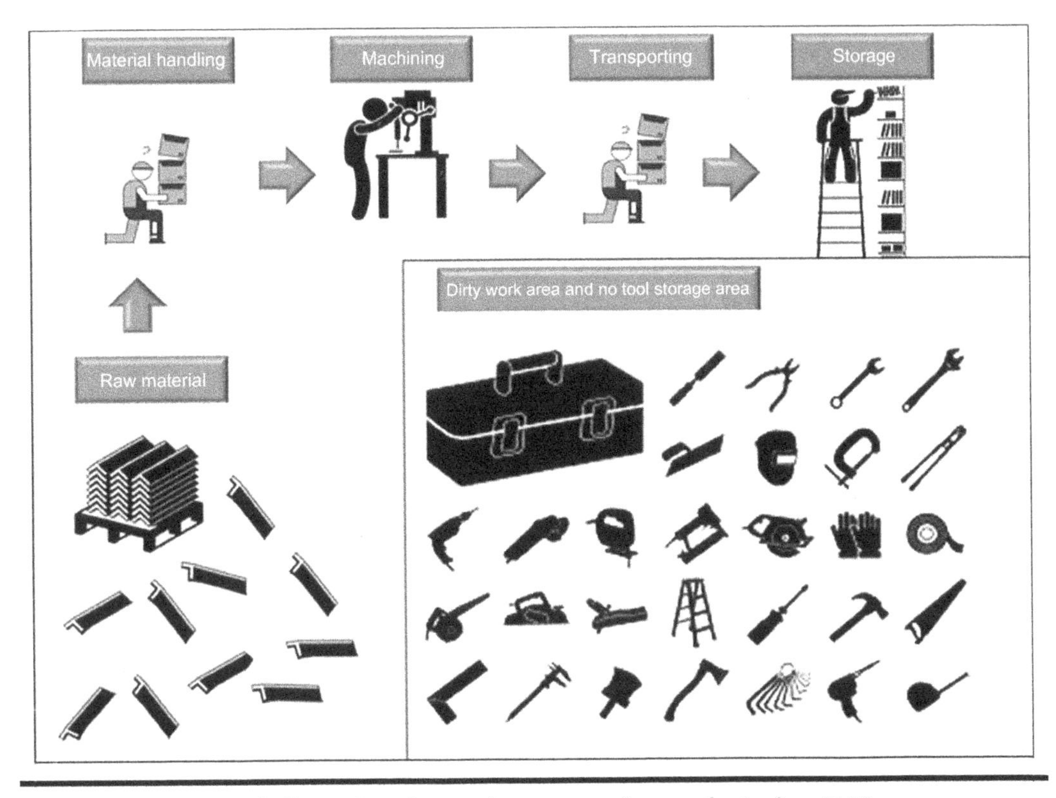

Figure 5.1 Factors influencing the environment of manufacturing SMEs. (From Author, 2017.)

and these SMEs are faced with redundancies. All these problem areas need to be taken seriously to avoid manufacturing SMEs spending more and more money and failing to operate for economic growth of the country.

5.2.1 Noise

The first element to be discussed facing manufacturing SMEs is noise. As explained by Aurich et al. (2012:337), noise is an unpleasant sound emitted from machinery, powered tools, or other activities such as air in manufacturing SMEs. Gunasekaran and Spalanzani (2012:40) explain that noise is sound produced from the manufacturing process through poor communication during the manufacturing operation. Kaljun and Dolšak (2012:168) refer to noise as the result of vibrations transferred from the tool used by the employee into the air and then into the employees' ears. Attarchi et al. (2013:243–4) report that noise in manufacturing SMEs is considered as noise loud enough to expose employees to health risks such as blood pressure, poor hearing, and stress in the workplace. As is reported by O'Driscoll and O'Donnell (2013:59), as most manufacturing SMEs depend on machinery for productivity enhancement, noise is generated during production, which, in turn, results in harm to employees. "Health is a state of complete physical, mental and social wellbeing and not merely the absence of diseases or infirmity" (Mehta, 2012:1).

Singh, Bhardwaj, and Deepak (2013:412–4) comment that noise is a continuous undesired sound of hammering, cutting, and punching press; grinding machinery; and barreling, which results in hearing disability and noise-induced syndromes such as speech interference, noise annoyance, headache, anxiety, and sleep disturbance and which, in turn, decreases the productivity of manufacturing SMEs. Amponsah-Tawiah et al. (2013:76) assert that noise is a manufacturing element stemming from a variety of increasingly mechanized practices, including boring, drilling, blasting, cutting, materials handling, ventilation, crushing, conveying, and product processing in manufacturing SMEs. All these mechanized practices result from the machinery used, which result in employee hearing impairment or traumatic injuries that force an employee to take excessive absenteeism or suffer loss of consciousness or death. The consequence of excessive absenteeism, loss of consciousness, or death represents a loss in productivity for manufacturing SMEs.

According to Sinay and Balážikova (2014:301–6), noise in manufacturing SMEs is regarded as a continuous loud sound emitted from different work stations utilizing machinery, operational processes, and procedures, resulting in employee safety risks such as hearing loss, high blood pressure, work-related stress, and accidents. Noise is an unpleasant sound occurring from machinery, which results in occupational disease such as hearing loss and, in turn, productivity decline in manufacturing SMEs (Sadeghi, Bahrami, & Fatemi, 2014:284–5).

As reported by Yang and Bagheri (2015:3–6), noise is an unsettled sound resulting in irregular records being kept by employees, employee and machine errors, and, in turn, poor-quality products in manufacturing SMEs. Luka and Akun (2015:591) state that noise in manufacturing SMEs is a sound that is undesired from machinery, which results in hearing loss, high blood pressure, stress and head pains, irritation, fatigue, illness, and injuries. Noise is defined as "any unsolicited or unwanted sound that is uncomfortable mentally or the cause of physiological and psychological tension." This noise focuses on indirect effects on human performance, such as a reduction in productivity and the increased risk of accidents and errors due to machinery moving parts, high-speed flow of fluids, and the structural vibration of devices as a result of improper foundations in manufacturing SMEs (Dehaghi et al., 2016:1). As is discovered by Dehaghi et al. (2016:219), noise is undesired sound resulting in employees' failure due to fatigue caused by the noise level, which contributes to employee errors and poor production quality and, in turn, productivity decline in manufacturing SMEs.

5.2.2 Pollution

Driessen et al. (2012:144–9) report that pollution comprises various elements influencing the environment, such as noise, air, and water pollution. Noise pollution is harmful, and annoying noise affects employees in the workplace. Air pollution involves the presence of poisonous effects in the air. Finally, water pollution is water that is not clean, which impacts on the health of employees in the workplace. All these elements of pollution occur in manufacturing SMEs and result in productivity decline in manufacturing industries both small and big businesses. Pollution in manufacturing SMEs focuses on water that contains chemical extracts and air that contains gas that can be harmful to employees in the workplace, which results in a decrease in productivity in SMEs' businesses (Nicot & Scanlon, 2012:3580).

Muduli et al. (2013:336) report that pollution results from acid drainage structure, toxic substances, emissions to air and dust emissions, noise, and vibrations in manufacturing SMEs, which causes many health problems in the manufacturing process. Pollution involves the production of air and noise pollution in the manufacturing process, which are harmful to employees' health in manufacturing SMEs (Tseng et al., 2013:4).

Jiang et al. (2014:119) comment that pollution includes noise, which is a sound that is causing disturbance; dirty water, which contains chemical oxygen and nitrogen pollutants; and air pollution, involving sulfur dioxide and burned dust in the workplace. All these pollutants negatively influence the productivity levels of manufacturing SMEs. Pollution in manufacturing SMEs includes noise occurring from machinery in the manufacturing process that is harmful to employee health; air pollution involving nitrogen and carbon dioxide; and water pollution of water consumed in the workplace (Li et al., 2014:205–9)

As is indicated by Ocampo (2015:40–8), pollution focuses on elements such as toxic substances, gas emissions, noise, and acid leakage, resulting in employee health problems in manufacturing SMEs. As pointed out by Hasanbeigi and Price (2015:38), pollution in manufacturing SMEs consists of water containing chemical components and suspended solid particles as well as air containing nitrogen oxide and carbon dioxide. Židonienė (2016:9–11) describes pollution as being categorized by two types of elements influencing the environment in manufacturing SMEs. These elements are air and water pollution. Air pollution involves harmful material causing illness, which impacts on the manufacturing environment of manufacturing SMEs, whereas water pollution is water that is contaminated, negatively affecting the productivity of manufacturing SMEs. According to Tanioka and Takahashi (2016:3760–73), pollution in manufacturing SMEs encompasses elements such as air pollution, water, and oil. Air pollution involves gas that is harmful to employee health in the workplace, whereas a mixture of both oil and water pollution results in harm among employees in the workplace.

5.2.3 Vibration

Spillane et al. (2012:412–14) explain that vibration in manufacturing SMEs is noise associated with machine operation during drilling, cutting, and driving on manufacturing sites or other manufacturing-related activities whereby employees are exposed to risk factors such as ill-health or disease, fatal accidents, or severe injuries. As noted by Kaljun and Dolšak (2012:167–9), vibration in manufacturing SMEs is an element of an environment causing circulation disorders, white fingers, or other serious cumulative trauma disorders in employees of manufacturing SMEs. White fingers, in terms of vibration, involves an industrial injury triggered by continuous use of vibrating hand-held movement of machinery by employees in their specific workplaces in manufacturing SMEs.

Vibration refers to the cumulative effect of manufacturing process activities that produce noise in manufacturing SMEs. These unpleasant sound conditions result in hearing loss, various health-related problems, and a loss of productivity in manufacturing SMEs (Barve & Muduli, 2013:1105). Amponsah-Tawiah et al. (2013:75–6) advise that the vibration leads to challenges such as shaking hand tools, the use of hammers, the application of machine tools, work on vibrating platforms, manual handling activities, uncomfortable postures, an employee pace of work beyond the normal pace, and working in adverse climates (hot or cold) in manufacturing SMEs. The consequences of these challenges in the working environment include employee injury, fatigue, and disability (hearing loss and other work-related disabilities) and lead to a decrease in productivity in manufacturing SMEs.

According to Sinay and Balážikova (2014:298–302), vibration involves negative effects resulting from adverse conditions endangering the health of the employee, such as machinery, tools, and manufacturing processes, and which end up leading to subjective disturbance, interference with human activity, work and health risks, and an inadequate manufacturing environment that affects the health and ecological balance of employees. As a result of negative effects, the productivity of manufacturing SMEs is affected. Koukoulaki (2014:199–209) reports that vibration includes physical and mechanical risk factors such as repetitive handling of tools and the use of machinery involving high frequency of abnormal postures. These postures negatively contribute to employee musculoskeletal disorders, stress, mental workload and disorders, time pressure, inflammation, and, ultimately, to productivity decline in manufacturing SMEs.

Moatari-Kazerouni et al. (2015:4459–75) comment that vibration occurs as a result of cutting items using a saw blade or machinery. The consequences of vibration from a saw blade are the physical factor such as hand shivering or stress and mental disorder in the workplace within the manufacturing SMEs. According to Chaudhary et al. (2015:268), vibration entails risk factors such as musculoskeletal disorders, lower back pain and spine disorders, and whole-body vibration (WBV) resulting from hand and arm vibration such as machine handling, the seat of a mobile machine, heavy machine driving and loading, as well as unleveled surfaces in manufacturing SMEs.

As mentioned by Dejanovi and Heleta (2016:440–46), vibration is considered as a risk factor causing a sound resulting in physical harm and stress, disability, and musculoskeletal and mental disorders caused by inappropriate manufacturing processes and the continuous shaking of machinery, equipment, and tools for material handling and production in manufacturing SMEs. As described by Pacaiova (2016:15–7), vibration in manufacturing SMEs is exposure to machinery, equipment, tools, and manufacturing processes repeatedly; abnormal temperature and abnormal working hours that expose the employee to the risk of harm, muscular arm and finger disorders, whole-body pain, disease, and spinal injury.

5.2.4 Hazardous Working Conditions

As is explained by Spillane et al. (2012:412–14) hazardous working conditions are conditions in the manufacturing environment whereby employees of manufacturing SMEs are exposed to risk factors such as hazardous working conditions, hazardous materials and equipment on manufacturing sites, weather conditions in the workplace, and spatial constraints. Working conditions in manufacturing SMEs can be bad or good. Working conditions that are bad for the manufacturing process involve an aging workforce, the number of hours worked, the nature of the job, the work schedule, and cigarette smoking. Working conditions that positively contribute to the effectiveness and efficiency of manufacturing SMEs involve exposing employees to exercise (Attarchi et al., 2013:244). As pointed out by Amponsah-Tawiah et al. (2013:75–6), hazardous working conditions in manufacturing SMEs entail conditions such as employee exposure to physical risk factors such as excessive noise; falls from height; mobile equipment accidents; injuries to employees' legs, hands, and head; gases; dusty conditions; fires; faulty machinery; electrocution and vehicular accidents on site; restricted space; harmful gases; awkward posture; poor infrastructure such as a low roof; inappropriate entrance or exit in case of emergencies; weather (heat and cold); poor visibility; and climate.

As noted by Pagell et al. (2014:1161–72), hazardous working conditions result in ineffective safety practices, illness, fatal accidents, poor quality, and high costs, which result in poor productivity in manufacturing SMEs. Da Silva et al. (2014:416–8) consider hazardous working conditions as risk factors such as disease, occupational injuries, working days lost, manual maintenance, repetitive stress, falling, and knocks. These risk factors result in poor planning, quality problems, and absenteeism, which ultimately impact on financial results through business loss and productivity decline in manufacturing SMEs. Moatari-Kazerouni et al. (2015:4461) refer to such hazardous working conditions as malfunctioning of employees due to lack of knowledge, training, and experience in the workplace, exposing employees to harm, injuries, accidents, and death. Hazardous working conditions involve such things as the exposure of employees to chemicals; poor employee health; inappropriate ventilation, preventing employees from breathing properly; work-related injury; illness; and fatality, which impact negatively on the quality and productivity of manufacturing SMEs (Scanlon et al., 2015:28).

As noted by Dejanovi and Heleta (2016:441–42), hazardous working conditions encompass the identification of hazards such as harm or injury at work, employee illness and health, and environmental damage caused by inadequate and unprescribed personal protective equipment and devices in the workplace; limited working space for the quantity of products produced; shortage of employees and machines; poor lighting and ventilation; poor planning; unhealthy environment; unskilled employees; lack of safety; gases; excessive temperature; and dusty working environment in manufacturing SMEs. Prasad, Khanduja, and Sharma (2016:415–24) explain that hazardous working conditions involve dangerous waste, poor employee morale, and industrial accidents, which lead manufacturing SMEs to poor quality and increased inefficiency in their businesses.

5.2.5 Excessive Temperature

Kolb et al. (2012:621–27) refers to excessive temperature as abnormal temperature such as heat and cold and thermal temperatures resulting in employee stress in manufacturing SMEs. These abnormal temperatures generate employee discomfort in the workplace and customer needs are not satisfied. So, lack of service toward customers is a result of poor employee performance in manufacturing SMEs. Temperature is the degree or intensity of heat present in a manufacturing working environment (Kolb et al., 2012:621–33).

As is reported by Lee et al. (2012:138–9), excessive temperature in manufacturing SMEs involves high temperatures or humidity due to the workplace being very hot and wet during summer, whereby employees cannot avoid this type of temperature. These risk factors, such as hot or wet temperatures, contribute to employee heat stress and heart disease. So, manufacturing SMEs are then faced with a poor working environment in their businesses. Amponsah-Tawiah et al. (2013:76–8) state that excessive temperature in manufacturing SMEs involves whether it is too hot or cold in the working environment. These temperatures are related to risk factors such as heat stress, heat stroke, and coldness in manufacturing SMEs. As a result, productivity decreases among these SMEs. Min et al. (2013:1298–1300) report that excessive temperature in manufacturing SMEs involves physical risks such as continuous high temperatures, making employees sweat whether working or not in the workplace, as well as nonstop low temperatures, whether indoors or outdoors, exposing employees to coldness. These risk factors contribute to employee tiredness and physical pain, resulting in poor task performance.

As advised by Sharma and Sidhu (2014:620–1), excessive temperature is heat, thermal, and coldness negatively affecting the performance of employees in manufacturing SMEs. Firstly, heat causes disturbance on the work piece and steadily damages tools used by employees in the workplace. Secondly, excessive thermal temperature occurs due to increases in chemical reactivity in the workplace. Finally, coldness makes it difficult for employees to attend to the manufacturing process in manufacturing SMEs. All these types of temperature lead manufacturing SMEs to poor quality of running a business. Singh et al. (2014:55–61) explain that excessive temperature focuses on poor temperature such as heat, thermal, and cold hindering the employees' operations during loading and standing work in manufacturing SMEs. Furthermore, oxygen uptake is hard for employees due to ambient temperature. Both cold and ambient conditions also cause greater energy of the work force. Thirdly, the results showed that environmental heat causes a fluctuation of heat in the employees for both light and heavy tasks. These risk factors end up resulting in muscle fatigue, metabolism, and heart disease affecting employees' health.

Lee et al. (2015:20–4) note that excessive temperature is temperature beyond control, either high or low, contributing to illness and disease among employees of manufacturing SMEs. The illness and disease incurred negatively affects the performance of these employees in their workplace. Ceballos et al. (2015:217) report that excessive temperature involves low temperature in cold rooms, causing employee discomfort and resulting in injury to body tissues caused by exposure to extreme cold, typically affecting the nose, fingers, or toes. This risk factor ultimately leads the employee to incur infection, heart disease, metabolism disorder, or musculoskeletal problems and result in a drop in employee performance.

Ozturkoglu et al. (2016:74) note that excessive temperature in manufacturing SMEs is bad temperature, either high or low, contributing to excessive coldness in manufacturing SMEs. This type of temperature leads to long-term sickness, resulting in workers being absent from work on a continuous basis. As noted by Reinhold et al. (2015:288), excessive temperature is poor temperature affecting employees due to deficiencies in ventilation systems, open doors, and insulation systems in the industrial building in manufacturing SMEs. The impact of such deficiencies in the workplace results in employee' eye and skin problems and dryness of mucus membranes, which in turn affect employee' performance in manufacturing SMEs.

5.2.6 Dust

Gupta et al. (2012:195–201) state that dust is tiny particles that are formed when material is heated and which contaminate the working environment in the form of vapors and fumes in manufacturing SMEs. The contamination of the working environment then contributes to inefficiency and ineffectiveness. Dusts include tiny and loose particles of material waste, causing employees allergies, which, in turn, lead to respiratory system problems and hearing and eye strain in the workplace. The employee ends up being in a situation whereby fatigue is incurred, resulting in poor employee productivity (Mehta, 2012:1–3).

Amponsah-Tawiah et al. (2013:75–82) report that dust comprises dusty particles harmful to employees' chests, resulting from poor safety planning by management, which in turn leads to short- and long-term effects of dust on the employees' respiratory systems. This dust creates poor visibility in the manufacturing environment, making it impossible for employees to see moving equipment such as forklifts and other moving work-related equipment. The nature of poor visibility then results in poor quality and ultimately productivity deterioration in manufacturing SMEs. As discovered by Magagnotti et al. (2013:784–5), dust includes material particles occurring during manufacturing operations for producing products using machinery, which negatively contribute to risk factors such as noxious effect (nasal), inhalation discomfort, poor respiratory performance, and serious occupational disease such as chest and eye symptoms, skin problems, asthma, and cancer. These risk factors lead to impaired employee performance in manufacturing SMEs.

As is advised by Taiwo (2014:71), dust is small particles of various metals contained in gaseous substances, such as fluorides, sulfur dioxide, and other contaminants floating in the air used by manufacturing SMEs, which results in lung disease, coughing, mucous generation, and respiratory symptoms. Chiarini (2014:226–31) reports that dust involves small particles resulting from oils and the emission of fumes into the atmosphere. These fumes come from the operation and leakage of machinery during the manufacturing of products. This dust-formation process also happens in cases of emissions through chimneys in the form of gases, and all these particles result in employee harm and ultimately employee poor performance in manufacturing SMEs.

Reinhold et al. (2015:282–9) declare that dust is a substance harmful to the respiratory system of the employee when manufacturing products in manufacturing SMEs. The risk of inhaling dusts ultimately results in cancer problems affecting the employees exposed to dust in the working environment. Employees affected by cancer do not perform well in the workplace. Dust is considered as particles from material affecting the employee while carrying out tasks in batch production in manufacturing SMEs. These particles enter organs such as the respiratory system, making it hard for employees to breathe. Ineffective breathing results in poor employee performance in the workplace (Dubey, Verma, & Khandelwal, 2015:4).

According to Echt and Mead (2016:1–2), dust is a substance made up harmful particles affecting the employee in manufacturing SMEs during manufacturing operations using machinery. These harmful substances endanger employees' respiratory health, which negatively affects the efficiency of these SMEs. Ozturkoglu et al. (2016:74) advise that dust is a harmful substance affecting employees during the loading and lifting operation in preparation for a manufacturing process involving machinery in manufacturing SMEs. These substances contribute to long-term sickness, resulting in workers being absent from work for a long period, negatively affecting manufacturing SMEs' performance.

5.2.7 Unavailability of Personal Protective Clothing

Silvestri, De Felice, and Petrillo (2012:4807) explain that the unavailability of personal protective clothing is a process of exposing employees to inappropriate safety standards and to accidents in manufacturing SMEs. Personal protective clothing is protective work clothes such as respirators, nuisance dust masks, surgical masks, gloves, safety glasses, lab coats, full-body protective suits, and protective smocks (Dahm et al., 2012:552).

Arcury et al. (2012:320–4) state that the unavailability of personal protective clothing entails exposure of employees to lifting, repetitive motion, and cutting in the workplace without protective work clothes, which in the long run leads to occupational injury and illness being frequently experienced by manufacturing processing employees. These injuries include musculoskeletal problems, skin infections, mental ill-health and inflammation caused by hazardous toxic substances, eye injury, hearing and footwear issues, head injury, and hand and finger injury.

By the same token, Arcury et al. (2013:182) emphasize that the unavailability of personal protective equipment involves exposing employees to the workplace without protective work clothes, which, in the long run, leads to occupational injury and illness being frequently experienced by manufacturing processing employees in manufacturing SMEs. The outcome of the literature not only refers to musculoskeletal problems, skin infection, mental ill-health and inflammation caused by hazardous toxic substances, eye injury, hearing and footwear problems, head injury, and hand and finger injury but also embraces employee dust masks, hard headgear or plastic helmets, safety goggles, glasses, overalls, jackets, and ear plugs. The absence of these components results in serious injuries and accidents in employees' sphere of operation in manufacturing SMEs.

Fleury et al. (2013:34) assert that the unavailability of personal protective clothing is a process of exposing employees in manufacturing SMEs to the workplace without using protective clothes such as dust masks, gloves, overalls, and overshoes, resulting from inappropriate standard safety procedures.

As explained by Potter and Hamilton (2014:11), the unavailability of personal protective clothing in manufacturing SMEs focuses on the exposure of employees to elements hazardous to the skin, the smell of chemicals, breathing difficulties, high temperatures, nausea, headaches, tiredness, injuries, and accidents. These hazardous elements contribute to abnormal conditions and incorrect standards, resulting in productivity decline in manufacturing SMEs. As pointed out by Sinay and Balážiková (2014:301–6), the unavailability of personal protective clothing exposes employees to danger in the workplace due to failure to use work clothes to carry out their tasks. The employees do not get access to clothes such as ear plugs, masks, and overall, resulting in employee discomfort and work-related stress, leading to accidents and high blood pressure.

Ceballos et al. (2015:217) assert that the absence of protective clothing in manufacturing SMEs involves exposure to cold rooms resulting in skin problems, infections, heart disease, metabolism disorders, or musculoskeletal problems, resulting in a decline in employee productivity. Yu et al. (2015:325) comment that the nonexistence of protective clothing in manufacturing SMEs results in employee exposure to light in the workplace, which results in radiation damage and infection. As discovered by Dababneh and Schwab (2015:338–9), the unavailability of personal protective clothing such as gloves in manufacturing SMEs results in exposure of employees to mechanical, electrical, chemical, and biological hazards; extreme temperatures; and vibration in the manufacturing environment.

As noted by Reinhold et al. (2015:288), the unavailability of protective clothing in manufacturing SMEs involves exposure of employees to toxic substances through inhalation by the employee during the manufacturing process. Risk associated with toxic substances causes health problems such as allergies, cancer, and accidents in manufacturing SMEs. Top et al. (2016:377–80) state that the absence of personal protective clothing makes the employees vulnerable to accidents, irritant chemicals, and dust particles that gradually cause skin disease. When employees are exposed to skin disease, their performance declines and negatively affects the productivity of manufacturing SMEs.

5.2.8 Poor Lighting

Lighting includes the use of both artificial light sources like lamps and light fixtures as well as natural illumination by capturing daylight.

Hemphälä and Eklund (2012:217–23) refer to poor lighting in manufacturing SMEs as causing eye discomfort, eye fatigue, blindness, and eye strain to employees in the working environment. These risk factors lead to employee visual disorders and wasted time, which contributes to productivity decline in manufacturing SMEs. According to Dianat et al. (2013:1536), lighting in manufacturing SMEs focuses on the visibility of the areas from a connected electricity using bulbs showing various characteristics such as light level, type of light sources, light color, and use of daylight. Whereas on the other hand, lighting in manufacturing SMEs involves the use of light sources by capturing visibility in the workplace for the employee to carry out the job in a proper manner (Zhang et al., 2016:451).

According to Motamedzade and Moghimbeigi (2012:231), poor lighting in manufacturing SMEs is lighting that prevents employees from doing the job productively. This type of lighting results in eye strain, awkward posture, and employees inclining their heads, necks, and backs to be able to look closer at their work, which contributes to headaches.

As is indicated by Rose, Orrenius, and Neumann (2013:371), poor lighting involves inadequate lighting that results in poor performance as a result of fatigue, injury or accident, pain, boredom, poor posture, and product errors. Poor quality lighting and the inability of the employees to perform work activities causes quality errors and financial loss. These outcomes contribute to reduced productivity in manufacturing SMEs. Dianat et al. (2013:1535–6) state that poor lighting in manufacturing SMEs involves insufficient illumination affecting employees in the working environment whereby these employees end up suffering from risk factors such as eye strain and eye irritation, impaired visual performance, vision breakdown, headaches, dizziness, and discomfort. These risk factors negatively impact on the productivity of manufacturing SMEs.

Mak et al. (2014:1155–6) explain that poor lighting in manufacturing involves the interruption of the manufacturing process, from preparation of components through assembly up to the testing of the product for quality. This interruption leads to employee fatigue and poor accuracy of results contributing to productivity problems such as defects and high costs. As noted by Potter and Hamilton (2014:7–12), poor lighting in manufacturing SMEs entails lighting affecting employees' working patterns in the workplace during long working hours, resulting in various risk factors such as ill-health, back pain, aching legs, headaches, fatigue, and accidents. These risk factors contribute to high cost and ultimately a decline in productivity in the manufacturing SMEs.

According to Reinhold et al. (2015:287), poor lighting in manufacturing SMEs involves improper lighting conditions affecting eyes and health, accidents, injuries, fatigue, and poor colors resulting from poor visibility in the working area. Manufacturing SMEs end up risking poor performance of their businesses, that is, productivity decline. Schippers, West, and Dawson (2015:773–82) report that poor lighting in manufacturing SMEs involves inadequate lighting, hindering employees from having easy physical access and meaning that they must work in a premises of poor quality. When employees are subjected to a poor quality working area, cost ineffectiveness and poor employee performance prevail.

Top et al. (2016:377–80) refer to poor lighting in manufacturing SMEs as bad lighting that causes accidents in the workplace, ultimately contributing to poor efficiency. This type of lighting hinders employees in that they are prevented from realizing the extent of the accumulation of dirt and dust as well as pollution, which impacts on employees' health and causes accidents in manufacturing SMEs.

As is pointed out by He et al. (2016:1–12), poor lighting in manufacturing SMEs involves insufficient lighting that affects employees both during the day and night. During the day, an employee can perform the job unless the employee is exposed to an area where light is needed but is not sufficient to do the job properly. When working during the night, serious cases happen whereby there is no environmental light source such as a ceiling lamp or infrared night vision in order to see. When these components are not available, material becomes invisible and gets damaged, employee discomfort is incurred, and accidents happen, which negatively impact on the productivity of manufacturing SMEs.

5.2.9 Poor Employee Posture

Halim and Omar (2012:85) assert that poor employee posture involves an employee standing while working in the assembly process, producing products for an extended period of time and thus being exposed to risk factors associated with standing jobs, such as working pose, muscle activity, standing duration, holding time, and whole-body vibration. These risk factors contribute to discomfort and muscle fatigue in the employee's back and legs, which results in impaired performance. Employee posture focuses on a specific position of the employee body whereby a normal operation of repetitive or recurring work is done by an employee at a reduced number of repetitive movements (Wang et al., 2014:161). As defined by Story and Neves (2016:114), employee performance is "the behaviour directed toward formal tasks, duties, and responsibilities that are written in their job description."

Mehta (2012:1) contends that poor employee posture in manufacturing SMEs involves bad positioning of or discomfort (seated working at an abnormal angle) to the employee, with lifted upper arm and forward inclined positioning of head and trunk and relatively unfavorable ankle and knee angles, respectively. All this bad positioning results in continuous handling of loads, prolonged standing, repetitive movement of both hands and wrists, and awkward positioning in relation to the job or working environment whereby the employees ends up being exposed to musculoskeletal disorders affecting areas such as muscles (tendons and ligaments), joints, and bones, causing back problems; stiffness at neck, shoulder, arm, wrist, and leg; nerves; and stress. The employee ultimately incurs a fatigue problem, which prevents this employee from working at standard performance.

According to Lin, Chen, and Pan (2013:223–7), poor employee posture involves poor positioning of the employee in relation to the working environment, with risk factors related to awkward positioning, high production demands associated with long working hours, prolonged standing and frequent walking, poor equipment design, awkward positioning, and repetitive handling activities. These risk factors put employees at risk of muscular injuries and irritation resulting in poor performance of the employee. Dhande (2013:2) explains that poor employee posture encompasses employee positioning that causes sickness, muscular pain, and other physical discomforts in areas such as the neck, shoulder, elbow, wrist/hand, upper back, lower back, knee, and eye. This harm results in the employee getting insufficient sleep, feeling tired, and not sleeping normal hours, which leads to poor performance and absenteeism. The harm takes place in cases of manual material handling by employees and overwork in their sphere of operation, in particular the manufacturing process.

As stated by Gerr et al. (2014:178), poor employee posture in manufacturing SMEs is a working posture that is awkward and repetitive in motion, resulting in employee muscular and skeletal disorders, injuries, illness, stress, and disease. According to Wang et al. (2014:161–3), poor employee posture involves failure to allow normal posture by exposing the employee to risk of excessive movement in the workplace. This excessive movement then contributes to musculoskeletal disorders occurring in the upper limb, neck, and trunk, which hinder the employee from performing well.

As noted by Basahel (2015:4643–45), poor employee posture is posture that is unfavorable in the workplace, leading to employee problems such as musculoskeletal disorders, physiological stress, and other risk factors such as pain in the lower back, shoulder, lower arm, and trunk (chest), which results from manual heavy material handling (lifting and pulling) and workload in manufacturing SMEs. These risk factors contribute to illness and injuries and prevent the employee from performing well. Rivero et al. (2015:4817) point out that poor employee posture refers to posture that is detrimental to employees' bodies, causing negative impacts such as disorders in the upper limbs and a weakened musculoskeletal system. This poor posture results from a poor working environment due to such factors as lighting, noise, temperature, and overloading. The result of poor posture is poor employee performance and absenteeism, negatively contributing to the productivity of manufacturing SMEs.

Poor employee posture is posture taking place on the assembly line, whereby the employee is moving products from a position of squatting to standing with the load and then placing it on top of a shelf for long and uncomfortable working hours. The squatting and standing of employees carrying heavy loads causes pain in the leg, back, and neck muscles as well as stress, resulting in employee fatigue and poor performance (Gonen et al., 2016:616). As indicated by Lu et al. (2016:41–3), poor employee posture is poor and prolonged positioning demanding assembly, visual inspection, and packaging, moving from one manufacturing process to another. This type of posture results in pain in employees' shoulders, ankles, and feet; lower back; hand and wrists; elbows; neck; and legs, as well as injuries from the job being carried out through manual material handling, heavy workload, and machine operation, causing fatigue and cumulative trauma disorders. This dilemma of poor posture causes increased costs and impaired business performance for manufacturing SMEs.

5.2.10 Poor Air Quality

Poor air quality is air processed by insufficient ventilation that threatens employee health and increases the risk of such things as respiratory disease to the employees carrying out work activities in manufacturing SMEs. This type of air impairs the productivity of manufacturing SMEs (Motamedzade & Moghimbeigi, 2012:232). Dahlman-Höglund et al. (2012:628–9) regard poor air quality as air loaded into a ventilation system that jeopardizes employee health as a result of gas entering the ventilation in the workplace. Poor ventilation in manufacturing SMEs causes risk elements such as employee respiratory disease, in this case asthma.

As pointed out by Khrais et al. (2013:233), poor air quality in manufacturing SMEs is low fresh air circulation in a confined space, affecting the employee health in the workplace. Meo et al. (2013:390–1) state that poor air quality in manufacturing SMEs is air processed through inappropriate ventilation systems, impacting on employees' lung function and resulting in chronic respiratory disease.

Pietroiusti and Magrini (2014.321) report that poor air quality in manufacturing SMEs consists of poor ventilation systems, causing respiratory distress. This respiratory disease results from inhalation of toxic substances such as chemicals, and these chemicals end up causing illness to employees. Trianni et al. (2014:211–7) explain that poor air quality in manufacturing SMEs involves risk factors such as high indoor air pollution and increased thermal discomfort from intensive heat resulting from insufficient ventilation in the area where employees are carrying out work activities. These risk factors have a negative impact on the health of employees in manufacturing SMEs.

As is indicated by Akram (2015:41–9), poor air quality results from an inadequately ventilated manufacturing working area where the air is unhealthy and creates an uncomfortable environment for employees in manufacturing SMEs. This environment is exposed to risk factors such as little air flow, no daylight, and high temperatures. Reinhold et al. (2015:285–9) state that poor air quality refers to air deficiencies in ventilation systems, causing temperatures outside the standards required based on the international standard organization (ISO). The absence of ISO in terms of ventilation systems incurs employee ill-health in the workplace.

Antuña-Rozado et al. (2016:1) refer to poor air quality as unventilated roofs and poor indoor air quality whereby smoking is allowed everywhere in the workplace, resulting in a health risk to employees in manufacturing SMEs. According to Reinhold et al. (2015:288), poor air quality is air loaded into inappropriate and inadequate local ventilation systems that do not comply with the required air capacities in the workplace. Poor air quality in the workplace affects employees' eyes and skin and causes dryness of the mucus membranes, which in turn lowers employee performance. The next section to be discussed is the standard working procedures to be followed in ensuring a safe and healthy environment in manufacturing SMEs. These procedures focus on the understanding of ergonomics and good housekeeping.

5.3 Standard Working Procedures In Ensuring a Safe and Healthy Environment in Manufacturing SMEs

In this section, standard working procedures, ensuring a safe and healthy environment in manufacturing SMEs, are outlined and discussed. The aim of addressing these procedures is to provide a detailed understanding of standards with which manufacturing SMEs must comply in their operational environment. These standards are used in compliance with ISO certification applicable within work study techniques such as method study for simplification and work measurement applied in setting the appropriated standard time for the task being carried out.

5.3.1 Understanding of Ergonomics

In this section, ergonomics as a concept is discussed and a detailed background of its role in manufacturing SMEs is presented. According to Jin et al. (2012:328–30), ergonomics (also referred to as *employee environmental system*) as a concept involves employees being provided with proper work instructions through the managers' process planning so that they can work at ease in their workplace. Quintana and Leung (2012:766) state that ergonomics involves implementation of good housekeeping, which reduces accidents and decreases absenteeism and labor compensation, as well as improving productivity. This improvement takes place through better handling of tools and material, elimination of hazards, a lessening of employee fatigue, and the easy flow of material in the manufacturing process.

Ergonomics comprises comfortable job activities for accessing tools, good lighting, ease of handling equipment, no leakage of hazardous chemicals, and suitable standard procedures and maintenances manuals (Agyapong-Kodua, Darlington, & Ratchev, 2013:531–6). Ergonomics in manufacturing SMEs focuses on identifying and avoiding hazards as well as improving health and safety (such as eliminating accidents and injuries) at work (Carrillo-Castrillo, Rubio-Romero, & Onieva, 2013:423). Khrais et al. (2013:227–32) relate ergonomics to the elimination of hazardous chemicals, noise, accidents, extreme temperatures, dangerous elevations, confined work spaces, poor lighting, monotonous and repetitive jobs, excessive vibration, and uncontrolled air circulation through awareness of and adherence to appropriate safety rules. Motamedzade (2013:475–9) contends that ergonomics focuses on using training programs for employees to work as a team to avoid the occurrence of work-related musculoskeletal disorders in all parts of the body (neck, shoulder, elbow, wrist, upper back, lower back, hip, knee, and ankle), which lead to absenteeism, disability, and increased compensation for workers. By avoiding these work-related musculoskeletal disorders, manufacturing SMEs save costs, which ultimately results in the improvement of productivity.

According to Andani et al. (2014:273–9), ergonomics involves introducing employees to healthy and suitable temperatures in their workstation in manufacturing SMEs. Also noted by Dotoli et al. (2014:178–9), ergonomics embraces good housekeeping and the improvement of working conditions, with the focus on reducing accident risk. Jaca et al. (2014:1474–5) define ergonomics as safety training in which employees are introduced to advanced creative thinking in identifying hazardous situations, at the same time ensuring a productive workplace in manufacturing SMEs.

Longoni and Cagliano (2015:1334–49) explain ergonomics as way to ensure employees in manufacturing SMEs work in a safe, healthy, comfortable environment.

Dubey, Gunasekaran, and Chakrabarty (2015:5216) report that ergonomics in manufacturing SMEs necessitates a well-defined environment through the use of ISO 14001, whereby employees are made aware of areas where pollution can affect their well-being in the workplace. According to Cheluszka (2015:62–79), ergonomics includes the process of using machine automation effectively to improve safety and operational processes in manufacturing SMEs. García-Hernández et al. (2016:116–21) argue that ergonomics is about improving safety protective clothing and exposure to normal loading conditions (such as forklift) in order to allow employees to work comfortably, reducing fatigue in the workplace in manufacturing SMEs. Guadix et al. (2015:1475) define ergonomics as a process of ensuring employees are aware of the health (hygiene) and safety (avoiding accidents) of the environment as well as managers' motivation and satisfaction with employees who comply with health and safety measures in their workstations. According to Gupta and Jain (2015:76–82), ergonomics involves the implementation of good housekeeping by adapting employees to the environment to ensure that they are trained to store material in the proper areas, to maintain machinery (lubrication of oil and other substances), and understand efficient ways of carrying out their tasks in manufacturing SMEs.

Harik et al. (2015:4118–25) consider ergonomics to embrace safety policies, procedures, and measures implemented to ensure a comfortable and motivational workplace to improve productivity in manufacturing SMEs. Ergonomics encompasses the promotion of employee safety and well-being and the eradication of physical and chemical hazardous substances in the workplace (Koho et al., 2015:13–14). Moatari-Kazerouni et al. (2015:4470–1) define ergonomics as the application of health and safety occupational tools to prevent exhaustion and tiredness (fatigue) in physical and mental terms, resulting from dangerous posture, as well as the removal of hazardous chemicals to avoid injuries in the workplace.

The next qualitative technique to be described is good housekeeping. As discussed earlier in various literature sources on how ergonomics works hand in hand with good housekeeping, there is a difference between the two techniques (Quintana & Leung, 2012:766; Dotoli et al., 2014:178–9; Gupta & Jain, 2015:76). Ergonomics focuses on the entire manufacturing process of SMEs, whereas good housekeeping puts an emphasis on specific areas within the manufacturing SME (Gupta & Jain, 2015:76).

5.3.2 Understanding of Good Housekeeping

In this section, good housekeeping is studied and defined to provide a detailed understanding of the concept in manufacturing SMEs. Based on the literature studied by various authors such as Srinivasan, Ikuma, Shakouri, Nahmens and Harvey, good housekeeping was first introduced in Japan as 5S. This 5S uses S as an abbreviation, which stands for five Japanese words. These word are provided in sequence as *Seiri* (sort), *Seiton* (set in order), *Seiso* (shine), *Seiketsu* (standardize), and *Shitsuke* (sustain). 5S was developed in the post-Second World War era to eliminate barriers facing the manufacturing process in manufacturing industries, whether they be small or big businesses. This tool was used to reduce waste and optimize productivity and quality through inspection of an orderly work area (Srinivasan et al., 2016:365).

Good housekeeping in manufacturing SMEs is regarded as the foundation of accident and fire prevention, through cleaning, maintenance of machines, and removal of waste on the shop floor. Good housekeeping simply means keeping work areas orderly; maintaining halls and floors free of slip and trip hazards; removing waste materials and other fire hazards from work areas; paying attention to the layout of the whole workplace; checking aisle marking; and ensuring the adequacy of storage facilities, as well as taking care of maintenance (Tezel & Aziz, 2017:129–44). According to the literature studied, good housekeeping involves adhering to standards such as ISO 9001 for quality (Aschehoug et al., 2012:6–8) to minimize employees' exposure to hazardous substances and to reduce material input so that the environment is employee friendly according to ISO 14001 and ISO 14040 (Aschehoug et al., 2012:6–8; Ortolano et al., 2014:125; Wolfgang Thurner & Roud, 2016:2854).

Suárez-Barraza and Ramis-Pujol (2012:78) refer to good housekeeping in manufacturing SMEs as reducing costs through the establishment and maintenance of a high-quality and clean working environment to maximize productivity. According to Al-Najem et al. (2013:292), good housekeeping involves a tidy workplace with well-organized items and equipment to ensure that they are located in the right zone to avoid unnecessary movement. Amin and Karim (2013:1156) explain that good housekeeping in manufacturing SMEs is perceived as ensuring that the area is clean and tools are in order and easier to find. As reported by Ene et al. (2013:124) and Klewitz and Hansen (2014:64–6), good housekeeping is ensuring that tools are washed and cleaned after use.

Sangwan et al. (2014:581–4); Gupta and Jain (2015:72–81); and Prasad et al. (2015:276–82) describe housekeeping as the incorporation of the 5Ss—sort, straighten, sweep, standardize, and self-discipline—making it possible to sort until standardization through standard operating procedures in the manufacturing process for safety and continuous improvement in manufacturing SMEs.

O'Neill et al. (2016:383) explain that good housekeeping involves storing material in the right place for inventory control and expedition (speed) of delivery to customers from the manufacturing SMEs.

Srinivasan et al. (2016:364–5) state that good housekeeping in manufacturing SMEs improves employee morale in the workplace, eliminating safety hazards, placing resources such as tools and products in order, thereby improving productivity.

Petek, Glavič, and Kostevšek (2016:2818) argue that good housekeeping not only focuses on the washing and cleaning of tools but also on washing the floor for cleanliness and for pollution prevention in manufacturing SMEs.

So, for companies in Gauteng, South Africa, as in other manufacturing industries worldwide, to manage the environment efficiently and effectively, ergonomics and good housekeeping are essential tools to improve the manner in which employees should perform and the companies reach their competitive edge. An example of a desired working environment in a manufacturing SME is illustrated in the form of a diagram in Figure 5.2.

A case study on identifying the environment for manufacturing SMEs in South Africa is the subsequent section to be addressed.

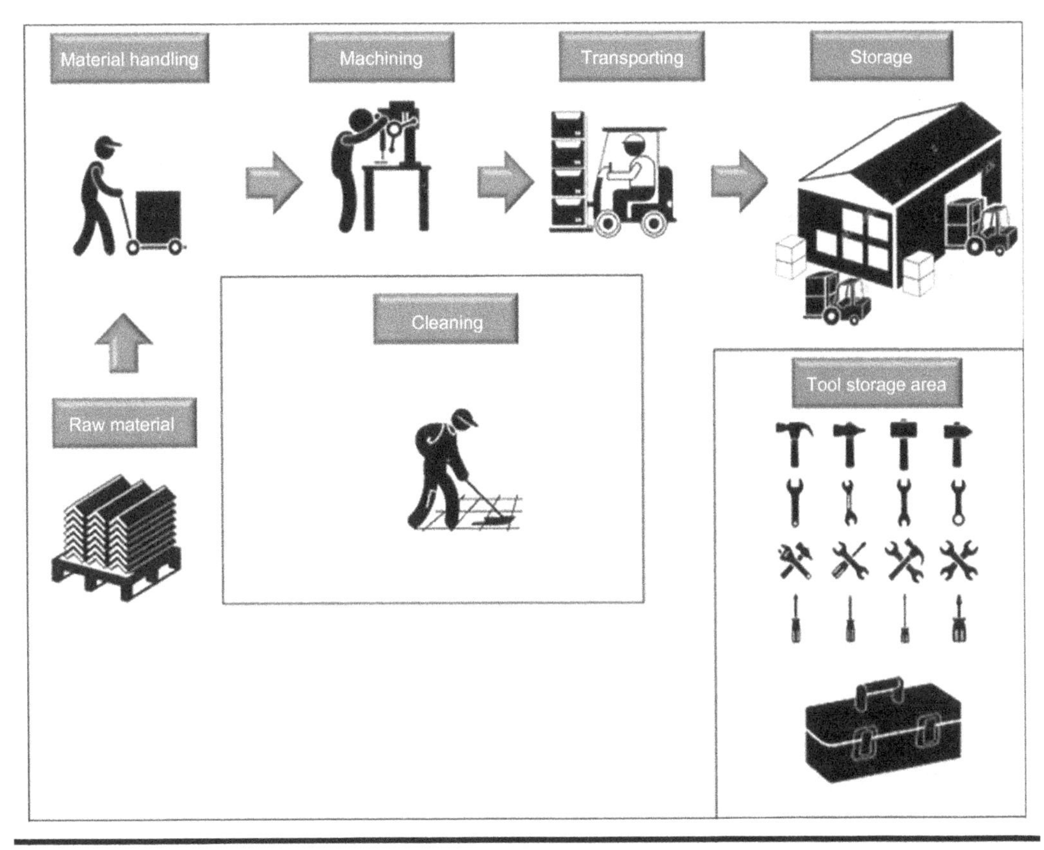

Figure 5.2 Standard working procedures using ergonomics and good housekeeping. (From Author, 2017.)

5.4 Case Study on Identifying the Environment for Manufacturing SMEs in South Africa

Even though the environmental factors facing manufacturing SMEs have been identified, such as noise, pollution, vibration, hazardous working conditions, excessive temperatures, dust, unavailability of personal protective clothing, poor lighting, poor employee posture, and poor air quality, it can been envisaged that these problem areas can be resolved through the guidance of work study in manufacturing SMEs.

Though Company A provides employees with the necessary protective clothing, employees do not comply with the safety measures in the workplace. Company B is exposed to poor lighting. The concept of ergonomics in Company C is not applicable. No protective clothing is used to ensure the safety for employees in terms of dust, noise, and temperature. Chairs used by employees are not suitable to adjust according to the work activities being carried out. Company B emphasizes the importance of ergonomics and good housekeeping but has no standard in places for ISO 14001 and ISO 18001, which is initiated by the company for the well-being of employees in the workplace. Work study in this case can be applied through the discourse of the standard working procedures involving ergonomics and good housekeeping to guide management in all companies visited during the research conducted in ensuring that employees in manufacturing SMEs work in a safer, healthier, and more comfortable workplace.

5.5 Summary

The purpose of this chapter is to identify how the environment affects employees and the standard procedures that should be followed to ensure a safe and healthy environment in manufacturing SMEs. These factors include noise, pollution, vibration, hazardous working conditions, excessive temperatures, dust, unavailability of personal protective clothing, poor lighting, poor employee posture, and poor air quality. Firstly, the implication of noise results in employees being exposed to vibrations, raised blood pressure, poor hearing, stress, speech interference, headache, anxiety, sleep disturbance, and traumatic injuries, which lead employees to loss of consciousness or death, compelling them to take excessive absenteeism. Secondly, pollution subjects the workplace environment to poisonous effects in the air, unclean water, chemical extracts, gas, dust, and noise occurring from machinery, resulting in employee ill-health. Thirdly, vibration is also critical to employee health in the sense that employees incur diseases, fatal accidents, severe injuries, serious cumulative trauma disorders, hearing loss, shaky hands, disability, musculoskeletal disorders, stress, mental workload and disorders, time pressure, inflammation, lower back pain, spine disorders, muscular arms, finger disorders, and whole-body pain, occurring during work but taking place after a long period of work activity. Fourthly, hazardous working conditions represent another factor negatively affecting the well-being of the employees in the workplace.

The factors involved are hazardous materials and equipment, weather conditions, aging, the number of hours worked, the nature of the job, work schedule and cigarette smoking, excessive noise, falls from height, mobile equipment accidents, injuries, gases, dusty conditions, fires, faulty machinery, electrocution and vehicular accidents, awkward posture, poor infrastructure such as a low roof, inappropriate entrance or exit in case of emergencies, inappropriate ventilation structure, dangerous waste, poor employee morale, and industrial accidents, which lead to poor quality and increased inefficiency in manufacturing SMEs. The fifth factor is excessive and abnormal temperatures, such as heat, cold, thermal, and dry temperatures resulting in employee discomfort in the workplace. Sixth, employees in manufacturing SMEs are unprotected from dust like tiny and loose particles of material waste, causing employees allergies, which, in turn, lead to respiratory difficulties and hearing and eye strain in the workplace. Seventh, unavailability of personal protective clothing in manufacturing SMEs, such as masks, safety glasses, lab coats, full-body protective suits, and protective smocks, leads employees to suffer from musculoskeletal problems, skin infections, mental ill-health, and inflammation caused by hazardous toxic substances; eye injuries; hearing and footwear problems; head injuries; and hand and finger injuries. Eighth, poor lighting such as lamps and light fixtures, as well as inadequate natural illumination, results in employees being exposed to danger of areas that cannot be seen. Ninth, poor employee working posture, muscle activity, standing duration, holding time, and whole-body vibration, repetitive movements, and prolonged standing result in poor body muscles (tendons and ligaments); joints, bone, and back problems; stiffness at the neck, shoulder, arm, wrist, and leg; nerves; and stress in manufacturing SMEs. Lastly, poor air quality is also a serious environmental factor negatively impacting the workplace environment. Insufficient ventilation threatens employee health and increases the risk factor of such things as respiratory diseases for employees carrying out work activities in manufacturing SMEs. The above-mentioned environmental factors can be improved by focusing on tools such ergonomics and good housekeeping in order to ensure that employees are working in a healthy, safe, and comfortable environment.

References

Agyapong-Kodua, K., Darlington, R., and Ratchev, S. 2013. Towards the derivation of an integrated design and manufacturing methodology. *International Journal of Computer Integrated Manufacturing*, 26(6): 527–539.

Akram, O. 2015. Occupational health, safety and extreme poverty: A qualitative perspective from Bangladesh. *International Journal of Occupational Safety and Health*, 4 (1): 41–50.

Al-Najem, M., Dhakal, H., Labib, A., and Bennett, N. 2013. Lean readiness level within Kuwaiti manufacturing industries. *International Journal of Lean Six Sigma*, 4(3): 280–320.

Amin, M.A. and Karim, M.A. 2013. A time-based quantitative approach for selecting lean strategies for manufacturing organisations. *International Journal of Production Research*, 51(4): 1146–1167.

Amponsah-Tawiah, K., Jain, A., Leka, S., Hollis, D., and Cox, T. 2013. Examining psychosocial and physical hazards in the Ghanaian mining industry and their implications for employees' safety experience. *Journal of Safety Research*, 45: 75–84.

Andani, A.M., Golbabaei, F., Shahtaheri, S.J., and Foroushani, A.R. 2014. Evaluating workers' exposure to metalworking fluids and effective factors in their dispersion in a car manufacturing factory. *International Journal of Occupational Safety and Ergonomics (JOSE)*, 20(2): 273–280.

Antuña-Rozado, C., García-Navarro, J., Reda, F., and Tuominen, P. 2016. Methodologies developed for EcoCity related projects: New Borg El Arab, an Egyptian case study. *Energies*, 9(631): 1–22.

Arcury, T.A., Grzywacz, J.G., Anderson, A.M., Mora, D.C., Carrillo, L., Chen, H., and Quandt, S.A. 2012. Personal protective equipment and work safety climate among Latino poultry processing workers in Western North Carolina, USA. *International Journal of Occupational and Environmental Health*, 18(4): 320–328.

Arcury, T.A., Grzywacz, J.G., Anderson, A.M., Mora, D.C., Carrillo, L., Chen, H., and Quandt, S.A. 2013. Employer, use of personal protective equipment, and work safety climate: Latino poultry processing workers, *American Journal of Industrial Medicine*, 56: 180–188.

Aschehoug, S.H., Boks, C., and Støren, S. 2012. Environmental information from stakeholders supporting product development. *Journal of Cleaner Production*, 31: 1–13.

Attarchi, M., Golabadi, M., Labbafinejad, Y., and Mohammadi, S. 2013. Combined effects of exposure to occupational: Noise and mixed organic solvents on blood pressure in car manufacturing company workers. *American Journal of Industrial Medicine*, 56: 243–251.

Aurich, J.C., Yang, X., Schröder, S., Hering-Bertram, M., Biedert, T., Hagen, H., and Hamann, B. 2012. Noise investigation in manufacturing systems: An acoustic simulation and virtual reality enhanced method. *CIRP Journal of Manufacturing Science and Technology*, 5: 337–347.

Barve, A. and Muduli, K. 2013. Modelling the challenges of green supply chain management practices in Indian mining industries. *Journal of Manufacturing Technology Management*, 24(8): 1102–1122.

Basahel, A.M. 2015. Investigation of work-related musculoskeletal disorders (MSDs) in warehouse workers in Saudi Arabia. *Procedia Manufacturing*, 3: 4643–4649.

Carrillo-Castrillo, J.A., Rubio-Romero, J.C., and Luis Onieva, L. 2013. Causation of severe and fatal accidents in the manufacturing sector. *International Journal of Occupational Safety and Ergonomics*, 19(3): 423–434.

Ceballos, D., Mead, K., and Ramsey, J. 2015. Recommendations to improve employee thermal comfort when working in 40°F refrigerated cold rooms. *Journal of Occupational and Environmental Hygiene*, 12(9): 215–221.

Chaudhary, D.K., Bhattacherjee, A., Patra, A.K., and Chau, N. 2015. Whole-body vibration exposure of drill operators in iron ore mines and role of machine-related, individual, and rock-related factors. *Safety and Health at Work*, 6(2015): 268–278.

Chavalala, B. 2013. Clean Technology Transition Potential in South Africa's Gold Mining Sector: Case of Harmony's Kusasalethu Mine. Available at: http://uir.unisa.ac.za/handle/10500/13601. [Accessed on 2016-10-23]

Cheluszka, P. 2015. Computer-aided design of robotised technology for manufacturing working units of mining machines. *International Journal of Mining, Reclamation and Environment*, 29(1): 62–81.

Chiarini, A. 2014. Sustainable manufacturing-greening processes using specific Lean production tools: An empirical observation from European motorcycle component manufacturers. *Journal of Cleaner Production*, 85: 226–233.

Dababneh, A.J. and Schwab, P.A. 2015. Impact of gloves on user performance and comfort in the semiconductor industry. *Human Factors and Ergonomics in Manufacturing and Service Industries*, 25(6): 638–645.

Dahlman-Höglund, A., Renström, A., Larsson, P.H., Elsayed, S., and Andersson, E. 2012. Salmon allergen exposure, occupational asthma, and respiratory symptoms among salmon processing workers. *American Journal of Industrial Medicine* 55: 624–630.

Dahm, M.M., Evans, D.E., Schubauerberigan, M.K., Birch, M.E., and Fernback, J.E. 2012. Occupational exposure assessment in carbon nanotube and nanofiber primary and secondary manufacturers. *Annals of Occupational Hygiene*, 56(5): 542–556.

Da Silva, M.P., Amaral, F.G., Mandagara, H., and Leso, B.H. 2014. Difficulties in quantifying financial losses that could be reduced by ergonomic solutions. *Human Factors and Ergonomics in Manufacturing and Service Industries*, 24(4): 415–427.

Dehaghi, B.F., Abadi, L.I.G., Rasi, H., and Angali, K.A. 2016. The feasibility of noise control in a soft drink plant. *Jundishapur Journal of Health Sciences*, 8(2): 1–5.

Dejanovi, D. and Heleta, M. 2016. An airport occupational health and safety management system from the OHSAS 18001 perspective. *International Journal of Occupational Safety and Ergonomics (JOSE)*, 22(3): 439–447.

Dhande, K.K. 2013. Practical approach towards issue on ergonomic training with respect to productivity improvement. *Journal of Ergonomics*, 3(2): 1–7.

Dianat, I., Sedghi, A., Bagherzade, J., Jafarabadi, M.A., and Stedmon, A.W. 2013. Objective and subjective assessments of lighting in a hospital setting: Implications for health, safety and performance. *Ergonomics*, 56(10): 1535–1545.

Dotoli, M., Fanti, M.P., Lacobellis, G., and Rotunno, G. 2014. An integrated technique for the internal logistics analysis and management in discrete manufacturing systems. *International Journal of Computer Integrated Manufacturing*, 27(2): 165–180.

Driessen, P.P.J., Dieperink, C., van Laerhoven, F., Runhaar, H.A.C., and Vermeulen, W.J.V. 2012. Towards a conceptual framework for the study of shifts in modes of environmental governance: Experiences from the Netherlands. *Environmental Policy and Governance*, 22: 143–160.

Dubey, R., Gunasekaran, A., and Chakrabarty, A. 2015. World-class sustainable manufacturing: Framework and a performance measurement system. *International Journal of Production Research*, 53(17): 5207–5223.

Dubey, A.D., Verma, M., and Khandelwal, A. 2015. Ergonomics concerns (OHS) to improve productivity in brick industry. *Journal of Applied Mechanical Engineering*, 4(2): 1–5.

Echt, A. and Mead, K. 2016. Evaluation of a dust control for a small slab-riding dowel drill for concrete pavement. *Annals of Occupational Hygiene*, 1–6.

Ene, S.A., Teodosiu, C., Robu, B., and Volf, I. 2013. Water footprint assessment in the winemaking industry: A case study for a Romanian medium size production plant. *Journal of Cleaner Production*, 43: 122–135.

Fleury, D., João, A.S., Bomfimb, J.A.S., Vignes, A., Girard, C., Metz, S. Muñoz, F., R'Mili, B., Ustache, A., Guiot, A., and Bouillard, J.X. 2013. Identification of the main exposure scenarios in the production of CNT-polymer nanocomposites by melt-moulding process. *Journal of Cleaner Production*, 53: 22–36.

Fore, S. and Mbohwa, C.T. 2010. Cleaner production for environmental conscious manufacturing in the foundry industry. *Journal of Engineering, Design and Technology*, 8(3): 314–333.

García-Hernández, C., Huertas-Talón, J., Sánchez-Álvarez, E.J., and Marín-Zurdo, J. 2016. Effects of customized foot orthoses on manufacturing workers in the metal industry. *International Journal of Occupational Safety and Ergonomics*, 22(1): 116–124.

Gerr, F., Fethke, N.B., Anton, D., Merlino, L., Rosecrance, J., Marcus, M., and Jones, M.P. 2014. A prospective study of musculoskeletal outcomes among manufacturing workers: Ill effects of psychosocial stress and work organization factors. *Human Factors*, 56(1): 178–190.

Gonen, D., Oral, A., and Yosunlukaya, M. 2016. Computer-aided ergonomic analysis for assembly unit of an agricultural device. *Human Factors and Ergonomics in Manufacturing and Service Industries*, 26(5): 615–626.

Guadix, J., Carrillo-Castrillo, J., Onieva, L., and Lucena, D. 2015. Strategies for psychosocial risk management in manufacturing. *Journal of Business Research*, 68: 1475–1480.

Gunasekaran, A. and Spalanzani, A. 2012. Sustainability of manufacturing and services: Investigations for research and applications. *International Journal of Production Economics*, 140: 35–47.

Gupta, V., Acharya, P., and Patwardhan, M. 2012. Monitoring quality goals through lean Six-Sigma insures competitiveness. *International Journal of Productivity and Performance Management*, 61(2): 194–203.

Gupta, S. and Jain, K.S. 2015. An application of 5S concept to organize the workplace at a scientific instruments manufacturing company. *International Journal of Lean Six Sigma*, 6(1): 73–88.

Halim, I. and Omar, A.R. 2012. Development of prolonged standing strain index to quantify risk levels of standing jobs. *International Journal of Occupational Safety and Ergonomics (JOSE)*, 18(1): 85–96.

Harik, R., EL Hachem, W., Medini, K., and Bernard, A. 2015. Towards a holistic sustainability index for measuring sustainability of manufacturing companies. *International Journal of Production Research*, 53(13): 4117–4139.

Hasanbeigi, A. and Price, L. 2015. A technical review of emerging technologies for energy and water efficiency and pollution reduction in the textile industry. *Journal of Cleaner Production*, 95: 30–44.

He, Q., Qiu., S., Fan, X., and Liu, K. 2016. An interactive virtual lighting maintenance environment for human factors evaluation. *Assembly Automation*, 36(1): 1–11.

Hemphälä, H. and Eklund, J. 2012. A visual ergonomics intervention in mail sorting facilities: Effects on eyes, muscles and productivity. *Applied Ergonomics*, 43: 217–229.

Jaca, C., Viles, E., Paipa-Galeano, L., Santos J., and Mateo, R. 2014. Learning 5S principles from Japanese best practitioners: Case studies of five manufacturing companies. *International Journal of Production Research*, 52(15): 4574–4586.

Jiang, L., Lin, C., and Lin, P. 2014. The determinants of pollution levels: Firm-level evidence from Chinese manufacturing. *Journal of Comparative Economics*, 42: 118–142.

Jin, Y., Curran, R., Burke, R., and Welch, B. 2012. An integration methodology for automated recurring cost prediction using digital manufacturing technology. *International Journal of Computer Integrated Manufacturing*, 25(4–5): 326–339.

Kaljun, J. and Dolšak, B. 2012. Ergonomic design knowledge built in the intelligent decision support system. *International Journal of Industrial Ergonomics*, 42: 162–171.

Khrais, S., Al-Araidah, O., Aweisi, A.M., Elias, F., and Al-Ayyoub, E. 2013. Safety practices in Jordanian manufacturing enterprises within industrial estates. *International Journal of Injury Control and Safety Promotion*, 20(3): 227–238.

Klewitz, J. and Hansen, E.G. 2014. Sustainability-oriented innovation of SMEs: A systematic review. *Journal of Cleaner Production*, 65: 57–75.

Koho, M., Tapaninaho, M., Heilala, J., and Torvinen, S. 2015. Towards a concept for realizing sustainability in the manufacturing industry. *Journal of Industrial and Production Engineering*, 32(1): 12–22.

Kolb, P., Gockel, C., and Werth, L. 2012. The effects of temperature on service employees' customer orientation: An experimental approach. *Ergonomics*, 55(6): 621–635.

Koukoulaki, T. 2014. The impact of lean production on musculoskeletal and psychosocial risks: An examination of sociotechnical trends over 20 years. *Applied Ergonomics*, 45: 198–212.

Ledwaba, P.F. 2015. The use of renewable energy in small scale mining. Available at: http://146.141.12.21/handle/10539/17683. [Accessed on 2016-10-23]

Lee, H., Chae, D., Yi, K.H., Im, S., and Cho, S.H. 2015. Multiple risk factors for work-related injuries and illnesses in Korean-Chinese migrant workers. *Workplace Health and Safety*, 63(1): 18–26.

Lee, N., Lee, B., Jeong, S., Yi, G.Y., and Shin, J. 2012. Work environments and exposure to hazardous substances in Korean tire manufacturing. *Safety and Health at Work*, 3: 130–139.

Li, C. Nie, Z., Cui, S., Gong, X., Wang, Z., and Meng, X. 2014. The life cycle inventory study of cement manufacture in China. *Journal of Cleaner Production*, 72: 204–211.

Lin, Y., Chen, C., and Pan, Y. 2013. The suitability for the work-related musculoskeletal disorders checklist assessment in the semiconductor industry: A case study. *Human Factors and Ergonomics in Manufacturing and Service Industries*, 23(3): 222–229.

Longoni, A. and Cagliano, R. 2015. Cross-functional executive involvement and worker involvement in lean manufacturing and sustainability alignment. *International Journal of Operations and Production Management*, 35(9): 1332–1358.

Lu, J., Twu, L., and Wang, M.J. 2016. Risk assessments of work-related musculoskeletal disorders among the TFT-LCD manufacturing operators. *International Journal of Industrial Ergonomics*, 52: 40–51.

Luka, M.F. and Akun, E. 2015. Occupational noise management in small and large scale industries in north Cyprus. *International Journal of Scientific and Engineering Research*, 6(4): 591–593.

Magagnotti, N., Nannicini, C., Sciarra, G., Spinelli, R., and Volpi, D. 2013. Determining the exposure of chipper operators to inhalable wood dust. *Annals of Occupational Hygiene*, 57(6): 784–792.

Majid Motamedzade, M. 2013. Ergonomics intervention in an Iranian tire manufacturing industry. *International Journal of Occupational Safety and Ergonomics*, 19(3): 475–484.

Mak, C.W., Afzulpurkar, N.V., Dailey, M.N., and Saram, P.B. 2014. A Bayesian approach to automated optical inspection for solder jet ball joint defects in the head gimbal assembly process. *IEEE Transactions On Automation Science and Engineering* 11(4): 1155–1162.

Mehta, R. 2012. Major health risk factors prevailing in garment manufacturing units of Jaipur. *Journal of Ergonomics*, 2(2): 1–3.

Meo, S.A., Al-Drees, A.M., Al Masri, A.A., Al Rouq, F., and Azeem, M.A. 2013. Effect of duration of exposure to cement dust on respiratory function of non-smoking cement mill workers. *International Journal of Environmental Research and Public Health*, 10: 390–398.

Min, K.B., Park, S.G., Song, J.S., Yi, K.H., Jang, T.W., and Min, J.Y. 2013. Subcontractors and increased risk for work-related diseases and absenteeism. *American Journal of Industrial Medicine*, 56: 1296–1306.

Moatari-Kazerouni, A., Chinniah, Y., and Agard, B. 2015. A proposed occupational health and safety risk estimation tool for manufacturing systems. *International Journal of Production Research*, 53(15): 4459–4475.

Motamedzade, M. 2013. Ergonomics intervention in an iranian tire manufacturing industry. *International Journal of Occupational Safety and Ergonomics*, 19(3): 475–484.

Motamedzade, M. and Moghimbeigi, A. 2012. Musculoskeletal disorders among female carpet weavers in Iran. *Ergonomics*, 55(2): 229–236

Muduli, K., Govindan, K., Barve, A., and Geng, Y. 2013. Barriers to green supply chain management in Indian mining industries: A graph theoretic approach. *Journal of Cleaner Production*, 47: 335–344

Nicot, J. and Scanlon, B.R. 2012. Water use for shale-gas production in Texas U.S.. *Environmental Science and Technology*, 46: 3580–3586

Ocampo, L.A. 2015. A hierarchical framework for index computation in sustainable manufacturing. *Advances in Production Engineering and Management*, 10(1): 40–50.

O'Driscoll, E. and O'Donnell, G.E. 2013. Industrial power and energy metering e a state-of-the-art review. *Journal of Cleaner Production*, 41: 53–64.

O'Neill, P., Sohal, A., and Teng, C.W. 2016. Quality management approaches and their impact on firms' financial performance: An Australian study. *International Journal of Production Economics*, 171: 381–393.

Ortolano, L. Sanchez-Triana, E., Afzal, J., Ali, C.L., and Rebellón, S.A. 2014. Cleaner production in Pakistan's leather and textile sectors. *Journal of Cleaner Production*, 68: 121–129.

Ozturkoglu, O., Saygılı, E.E., and Ozturkoglu, Y. 2016. A manufacturing-oriented model for evaluating the satisfaction of workers: Evidence from Turkey. *International Journal of Industrial Ergonomics*, 54: 73–82.

Pacaiova, H. 2016. Risk assessment of vibration exposure for small and medium enterprises. *Theoretical Issues in Ergonomics Science*, 18(1): 14–23.

Pagell, M., Johnston, D., Veltri, A., Klassen, R., and Biehl, M. 2014. Is safe production an oxymoron? *Production and Operations Management Society*, 23(7): 1161–1175.

Petek, J., Glavič, P., and Kostevšek, A. 2016. Comprehensive approach to increase energy efficiency based on versatile industrial practices. *Journal of Cleaner Production*, 112: 2813–2821.

Pietroiusti A. and Magrini, A. 2014. Engineered nanoparticles at the workplace: Current knowledge about workers' risk. *Occupational Medicine*, 64: 319–330.

Potter, M. and Hamilton, J. 2014. Picking on vulnerable migrants: Precarity and the mushroom industry in Northern Ireland. *Work, Employment and Society*, 1–17.

Prasad, K.D., Jha, S.K., and Prakash, A. 2015. Quality, productivity and business performance in home based brassware manufacturing units. *International Journal of Productivity and Performance Management*, 64(2): 270–287.

Prasad, S., Khanduja, D., and Sharma, S.K. 2016. An empirical study on applicability of lean and green practices in the foundry industry. *Journal of Manufacturing Technology Management*, 27(3): 408–426.

Quintana, R. and Leung, M.T. 2012. A case study of Bayesian belief networks in industrial work process design based on utility expectation and operational performance. *International Journal of Productivity and Performance Management*, 61(7): 765–777.

Reinhold, K., Järvis, M., and Tint, P. 2015. Practical tool and procedure for workplace risk assessment: Evidence from SMEs in Estonia. *Safety Science*, 71: 282–291.

Rivero, L.C., Rodríguez, R.G., Pérez, M.D.R. Mar, C., and Juárez, Z. 2015. Fuzzy logic and RULA method for assessing the risk of working. *Procedia Manufacturing*, 3: 4816–4822.

Rose, L.M., Orrenius, U.E., and Neumann, W.P. 2013. Work environment and the bottom line: Survey of tools relating work environment to business results. *Factors and* Ergonomics *in Manufacturing and Service Industries*, 23(5): 368–381.

Sadeghi, F., Bahrami, A., and Fatemi, F. 2014. The effects of prioritize inspections on occupational health hazards control in workplaces in Iran. *Journal of Research in Health Sciences*; 14(4): 282–286.

Sangwan, K.S., Bhamu, J., and Mehta, D. 2014. Development of lean manufacturing implementation drivers for Indian ceramic industry. *International Journal of Productivity and Performance Management*, 63(5): 569–587.

Scanlon, K.A., Lloyd, S.M., Gray, G.M, Francis, R.A., and LaPuma, P. 2015. An approach to integrating occupational safety and health into life cycle assessment development and application of work environment characterization factors. *Journal of Industrial Ecology*, 19(1): 27–37.

Schippers, M.C., West, M.A., and Dawson, J.F. 2015. Team reflexivity and innovation: The moderating role of team context. *Journal of Management*, 41(3): 769–788.

Sharma, J. and Sidhu, B.S. 2014. Investigation of effects of dry and near dry machining on AISI D2 steel using vegetable oil. *Journal of Cleaner Production*, 66: 619–623.

Silvestri, A., De Felice, F., and Petrillo, A. 2012. Multi criteria risk analysis to improve safety in manufacturing systems, *International Journal of Production Research*, 50(17): 4806–4821.

Sinay, J. and Balážikova, M. 2014. Acoustic risk management. *Human Factors and Ergonomics in Manufacturing and Service Industries*, 24(3): 298–307.

Singh, R.P., Batish, A., Singh, T.P., and Bhattacharya, A. 2014. An experimental study to evaluate the effect of ambient temperature during manual lifting and design of optimal task parameters. *Human Factors and Ergonomics in Manufacturing and Service Industries*, 24(1): 54–70.

Singh, L.P., Bhardwaj, A., and Deepak, K.K. 2013. Occupational noise-induced hearing loss in Indian steel industry workers: An exploratory study. *Human Factors*, 55(2): 411–424.

Spillane, J.P., Oyedele, L.O., and von Meding, J. 2012. Confined site construction. *Journal of Engineering, Design and Technology*, 10(3): 397–420.

Srinivasan, S., Ikuma, L.H., Shakouri, M., Nahmens, I., and Harvey, C. 2016. 5S impact on safety climate of manufacturing workers. *Journal of Manufacturing Technology Management*, 27(3): 364–378.

Story, J. and Neves, P. 2016. When corporate social responsibility (CSR) increases performance: Exploring the role of intrinsic and extrinsic CSR attribution. *Business Ethics: A European Review*, 24(2): 111–124.

Suárez-Barraza, M.F. and Ramis-Pujol, J. 2012. An exploratory study of 5S: A multiple case study of multinational organizations in Mexico. *Asian Journal on Quality*, 13(1): 77–99.

Taiwo, O.A. 2014. Diffuse parenchymal diseases associated with aluminum use and primary aluminum production. *Journal of Occupational and Environmental Medicine*, 56(5): 71–72.

Tanioka, A. and Takahashi, M. 2016. Highly productive systems of nanofibers for novel applications. *Industrial and Engineering Chemistry Research*, 55: 3759–3764.

Tezel, A. and Aziz, Z. 2017. Benefits of visual management in construction: Cases from the transportation sector in England. *Construction Innovation*, 17(2): 125–157.

Top, Y., Adanur, H., and Öz, M. 2016. Comparison of practices related to occupational health and safety in microscale wood-product enterprises. *Safety Science*, 82: 374–381.

Trianni, A., Cagno, E., and De Donatis, A. 2014. A framework to characterize energy efficiency measures. *Applied Energy*, 118: 207–220.

Tseng, M., Chiu, S.F., Tan, R.R., and Siriban-Manalang, A.B. 2013. Sustainable consumption and production for Asia: Sustainability through green design and practice. *Journal of Cleaner Production*, 40: 1–5.

Wang, H., Hwang, J., Lee, K., Kwag, J., Jang, J., and Jung, M. 2014. Upper body and finger posture evaluations at an electric iron assembly plant. *Human Factors and Ergonomics in Manufacturing* and *Service Industries*, 24(2): 161–171.

Wolfgang Thurner, T.W. and Roud, V. 2016. Greening strategies in Russia's manufacturing: From compliance to opportunity. *Journal of Cleaner Production*, 112: 2851–2860.

Yang, S. and Bagheri, B. 2015. A unified framework and platform for designing of loud-based machine health monitoring and manufacturing systems. *Journal of Manufacturing Science and Engineering*, 137: 1–6.

Yu, Z., Zhang, J., Lou, C., and Lin, J. 2015. Processing and properties of multifunctional metal composite yarns and woven fabric. *Materials and Manufacturing Processes*, 30(3): 320–326

Zhang, J., Chen, S., Kim, J., and Cheng, S. 2016. Mercury flow analysis and reduction pathways for fluorescent lamps in mainland China. *Journal of Cleaner Production*, 133: 451–458.

Židonienė, S. 2016. Environmental impact assessment of manufacturing industry projects. *Journal of Environmental Research, Engineering and Management*, 72(1): 8–17.

Chapter 6

Work Study Techniques: Method Study

6.1 Introduction

In this chapter, work study techniques are studied and addressed with the aim of improving the input resource factors used in manufacturing small and medium enterprises (SMEs). These techniques, as mentioned earlier, involve method study and work measurement. The following various areas of work study (WS) relations in manufacturing SMEs are addressed in this section: traits of WS specialists, WS and relations with management, WS and relations with supervisors, WS and relations with workers, and, finally, WS and relations with trade unions. These areas are followed by method study as the next section to be discussed, including analysis of the current applications of method study in South Africa as well as the summary.

6.2 WS Characteristics in Manufacturing SMEs

6.2.1 Traits of WS Specialists in Manufacturing SMEs

A WS expert must have various qualities, such as being analytical, acting with integrity and tact, being enthusiastic (customer focused), and being cooperative in order to ensure productivity progress in manufacturing SMEs (Kulkarni, Kshire, & Chandratre, 2014:432–3).

Magu, Khanna, and Seetharaman (2015:6475–80) explain that the traits of WS specialists include being reliable, having different skill sets (tact), and occupying the position of a WS specialist in manufacturing SMEs. According to Jain et al. (2016:486–506) the traits of the WS specialist include enthusiasm, cooperation, and the ability to be customer focused (tact). The following traits are identified in the literature sources studied in terms of WS specialists: insight (tact), belief in oneself, having an interest in what one is doing (enthusiasm), and having the educational background to carry out a WS investigation in manufacturing SMEs (Vijai et al., 2017:446–66).

This section goes on to identify the expertise of WS specialists in manufacturing SMEs.

6.2.2 Expertise of the WS Specialists in Manufacturing SMEs

A WS expert must have expertise in the areas of training and practice of WS; knowledge of manufacturing SMEs; know-how of capabilities such as process, product, and service design, standards, inventory management, just-in-time (JIT), employee ability, location, layout, quality, and scheduling; familiarity with management operations in terms of regulations, policies, and standards; information about the functioning of other rivals; and how to simplify employees' activities through standardizing performance in order to eliminate waste. Eliminating waste enhances the productivity of manufacturing SMEs (Kulkarni et al., 2014:429–33).

As described by Magu et al. (2015:6475–80), the expertise of WS specialists consists of competencies such as knowledge of WS and the workplace processes, as well as the ability to guide management in ensuring employees work in a safe, healthy, and comfortable manner in the manufacturing processes of the manufacturing SMEs.

The expertise of WS specialists is based on knowledge of WS, know-how concerning manufacturing SMEs' function, and information regarding capabilities such as products, process, service, employee abilities, scheduling, machine maintenance, work space, layout, inventory improvement, production planning and control, and JIT as well as being able to determine the standard time for carrying out tasks as required for productivity improvement (Jain et al., 2016:486–506).

As pointed out by Vijai et al. (2017:446–66), the expertise of WS specialists entails having information regarding manufacturing SMEs; and the ability to make decisions on problem solving in the workplace, taking into consideration management capabilities such as process, product, services, scheduling, shop floor, plant, standardized procedures, JIT, TQM, production planning and control, employee abilities and job design, supply chain management, and operating systems.

The following traits are identified in the literature sources for WS specialists: insight (tact), belief in oneself, having an interest in what one is doing (enthusiasm), as well as the educational background to carry out WS investigation in manufacturing SMEs. The next issue to be discussed is WS relations with management in manufacturing SMEs.

6.2.3 WS and Relations with Management in Manufacturing SMEs

In this section, various relations are presented to ensure that management engages all parties in the business in order to ensure productivity improvement in manufacturing SMEs. The relation between WS expert and top management, WS expert and middle management, and WS expert and lower management are applicable in big manufacturing industries. So, the concept of the organizational structure of the chain of command in manufacturing SMEs differs from business to business depending on the size of these SMEs.

Based on the literature reviewed, this book advises manufacturing SMEs to be aware of the fact that there are consultancies out there that can assist SMEs to ensure progress in their businesses. The establishment of a WS unit within the manufacturing SME can also play an important role in ensuring that these SMEs improve productivity in their businesses (Muruganantham, Krishnan, & Arun, 2013:451–55; Bechar & Eben-Chaime, 2014:197–201; Odesola, Okolie, & Nnametu, 2015:1). For the purpose of this book, WS is the key tool that allows manufacturing SMEs to ensure that their productivity improves (Muruganantham et al., 2013:451–55; Ramish & Aslam, 2016:705–16; Vijai et al., 2017:447–60).

An example of manufacturing SMEs' organizational structure is provided in Figures 6.1 through 6.3. Various steps are followed by operations managers in interacting with WS experts or consultants for assistance in resolving complex problems facing manufacturing SMEs (see Figure 6.3).

The operational manager reads monthly production reports to identify problem areas. Corrective measures are taken in cases where the manufacturing SME needs assistance with the problem areas experienced, be it in the form of figures or areas of input resource factors, by consulting WS unit. The operations manager then waits for the WS section to arrange a meeting and

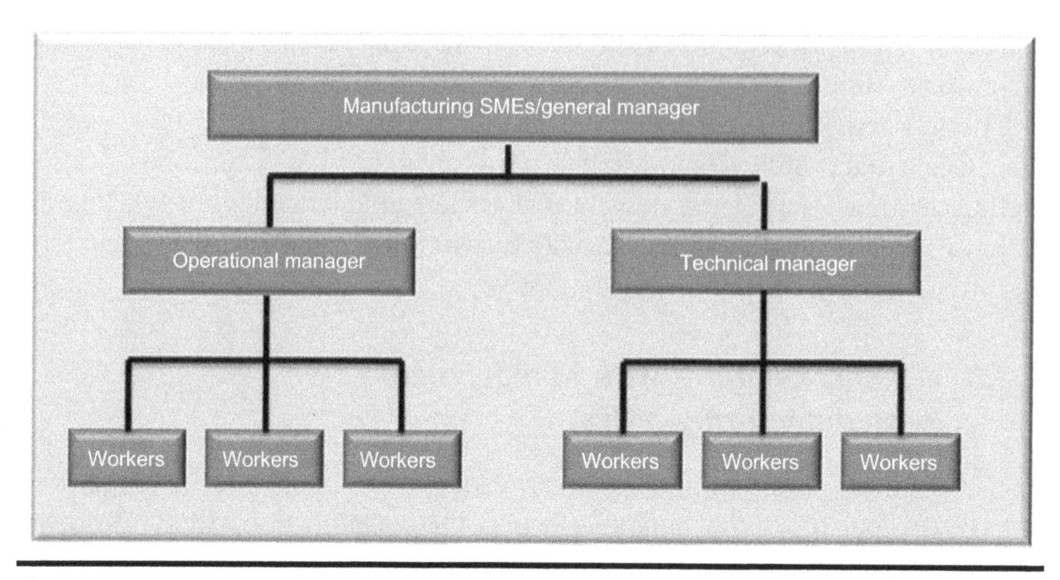

Figure 6.1 Current structure without WS unit. (From Author, 2017.)

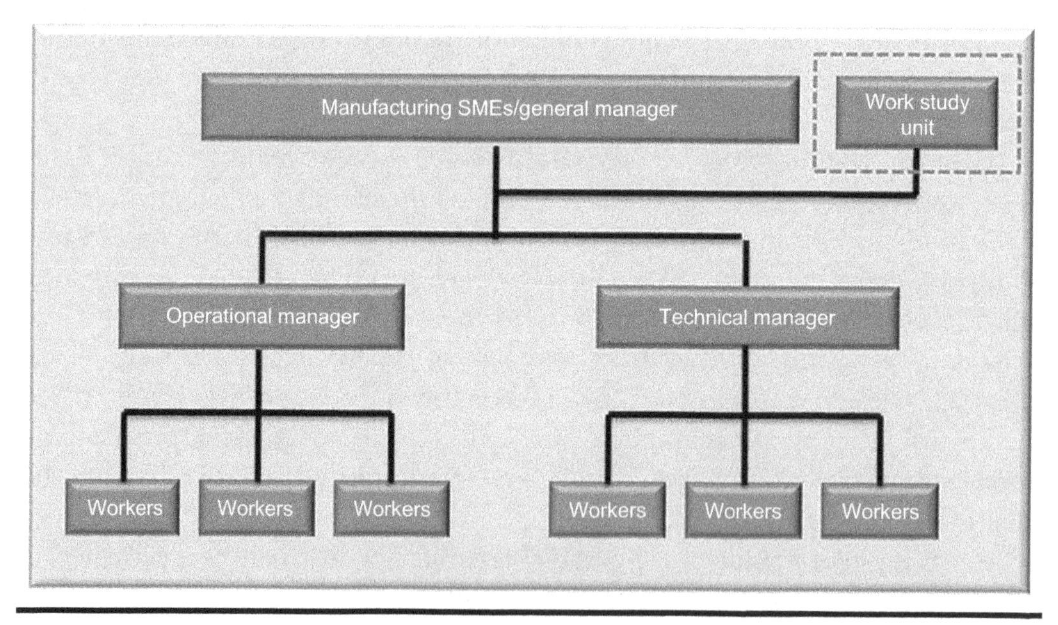

Figure 6.2 Current structure with WS unit. (From Author, 2017.)

an investigation begins. The following standardized procedure is followed to ensure WS is applied within the selected problem area to be studied. Managers firstly consult with the WS unit about a problem that has been identified in the manufacturing section. After the consultation, meetings take place between management, including supervisors, and the WS department so that agreement can be reached on how the problem is going to be resolved. An

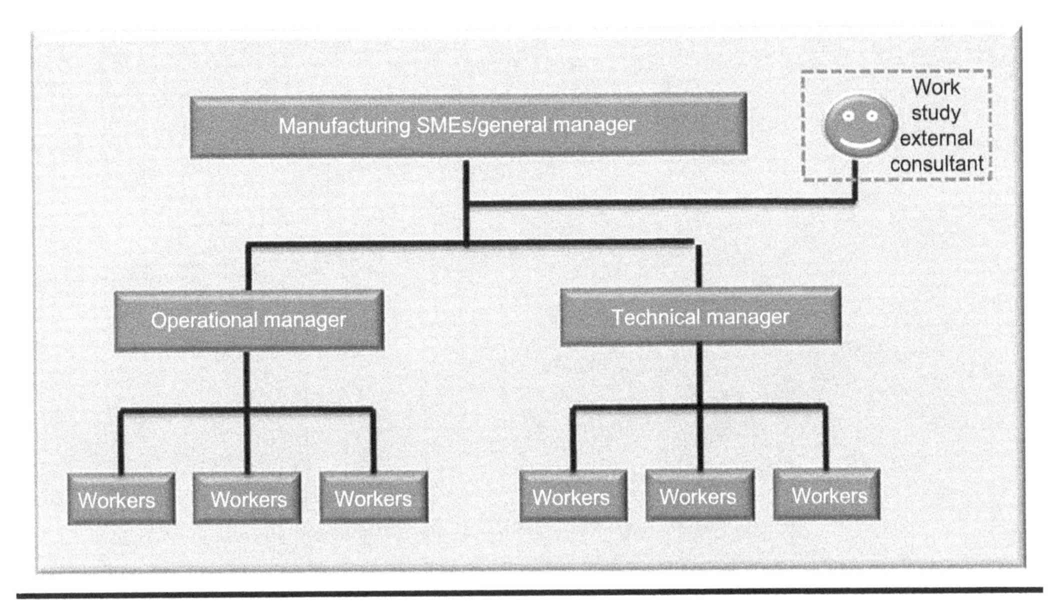

Figure 6.3 Proposed structure with external WS consultants. (From Author, 2017.)

authorization letter is provided to the WS department for approval of the investigation. Finally, a date is arranged for the WS experts to start the investigation (Jain et al., 2016:489–506; Vijai et al., 2017:448–66; Mehralian et al., 2017:113–20). An example of a WS and management relations procedure in the manufacturing SMEs is provided in Figure 6.4.

The information presented is a typical example of WS and management relations procedures in manufacturing SMEs as indicated in the diagram in Figure 6.4. After the preparation of the start date for the study as initiated by the operations departmental manager, the WS experts are introduced to operational supervisors in order to inform these supervisors on the WS invitation, the purpose of the investigation, its duration, and the area in which the investigation is going to occur. The supervisors will then introduce the WS expert to employees during the green area meeting/quality circle meeting. The aim of the supervisor is to clarify areas that will be affected during the investigation as well as to build trust and loyalty between the WS expert and the employees.

6.2.4 WS and Relations with Supervisors in Manufacturing SMEs

6.2.4.1 WS and Relations with Supervisors

In this section, the relations between WS and supervisors are examined and discussed as explained by different academic scholars with the intent to

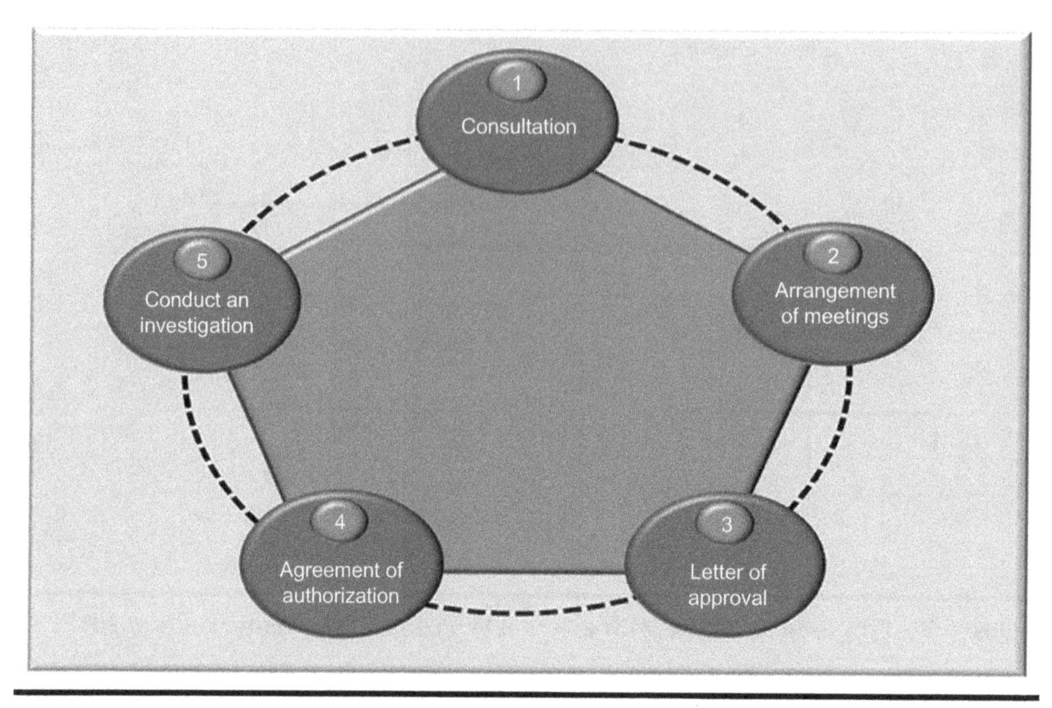

Figure 6.4 WS and management relations procedures in the manufacturing SMEs. (From Author, 2017.)

examine how WS impacts on supervisors in their workplace in manufacturing SMEs. Before a WS specialist can commence with an investigation of the respective workstations in manufacturing SMEs, the reasons and the procedures for carrying out an investigation need to be shared with supervisors (Ingvaldsen, Holtskog, & Ringen, 2013:280–1).

Standards set for the carrying out of the employees' tasks may be revealed by WS specialists when employees or supervisors experience defects, such as breakdowns, that negatively impact on time as well as causing poor product quality. So, rational WS specialists together with cooperative supervisors acquire results by enabling employees to get the job done efficiently (Ingvaldsen et al., 2013:280–1; Bechar & Eben-Chaime, 2014:195).

Based on the scientific literature sources used, continuous experiments by the WS specialist in introducing WS tools to prescribe the most efficient method and appropriate standard time for jobs being carried out by employees is essential if manufacturing SMEs are to ensure they improve the productivity of their businesses (Oeij et al., 2012:94–107; Bateman, Hines, & Davidson, 2013:553–64; Ingvaldsen et al., 2013:280–1; Jagoda, Lonseth, & Lonseth, 2013:383–99). Thus, if supervisors and employees are not guided by

WS specialists, both supervisors and employees will not realize the most efficient ways of working (Ingvaldsen et al., 2013:280–1; Bechar & Eben-Chaime, 2014:195).

By involving supervisors and employees in WS investigations, WS specialists make the parties feel part of the solution and enable these parties to accept the changes taking place in their workplace. By doing this, supervisors and employees end up being engaged in the effort to achieve results (Oeij et al., 2012:94–107; Bateman et al., 2013:553–64; Ingvaldsen et al., 2013:280–1; Jagoda et al., 2013:383–99; Bechar & Eben-Chaime, 2014:195). However, it is important to consider not only the relations between WS and supervisors but also the misconception of supervisors with regard to WS specialists' interference.

6.2.4.2 Misconception of Supervisors With Regard To WS Specialists Interference

The following are some points mentioned in terms of the misconception of supervisors with regard to WS specialists' interference.

As is indicated by Jagoda et al. (2013:390), it is imperative that supervisors associate themselves with employee behaviors with the founded activities from the initial stage. When these behaviors are attended, it will be difficult for manufacturing SMEs to achieve their results. The misunderstanding of supervisors can be seen when they interpret a WS as criticism. In order for supervisors to encourage WS application in their workplace, they must change their behaviors to allow constructive feedback in order to engage employees to perform well and continuously expand the productivity of manufacturing SMEs for growth in their businesses (Ingvaldsen et al., 2013:281–3; Jagoda et al., 2013:390).

Second, as reported by Ongkunaruk and Wongsatit (2014:685–90), in many manufacturing industries, be they large or small, activities such as production planning, development of work methods, work schedules, standard time setting, and piece work rate in terms of employee payment have been done by supervisors before. Thirdly, based on the literature discussed earlier, the WS expert's responsibility to advise these SMEs on productivity improvement is interpreted as a threat by supervisors. As a result, these supervisors lose confidence in their responsibilities to delegate employees on their work activities in manufacturing SMEs.

Finally, it is apparent that supervisors are the first managers on the shop floor responsible for resolving employee matters as well providing employees with the necessary information that will guide them to address disputes and make decisions on reaching a solution (Ingvaldsen et al., 2013:287–8; Lasrado, Arif, & Rizvi, 2015, 434–53).

6.2.4.3 Management Guidelines to Supervisors With Regard To WS Specialists' Interference

The following guidelines are presented by management as a supporting function to supervisors with regard to WS specialists' interference, since the supervisors are closely involved in the practical aspects of manufacturing and running shop floors in manufacturing SMEs (Ingvaldsen et al., 2013:281; Lasrado et al., 2015:434–53).

As advised by Ingvaldsen et al. (2013:280–89), the guidelines of top management to supervisors with regard to WS in manufacturing SMEs are as follows:

The supervisor needs to

- Resolve problems among the employees
- Directly participate in WS orientation
- Be knowledgeable in techniques of method study and work measurement (WS)
- Develop a friendly and respectful relationship with the WS specialist

6.2.4.4 Effective Communication by WS Specialists to Ensure That the WS Specialist Adheres to the Following Rules

According to Ingvaldsen et al. (2013:280–89) and Kafetzopoulos, Gotzamani, and Psomas (2014:508–14), WS specialists must comply with the following instructions:

- WS specialists must firstly contact the supervisor before beginning an investigation in the workplace where employees are carrying out their tasks.
- WS specialists should ensure that any tasks they perform do not counteract what the supervisor is doing.
- WS specialists' decisions in terms of the investigation must not be influenced by supervisors to make changes on the study conducted in the workplace.
- Any setback experienced by an employee that is outside the WS must be communicated by the employee to the supervisor.

6.2.5 WS and Relations with Workers in Manufacturing SMEs

The literature addressed points out that resistance to change has been a practice of employees, whereby these employees would not follow the instructions given under any program introduced by management, one of which is WS, in manufacturing SMEs. This resistance is attributed to failure and lack of effective communication and absence of feedback by management concerning employee accomplishments (Jagoda et al., 2013:392).

As indicated in Chapter 2, McGregor also stated the importance of employee behavior in his theory of Manager X; according to this theory, employees are lazy and reluctant to engage themselves in carrying out activities that will reach the goals of their organizations, in this case, manufacturing SMEs (Maskaly & Jennings, 2016:623).

The debate formulated to answer this is that WS as a technique is not used to focus directly on employee behavior but on the work activities that employees are doing for continuous improvement in manufacturing SMEs (Kafetzopoulos et al., 2014:503–4).

Tanvir and Ahmed (2013:53) emphasize that changes are essential before a WS as a technique is introduced to ensure efficient methods of working and appropriate standards for employees in the workplace in order to improve productivity in manufacturing SMEs. However, it is critical that these changes are communicated by management to employees. By the same token, Oeij et al. (2012:100–1) and Jagoda et al. (2013:394–7) recommend that when feedback is effectively communicated to employees on changes being made for improving work processes and thus improving productivity in manufacturing SMEs, employees become motivated and satisfied. As a result, these employees willingly commit themselves to engage their efforts to achieve the goals of manufacturing SMEs. Thus, WS will have a positive influence on the relationship between employee, supervisor, and top management due to various productivity improvement factors such as problem identification and improvement, employee training, and employee involvement in decision making.

6.2.6 WS and Relations with Trade Unions in Manufacturing SMEs

Since most of the manufacturing SMEs located in South Africa are represented by trade unions, negotiations play an important role in ensuring that the needs of employees are met. So, based on the global economy, unionists need to go to China for training in order for them to improve their negotiation skills and share this with employees for productivity improvement in manufacturing SMEs (Herman, 2011:124). During the negotiations between two parties such as the unions and management, WS plays a vital role in ensuring an efficient method and improved manufacturing process for productivity progress in manufacturing SMEs. So, the concepts of WS on communication with, consultation with, participation of, and empowerment of trade unions for employee support are apparent with the intent to ensure that employees are taken care of in manufacturing SMEs (Ingvaldsen et al., 2013:283–6).

As is indicated by Ingvaldsen et al. (2013:283–6) and James and Jones (2014:2184–5), the purpose of the use of WS in manufacturing SMEs does not always reflect negatively on employees as members of the trade unions but instead assists in eradicating bureaucracy in the business. Furthermore, WS specialists guide unions on various factors such as employee promotion, security, and better working conditions in manufacturing SMEs. There are two techniques of WS as indicated in Chapter 2, one of which is method study followed by the next technique, which is referred to as *work measurement*. In the case of this chapter, only method study will be discussed.

6.3 Method Study

Method study refers to the recording of factors and resources used in manufacturing processes and examining them analytically in order to make improvements to processes in manufacturing SMEs (Schulze et al., 2016:3693). Certain procedures need to be followed when conducting a method study investigation. This procedure involves selection of the task to be studied, recording of information regarding the task, analytical examination of the information collected, development and selection of an alternative solution, definition of a new method, and implementation and maintenance of the new method (Trianni, Cagno, & Farné, 2016:1539–51).

6.3.1 Method Study Procedure

The method study procedure in manufacturing SMEs focuses on various systematic procedures: select, record, examine, define, develop, install/implement, and maintain (SREDDIM). These procedures are used to firstly identify the problem being experienced by manufacturing SMEs at the business level in a particular department, section, or unit and then to ensure simplification of the job being carried out and improvement of the job for productivity enhancement in the business. The method of evaluation is added if both method study and work measurement (i.e., WS) are applied in the study to improve productivity. When the method of evaluation is added, the procedure for WS is abbreviated as SREDEDIM (Kulkarni et al., 2014:432).

Kulkarni et al. (2014:432) present the following method study procedures in sequence.

6.3.1.1 Select

Selecting the work activities to be studied in manufacturing SMEs means deciding on the task to be studied. Problem areas facing manufacturing SMEs are identified and defined to provide an understanding of the problems. As indicated earlier in the introduction in Chapter 1, manufacturing SMEs are faced with a drop in productivity whereby these SMEs are operating below expected production output levels (Coka, 2014:1; ILO, 2015:25; Statistics SA, 2015:7). Even though manufacturing SMEs are faced with this dilemma, such as productivity decline (Coka, 2014:1; ILO, 2015:25; Statistics SA, 2015:7), problems can still be solved in a department, section, or unit through the employment of WS experts. Problem areas of problem facing manufacturing SMEs are based on input resource factors such as physical capital (human resources, material, machinery, layout, location, finance, technology); technological capital (tangible and intangible elements); and management. As indicated in Chapter 1, these elements need to be taken into account if manufacturing SMEs are to ensure competitiveness in their market with the support of the government (ILO, 2015:61; Schwab, 2015:39).

6.3.1.2 Recording

The next procedure to be followed is to record information regarding the task being carried out. Recording of the job in manufacturing SMEs is done by using various method study techniques such as interviews, questionnaires, written notes, and/or observations. This recording procedure needs to have existing (current) and proposed (improved) information regarding the job being carried out. Even though the information is recorded for both current and improved information, improved information is exercised and the method "develops."

Before process charts are drawn up, process chart symbols are addressed based on the literature studied such as operation, inspection, transport, delay, and storage which can be utilized in these charts to address operations taking place in manufacturing industries such as manufacturing SMEs (van Niekerk, 19862:53). An example of these symbols is depicted in the form of a diagram in Figure 6.5.

Various types of recording methods are used, such as process charts (flow process chart, outline process chart, two-handed process chart, and multiple activity chart). The charts are show in Figures 6.6 through 6.9, respectively. Firstly, an example of the flow process chart symbol is used in Companies A, B, C, and D. The first company to be presented is Company A, and a flow process chart is provided in Figure 6.6. The aim is to show the flow of work taking place at the company from the cutting section to dispatch, where the last operation takes place in preparation for customer order.

Based on Figure 6.6, the number of operation, inspection, transport, delay, and storage stages need to be counted to show how productive work is as compared to nonproductive work. The resulting figure enables the company to realize problem areas facing the company and for corrective measures to be undertaken to ensure improvement. So, a proposed method can be prepared to assist the company in order to eliminate unnecessary delays and transport for productivity to improve operations.

The second case study is of Company B, whereby the flow of work starts with the collection of material from a warehouse and ends with a product being stored in a warehouse. The chart used for Company B is shown in Figure 6.7. The aim is to show the flow of work taking place at the company from the cutting section to dispatch, where the last operation takes place in preparation for customer order.

The third case study is of Company C, whereby the flow of work starts with the collection of material from a warehouse and ends with a product being stored in a warehouse for distribution. The chart used for Company C is shown in Figure 6.8. The aim is to show the flow of work taking place at the company from the pen-packaging section to the warehouse, where the last operation takes place in preparation for customer order.

The final case study is of Company D, whereby the flow of work starts with the collection of material from a warehouse and ends with a product being stored in a warehouse for distribution.

The chart used for Company D is shown in Figure 6.9. The aim is to show the flow of work taking place at the company from the cutting section to distribution, where the last operation takes place in preparation for customer order.

An example of a layout, also referred to as *flow diagram*, for Company A only is provided as a sketch to show how the flow of work can be carried out (Figure 6.10).

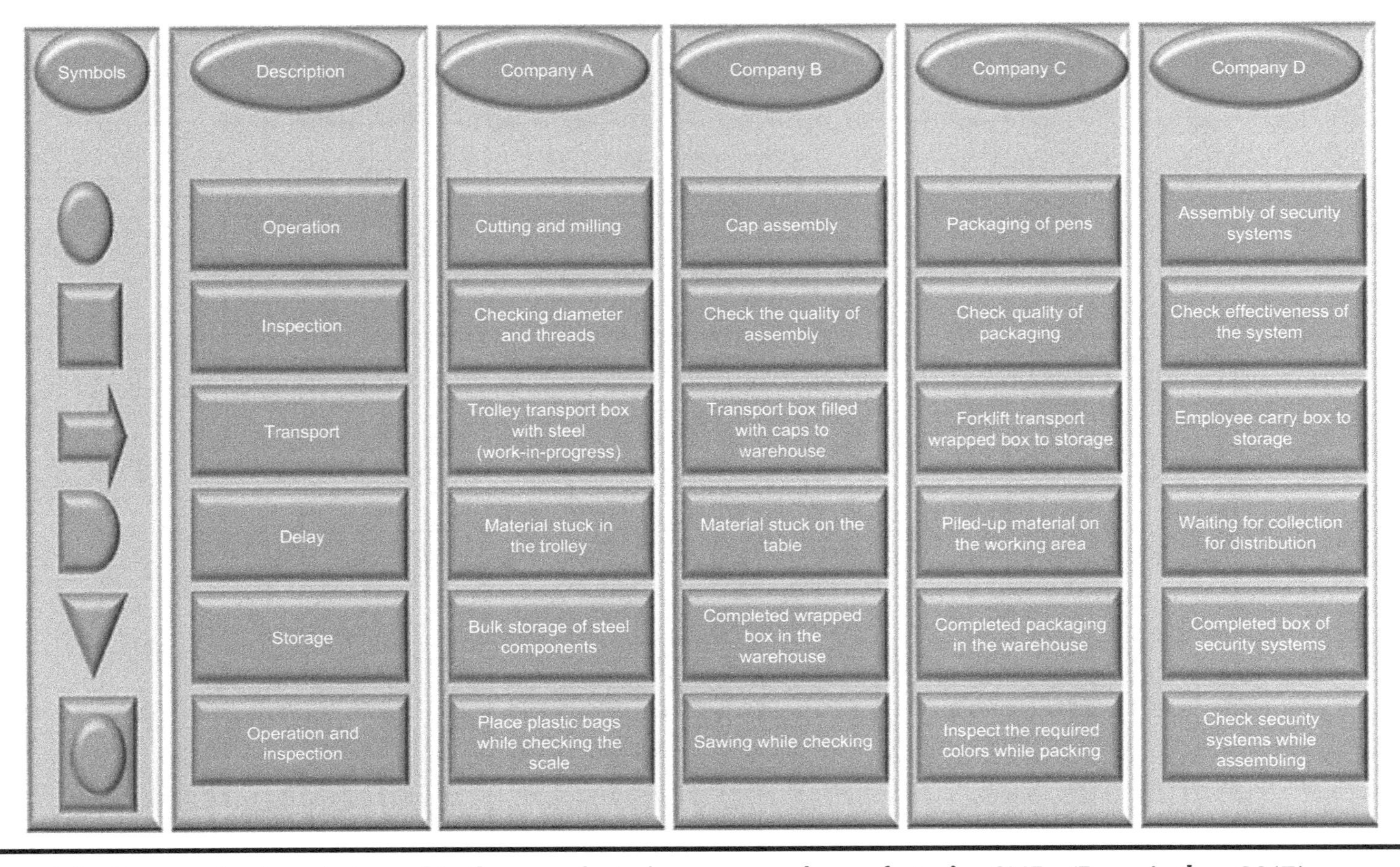

Figure 6.5 Process chart symbols used in the manufacturing process of manufacturing SMEs. (From Author, 2017.)

COMPANY A: PRESENT CHART

FLOW PROCESS CHART						
Name of the Company	Company A		Division/Department/ Section		Assembly Section	
Name of the work study specialists	Mr X		Name of the operator/ employee		Mr Y	
Study number	1		Sheet number		1	
Operator number	1		Job number		1	
Chart begins	Cutting Section		Chart finish		Despatch Section	
Method	Present	X	Proposed		Method	
Type	Worker		Material	Type	Worker	
Description of the task: Prepare rod for customers						

Step No.	Description of work activities	Symbols						Time (in min)	Distance (in meters)	Remarks
A	Cutting section									
1	Taking material from warehouse to cutting section									
2	Set machine									
3	Cut the steel rod to size									
4	Inspect rod									
5	Fill in trolley									
6	Transport steel rod to CNC/ milling section									
B	CNC/milling section									
7	Set machine									
8	Mill both side of steel rod									
9	Inspect rod									
10	Fill in trolley									
11	Transport steel rod to herbst/ threading section									

Figure 6.6 Flow process chart for Company A in Gauteng, South Africa. (From Author, 2017.)

Outcomes	Operation	Inspection	Transport	Delay	Operation and inspection	Storage
Present						
Proposed						

	Distance (in meters)	Time	Cost (Rands)

Figure 6.6 (Continued)

The next procedure to be addressed is the critical examination of the recorded information regarding the task being carried out.

6.3.1.3 Analytical Examination

Analytical examination of the information collected is the current information collected during the method study investigation. The information is examined analytically by the WS expert using questions such as "What is being done?" "Who is doing it?" "Where is it done?" "When is it done?" "How is it done?" and "Why is it done that way, by that person, in that place, and using that method?" The aim of asking these questions is for the WS expert to get clarity on the problem experienced by the manufacturing SME and to anticipate the improved method.

6.3.1.4 Development and Selection of Alternative Solution

Development and selection of an alternative solution is the fourth step, and this is addressed in this section so as to provide the reader with an understanding of what *development* means when conducting a method study procedure. This method study procedure focuses only on proposed (improved) information regarding the job being carried out in manufacturing SMEs with aim of developing improved methods. These methods are flow process charts, two-handed process charts, outline process charts, and multiple activity charts with the support of proposed flow diagrams to identify the layout of manufacturing SMEs.

The information that has already been examined by the WS expert, using questions such as "What is being done?" "Who is doing it?" "Where is it done?" "When is it done?" "How is it done?," and "Why is it done that way, by that person, in that place, and using that method?" is supported by using the proposed questions to ensure productivity improvement in terms of the problems facing manufacturing SMEs. In this method, proposed questions

COMPANY A: PRESENT CHART (CONTINUED)

FLOW PROCESS CHART			
Name of the Company	Company A	Division/ Department/ Section	Assembly Section
Name of the work study specialists	Mr X	Name of the operator/ employee	Mr Y
Study number	1	Sheet number	2
Operator number	1	Job number	1
Chart begins	Cutting Section	Chart finish	Despatch Section

Method	Present	X	Proposed	Method

Type	Worker		Material	Type	Worker	

Description of the task: Prepare rod for customers

Step number	Description of work activities	Symbols						Time (in min)	Distance (in meters)	Remarks
		●	■	➡	D	▣	▼			
C	Herbst/ threading section									
12	Set the machine									
13	Thread both sides									
14	Inspect threaded sides									
15	Fill in trolley									
16	Transport steel rod to slotting/ milling section									
D	Slotting/ milling section									
17	Place the rod on Jid									
18	Tighten the stand									
19	Slot both sides									
20	The rod (1st and 2nd operation)									
21	Inspect slotting									

Figure 6.6 (Continued)

Outcomes	Operation	Inspection	Transport	Delay	Operation and inspection	Storage
Present						
Proposed						
	Distance (in meters)		Time		Cost (Rands)	

Figure 6.6 (Continued)

are asked by the WS expert, such as “What else could be done?” “Who else could be doing it?” “Where else could it be done?” “When else could it be done?” “How else could it be done?” and “Could it be done in another way, by another person, in another place, and using another method?”

6.3.1.5 Definition

The fifth step is the definition of the new method. By defining the new method, the WS expert introduces the proposed method to management for approval, so that, when approved, it can be implemented to improve the productivity of manufacturing SMEs in their businesses.

6.3.1.6 Implementation

During the implementation stage, the qualified employee is introduced to the proposed method to ensure the appropriate standard procedure for the job, focusing on various standards such as ISO 9001, ISO 14001, and ISO 18001.

6.3.1.7 Maintenance

The last step in the method study procedure is maintaining the new method to avoid deviations from the proposed method.

6.3.2 Method Study (Qualitative Techniques)

In this section, qualitative techniques are studied and presented with the aim of providing insight into the techniques’ composition in manufacturing SMEs. Qualitative techniques are also referred to as method study under the umbrella of WS. A diagram of WS highlights method study techniques, including method study procedures and is shown in Figure 6.11.

COMPANY A: PRESENT CHART (CONTINUED)

FLOW PROCESS CHART					
Name of the Company	Company A	Division/ Department/ Section	Assembly Section		
Name of the work study specialists	Mr X	Name of the operator/ employee	Mr Y		
Study number	1	Sheet number	3		
Operator number	1	Job number	1		
Chart begins	Cutting Section	Chart finish	Despatch Section		
Method	Present	X	Proposed	Method	
Type	Worker		Material	Type	Worker
Description of the task: Prepare rod for customers					

Step number	Description of work activities	Symbols						Time (in min)	Distance (in meters)	Remarks
22	Fill in trolley									
23	Transport to despatch section									
	Despatch Section									
24	Final inspection									
25	Weigh and record									
26	Pack in plastic bags									
27	Packaging in dispatch warehouse									
28	Storage									

Figure 6.6 (Continued)

Outcomes	Operation	Inspection	Transport	Delay	Operation and inspection	Storage
Present						
Proposed						
	Distance (in meters)		Time		Cost (Rands)	

Figure 6.6 (Continued)

These techniques are presented next and include preliminary survey studies; benchmarking; brainstorming; method study process charts; time scale; diagrams; filming techniques; value stream mapping (VSM); cause-and-effect diagram; as well business process reengineering (BPR).

6.3.2.1 Method Study Preliminary Survey Studies

The following preliminary survey studies are addressed: interviews, observations, and questionnaires.

6.3.2.1.1 Interviews

As is explained by Gupta, Acharya, and Patwardhan (2013:635–45), interviews refer to consulting management in all divisions of the manufacturing SMEs on their operational background regarding setting operational improvement targets through waste elimination and ensuring flexibility, cost-effectiveness, product quality, and on-time delivery for productivity in businesses of these SMEs.

Wlazlak and Johansson (2014:264) outline the overview of the interviews used for management at all levels, whereby information is collected from the product manager, project manager, team members, design engineers, laboratory engineers, research and development (R&D) managers, manufacturing engineers, quality engineer, and buyers. The types of interviews used are semistructured and open-ended interviews based on projects, assembly, and continuous manufacturing processes for productivity improvement.

As pointed out by Gupta and Jain (2015:75–6), interviews are used to collect information to prepare method study reports for management through the channel of senior managers, supervisors, and employees, addressing disorganized input resource factors as human capital, location (shop floor), material, machinery, and layout (housekeeping and processes), with the aim being to improve these factors in manufacturing SMEs.

COMPANY B: PRESENT CHART

FLOW PROCESS CHART			
Name of the Company	Company B	Division/ Department/ Section	Assembly Section
Name of the work study specialists	Mr X	Name of the operator/ employee	Mr Y
Study number	2	Sheet number	1
Operator number	1	Job number	1
Chart begins	Collection Area	Chart finish	Warehouse Section

Method	Present	X	Proposed	Method

Type	Worker		Material	Type	Worker	

Description of the task: Prepare caps for the customers

Step No.	Description of work activities	Symbols						Time (in min)	Distance (in meters)	Remarks
	Cap assembly									
1	Taking material from warehouse to cutting section									
2	Cut material to size									
3	Prepare base									
4	Inspect base									
5	Pass inspected base to next operator									
6	Prepare sides and sawing									
7	Inspection									
8	Saw front side									
9	Transport completed cap for final inspection									
10	Final inspection									
11	Pack in plastic									
12	Pack in box									

Figure 6.7 Flow process chart for Company B in Gauteng, South Africa. (From Author, 2017.)

Outcomes	Operation	Inspection	Transport	Delay	Operation and inspection	Storage
Present						
Proposed						
	Distance (in meters)		Time		Cost (Rands)	

Figure 6.7 (Continued)

As reported by Dora, Kumar, and Gellynck (2016:6–11), interviews are based on searching for information to understand specific questions concerning the enabling and obstructing factors influencing the manufacturing process in manufacturing SMEs.

These interviews are conducted in the form of semistructured interviews seeking information from operators, operations managers, and general managers on the job being carried out. The operational processes used in this case are batch, continuous, project, and assembly processes.

6.3.2.1.2 Observations

Esan et al. (2013:266–8) explain that observations are made based on elements such as technology, procedures on how the instructions are followed, and manufacturing systems. In terms of technology, observations are made to see whether manufacturing SMEs update technological information. Secondly, observations are used to ensure that manufacturing SMEs comply with standard operating procedures to improve their manufacturing processes. Finally, the information collected identifies the availability of input factors such as education, training, and awareness. The aim of using observations to gather information regarding all these elements is to ensure there is maintenance of the manufacturing systems for continuous improvement in manufacturing SMEs.

According to Gupta and Jain (2014:28), observations involve the use of flow process charts to write information that identifies negative factors affecting the quality of productivity in manufacturing SMEs. These factors include material misplacement and loss, lost tools and equipment due to poor housekeeping, unnecessary transportation, poor layout on shop floors, unavailability of operating procedure manuals, and confusing processes due to poor planning and disorganized layout.

Pandey et al. (2014:113–4) state that observations are made using outline process charts, which provide an overall picture recording the information in

COMPANY B: PRESENT CHART (CONTINUED)

FLOW PROCESS CHART										
Name of the Company	Company A			Division/ Department/ Section			Assembly Section			
Name of the work study specialists	Mr X			Name of the operator/ employee			Mr Y			
Study number	2			Sheet number			2			
Operator number	1			Job number			1			
Chart begins	Cutting Section			Chart finish			Despatch Section			
Method	Present	X		Proposed			Method			
Type	Worker		Material	Type		Worker				
Description of the task: Prepare rod for customers										
Step number	Description of work activities	Symbols						Time (in min)	Distance (in meters)	Remarks
		●	■	➡	D	▣	▼			
13	Seal box									
14	Transport sealed box to warehouse									
15	Storage									

Figure 6.7 (Continued)

Outcomes	Operation	Inspection	Transport	Delay	Operation and inspection	Storage
Present						
Proposed						
	Distance (in meters)		Time		Cost (Rands)	

Figure 6.7 (Continued)

sequences using only operations and inspections, whereas with a flow process chart, observations show information for the setting out of the sequence of the flow of a product or a procedure using appropriate process chart symbols. Gupta and Jain (2015:76–80) refer to observations as the selection of work areas to be examined for the background of how input resource factors are utilized, such as shop floors. These observations aim to raise awareness of problem areas such as material waste and energy, which need to be avoided to improve efficiency in manufacturing SMEs for growth in their businesses.

According to Ohu et al. (2016:48–9), observations comprise the information collected through the use of two-handed process charts, determining how factors such as risk and capacity affect employees with respect to their body in manufacturing SMEs. From the observations of authors used in this literature, using automation in handling material plays an important role in ergonomic risks as well as in employee performance.

6.3.2.1.3 Questionnaires

As is stated by Kristianto, Ajmal, and Sandhu (2012:34–7) questionnaires focus on the action plan and contain various questions that cover the information needed for product and service quality for the continuous improvement of operational processes in manufacturing SMEs. Gupta et al. (2013:636–40) argue that not only interviews but also questionnaires should be used as tools to gather information from management in all divisions of the manufacturing SMEs on their operational background. Questions addressed are also based on how operational improvement targets are set in terms of waste elimination, flexibility, cost-effectiveness, product quality, and just-in-time delivery on the operational process for growth in businesses of these SMEs.

According to Granly and Welo (2014:195–7), a questionnaire is a form comprised of questions to be ticked, which is distributed to manufacturing SMEs that operate in their continuous processes. This questionnaire aims to check

COMPANY C: PRESENT CHART

FLOW PROCESS CHART			
Name of the Company	Company C	Division/ Department/ Section	Assembly Section
Name of the work study specialists	Mr X	Name of the operator/ employee	Mr Y
Study number	3	Sheet number	1
Operator number	1	Job number	1
Chart begins	Collection Area	Chart finish	Warehouse Section

Method	Present	X	Proposed	Method

Type	Worker		Material	Type	Worker	

Description of the task: Packaging pens for customer

Step No.	Description of work activities	Symbols						Time (in min)	Distance (in meters)	Remarks
		●	■	⇨	D	▣	▼			
	Pens Packaging Section									
1	Taking material from warehouse to cutting section									
2	Stock in the production area									
3	Inspection									
4	Run through production line									
5	Feed the small box									
6	Pack pens in the box following colors sequence									
7	Inspection									
8	Close the box									
9	Final inspection									
10	Pack in big box									
11	Place in the conveyor for sealing									

Figure 6.8 Flow process chart for Company C in Gauteng, South Africa. (From Author, 2017.)

12	Pack inbox									

Outcomes	Operation	Inspection	Transport	Delay	Operation and inspection	Storage
Present						
Proposed						
	Distance (in meters)		Time		Cost (Rands)	

Figure 6.8 (Continued)

whether these SMEs comply with ISO 14001 for health and safety of employees and with ISO 9001 to ensure quality management in their businesses.

Marodin and Saurin (2015:69–71) report that questionnaires in manufacturing SMEs focus on the background of risks associated with top, middle, and operational management as a well employees in mass production as well as projects focusing on short-term goals for continuous improvement. This risks having an impact on the management and operational practices exercised by all levels of management, which delay the operational process of manufacturing SMEs to ensure competitiveness in their market. These questionnaires aim to seek information regarding risks such as commitment from top, middle, and first-line management; employee morale and technical and job knowledge; communication within the business, both vertically and horizontally; and employee empowerment and safety in the workplace. These risk factors are inspected to ensure improvement in operational processes and ensure a gain in competitiveness in the market for manufacturing SMEs.

As indicated by Thomas et al. (2016:89), questionnaires involve various questions regarding financial background, organizational profile, business type, technology used, business attitude, and operational processes and systems as well as drivers for sustainability in manufacturing SMEs' growth. With financial background, manufacturing SMEs focuses on materials and labor costs, growth profile, operating costs, investment in processing technology, and major investments over the past 5 years. In terms of organizational profile, manufacturing SMEs look at the number of employees and direct and indirect staffing ratios. The questionnaire also captures the following information with regard to business type: growth profile, customer base, relationships with customers, types of products manufactured, methods of production employed, labor skills, and knowledge base. Technology in manufacturing SMEs is also addressed based on the technology currently employed, future plans for investing in new technologies, previous experiences in implementing new technology, and benefits of employing manufacturing technology,

COMPANY C: PRESENT CHART (CONTINUED)

FLOW PROCESS CHART										
Name of the Company	Company C							Assembly Section		
Name of the work study specialists	Mr X				Division/ Department/ Section			Mr Y		
Study number	3				Sheet number			2		
Operator number	1				Job number			1		
Chart begins	Cutting Section				Chart finish			Despatch Section		
Method	Present	X				Proposed			Method	
Type	Worker		Material			Type		Worker		
Description of the task: Packaging pens for customers										
Step number	Description of work activities	Symbols						Time (in min)	Distance (in meters)	Remarks
13	Pack sealed box in the pallet									
14	Wrap loaded pallet with plastic									
15	Transport to warehouse									
16	Storage									

Figure 6.8 (Continued)

as well as worker and management skills requirements based on technology. With regard to business attitudes in manufacturing SMEs, ambitions to grow the businesses based on market and new technologies is also focused on operational processes and systems, in terms of operational management

Outcomes	Operation	Inspection	Transport	Delay	Operation and inspection	Storage
Present						
Proposed						

	Distance (in meters)	Time	Cost (Rands)

Figure 6.8 (Continued)

style structure and approach as well as the development and management of manufacturing systems in manufacturing SMEs are also considered, respectively. Finally, the drivers for sustainability in manufacturing SMEs on how the business is growing is also dealt with, and this process is also done through the use of a questionnaire. All this information is gathered to ensure whether the manufacturing SME is experiencing growth.

6.3.2.2 Benchmarking

Benchmarking is a tool used by management in manufacturing SMEs to identify weakness and strengths in terms of product quality, cost, delivery, flexibility, and dependability against that of competitors through the application of ISO 9001 quality management system concept. The aim is to reduce the limitations (weaknesses) and optimize strengths in order to increase productivity with the aim of being competitive in the marketplace (Psomas, Kafetzopoulos, & Fotopoulos, 2012:66–71).

Panwar et al. (2013:777–8) define benchmarking as a process by management to learn best practices and incorporate and monitor successful capabilities such as employee effort, product design, just-in-time delivery, layout, and value-adding drivers for productivity improvement in manufacturing SMEs. Dai and Kuosmanen (2014:178–9) and Schulze et al. (2016:3701–6) emphasize benchmarking as feasible tool for energy consumption analysis and efficient use of building and machinery as compared to rivals. The monitoring of energy by management promotes government support at the same time as saving costs needed for the increase in productivity required in manufacturing SMEs.

Benchmarking involves determining the factors contributing to efficiency improvement, efficient manufacturing processes, and effort expended by managers in manufacturing SMEs. These factors are compared, looking at internal

COMPANY D: PRESENT CHART

FLOW PROCESS CHART			
Name of the Company	Company D	Division/ Department/ Section	Assembly Section
Name of the work study specialists	Mr X	Name of the operator/ employee	Mr Y
Study number	4	Sheet number	1
Operator number	1	Job number	1
Chart begins	Collection Area	Chart finish	Warehouse Section

Method	Present	X	Proposed	Method

Type	Worker		Material	Type	Worker	

Description of the task: Prepare security system for the customers

Step No.	Description of work activities	Symbols						Time (in min)	Distance (in meters)	Remarks
		●	■	➡	D	▣	▼			
	Security System Section									
1	Taking material from warehouse to cutting section									
2	Kitting components in downfloor (production line)									
3	Carry to quality section for testing									
4	Transport the completed kit									
5	Assemply (components)									
6	Soldering of security systems									
7	Fill in box									
8	Final inspection									
9	Place sticker (pass/ rejection)									

Figure 6.9 Flow process chart for Company D in Gauteng, South Africa. (From Author, 2017.)

10	Carry inspected box to packing bay									
11	Pack final box a with remote									
12	Store in warehouse									

Outcomes	Operation	Inspection	Transport	Delay	Operation and inspection	Storage
Present						
Proposed						

	Distance (in meters)	Time	Cost (Rands)

Figure 6.9 (Continued)

factors such as product cost, speed, safety, flexibility, and dependability, from one section to the other within the same manufacturing SME in order to improve efficient business processes. Also compared are the external effects between one manufacturing SMEs and another such as product quality, lower costs, and on-time delivery for customer service. This benchmarking method is followed by brainstorming to bring about change in terms of product, process, and services to customers for productivity improvement in manufacturing SMEs (Elaswad et al., 2015:2039–44).

As stated by Battagello, Cricelli, and Grimaldi (2016:324–44), benchmarking involves comparing the performance and outputs of manufacturing SMEs to others, measured in financial figures such as profitability and in terms of reliability such as customer expectations; these are obtained from the assets involved in the balance sheets and cash flow statements in relation to expenditure used. These assets are human capital (education), structural capital (such as building), social capital (teamwork), and management abilities (planning and control).

6.3.2.3 Brainstorming

Mathur, Mittal, and Dangayach (2012:754–8) and Chompu-inwai, Jaimjit, and Premsuriyanunt (2015:1358–9) maintain that brainstorming involves a team of employees recording challenges facing manufacturing with the intention of scaling down input resource factors such as employee, material, machinery, method, and standard operating procedures contributing to a decline in productivity. The proposals are then pursued as part of problem solving in the manufacturing process, resulting in incremental improvements in productivity.

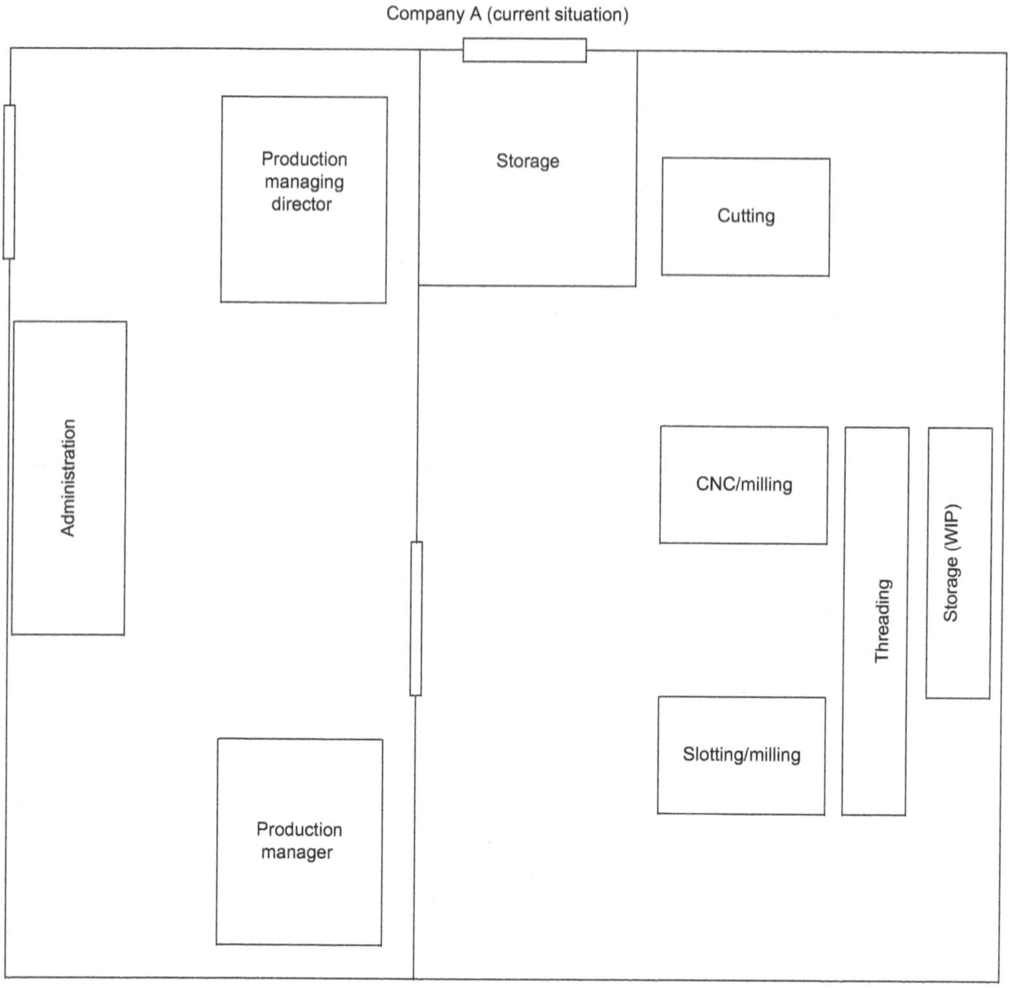

Figure 6.10 Flow diagram for Company B in Gauteng, South Africa. (From Author, 2017.)

Gupta et al. (2013:639–48) and Gupta and Jain (2014:22–28) state that brainstorming involves a group of dynamic employees together with management discussing matters relating to product quality, material requirement planning, and employees. The discussion group in the meeting utilizes a questionnaire in listing problems and solves problem through the elimination of seven types of waste, such as inventory, defect, overproduction, idling, overprocessing, motion, and underutilization. Silva et al. (2015:175–6) define brainstorming as the organized generation of creative ideas by the process team, outlining problems and addressing production problems.

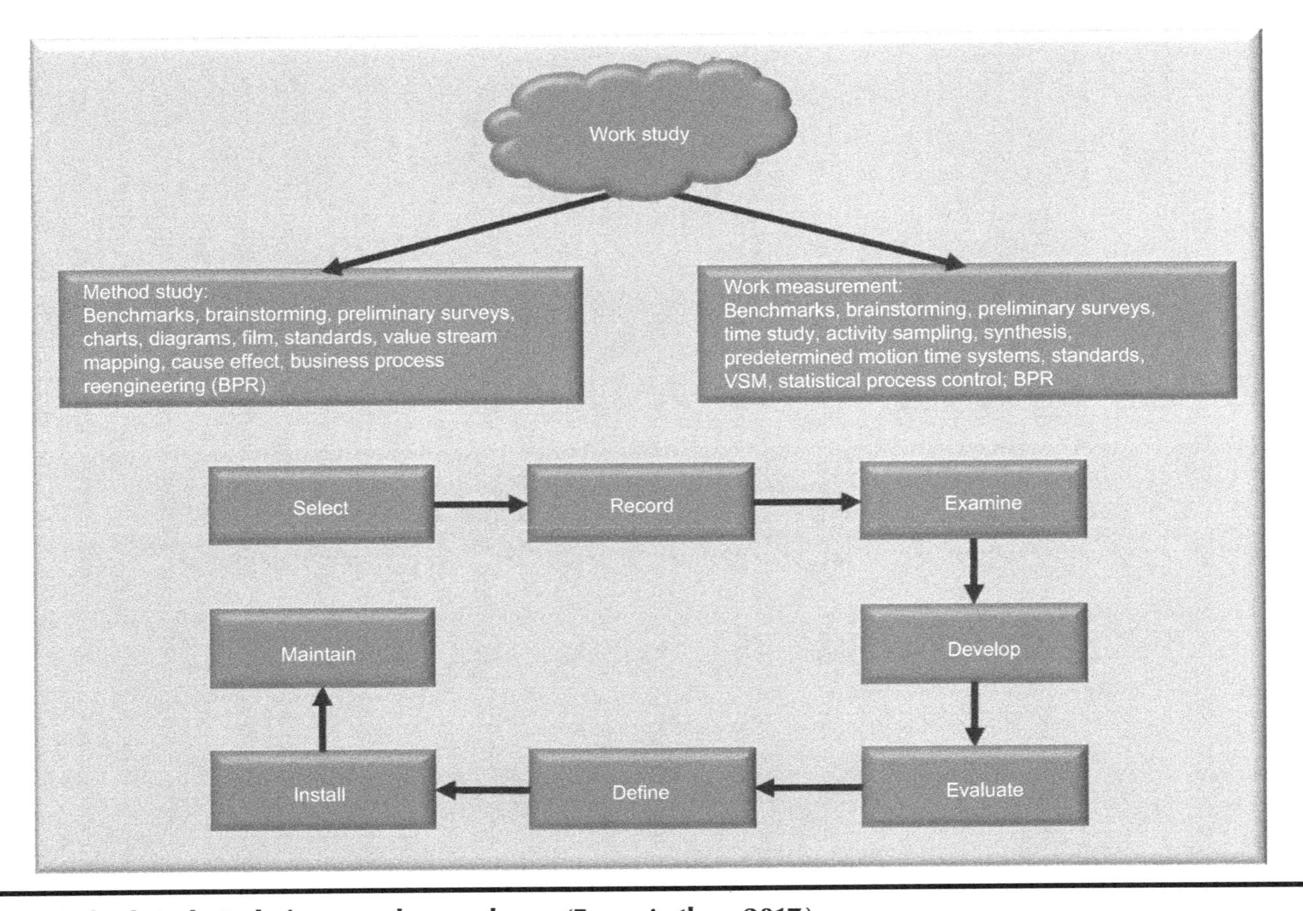

Figure 6.11 Method study techniques and procedures. (From Author, 2017.)

Belekoukias, Garza-Reyes, and Kumar (2014:5346–53) and Low, Kamaruddin, and Azid (2015:705–18) express brainstorming as techniques applied by a small team to improve manufacturing problem complexities. The techniques support the team in identifying a problem, determining the cause of the problem, developing an alternative solution, and devising an action plan for improvement. Sharma and Shah (2016:571–98) articulate brainstorming as involving shop-floor employees and management through meeting and sharing knowledge with regard to the input resource factors, process capabilities, and value-adding activities that are hindering productivity. The team uses a cause-and-effect diagram to identify problems and before-and-after value stream mapping to achieve continuous improvement in manufacturing SMEs so that they can gain a competitive edge.

6.3.2.4 Method Study Process Charts and Time Scale Diagrams and Films

Method study process charts consist of flow process, outline, and two-handed process charts.

6.3.2.4.1 Flow Process Chart

A flow process chart is a chart that represents activities, equipment, or flow of material in order to identify problem areas faced in the manufacturing process. This can be compared with an improved chart to improve productivity in manufacturing SMEs. This process chart uses chart symbols such as operation, inspection, transport, delay, and storage. The aim of comparing an improved chart with an existing chart is that management can improve processes by eliminating waste, rearranging and combining (synthesizing) work, and simplifying work activities in their businesses (Wajanawichakon & Srimitee, 2012:957).

Das, Ghosh, and Gangopadhyay (2013:247–9) explain that a flow process chart comprises process chart symbols such as operation, inspection, transport, delay, storage, and a combination of both operation and inspection with the aim of identifying problems facing manufacturing SMEs in their operation processes. Gupta and Jain (2014:22–34) state that a flow process chart involves information on operation, inspection, movement, delay due to material misplacement and loss, lost tools and equipment due to poor housekeeping, unnecessary transportation, poor layout on shop floors, and unavailability of operating procedure manuals, as well as confusing processes due to poor planning and disorderly layout as well as storage issues in manufacturing SMEs.

As found by Kumar et al. (2015:132–39), a flow process chart focuses on information regarding material flow from suppliers moving toward quality checks as input. When this material is checked and approved, this material goes to the manufacturing process via operational decisions or capabilities such as product design, process design, employee ability, supply chain relationship, employee activity time scheduling, capacity, maintenance, inventory flow and management, technology, just-in-time and customer service, and adding-value drivers (such as quality, costs, delivery, speed, reliability, and dependability). The information regarding storage at the completion of the product in manufacturing SMEs is also recorded on this process chart.

As indicated by Sanjog et al. (2016:33–40), a flow process chart compares information about employee activities at the proposed workstation with activities at the existing workstation to check physical ergonomic conditions of the workstations in manufacturing SMEs.

6.3.2.4.2 Outline Process Chart

Lyons et al. (2013:477–80) consider an outline process chart as a chart that entails operations and inspections, showing how operations are taking place using the standard operating procedures and how these operations are monitored in different manufacturing system types such as project, job shop, batch, and continuous systems.

Pandey et al. (2014:113–4) refer to an outline process chart as "a process chart giving an overall picture by recording in sequences only the main operations and inspections. In an outline process chart, only the principle operations are carried out and the inspections made to ensure the effectiveness. These process chart symbols are recorded, irrespective of who does them and where they are performed."

According to Gould and Colwill (2015:56–63), an outline process chart involves the use of process chart symbols such as operations to outline how machinery is operated and material flow as well the use of inspections to assess the efficiency of the system and to monitor the system for quality, cost, flexibility, dependability, and service so as to ensure continuous improvement in manufacturing SMEs.

As found by Manne, Ahmad, and Waterman (2016:32–40), an outline process chart comprises operations and inspections, showing operations activities and monitoring through the use of standard operating procedure in an assembly for improvement in manufacturing SMEs.

6.3.2.4.3 Two-Handed Process Chart

Ray and Teizer (2012:439–55) define a two-handed process chart in manufacturing SMEs as a chart using all process chart symbols such as operation, inspection, and transport to indicate movement of hands carrying out tasks such as lifting loads through the back and knee and working overhead using the neck and shoulder, which results in problems such as heavy loads, kneeling, contact stress, vibration, extreme temperatures, twisting hands or wrists, stretching to work overhead, or other awkward positions while performing job activities. A proposed two-handed chart is then used to assist manufacturing SMEs to determine the right employee posture to prevent these problems from happening in the workplace by improving the work-site environment in order to ensure employee safety and health.

According to Xia, Lopes, and Restivo (2013:69–73), a two-handed process chart in manufacturing SMEs involves information on using both hands for handling parts and tools and the free movement of the employee's hand toward the tool or any work-related object for assembly.

Sylla et al. (2014:475–7) advise that a two-handed process chart comprises information regarding the use of two hands simultaneously to carry out assembly line work at different workstations. The gathering of this information through this chart aims to ensure that a final product is completed without inappropriate posture, resulting in employee discomfort and injury and an unsafe working environment in manufacturing SMEs. De Macedo Guimarães et al. (2015:105) comment that a two-handed process chart in manufacturing SMEs involves information for using both hands for material handling and posture for the movement of hands toward a tool or any work-related object in the workplace. This process take place using all process chart symbols such as operation, evaluate (check), delay, and movement. This chart allows management to be aware of information shown in the existing chart indicating employee physical pain resulting from repetitive work, inappropriate postures/motion, and heavy materials handling. Then, the improved two-handed process chart is proposed by management to eliminate repetitive work for employees and for workers to be able to engage themselves in using proper posture and handling materials using new technology. As a result, this technology will reduce physical pain and also save time and costs for manufacturing SMEs.

According to Ohu et al. (2016:48–9) and Sanjog et al. (2016:36), a two-handed process chart involves using left- and right-hand operation charts to scrutinize and record activities performed by employees in the existing workstations.

6.3.2.5 Process Chart Based on Time Scale (Multiple Activity Chart)

Based on the literature studied, a process chart based on a time scale is referred to as a *multiple activity chart* (Dai et al., 2012:59; Naber & Kolisch, 2014:336; Dinis-Carvalho et al., 2015:235–44). Dai et al. (2012:51–62) state that a multiple activity chart is used in several manufacturing systems such as batch, mass, continuous, project, and assembly systems, whereby the information is gathered on parallel work activities of planning, inspection, operation, handling, and delivery for manufacturing SMEs.

Padhi et al. (2013:1886–7) clarify that a multiple activity chart generates information on work activities carried out by employees and/or machinery, such as job preparation, operation, and inspection of products during the assembly line process using a common time scale. As pointed out by Naber and Kolisch (2014:336), a multiple activity chart presents information on work activities carried out by employees and/or machinery in parallel.

Dinis-Carvalho et al. (2015:235–44) explain that a multiple activity chart presents information on work activities executed by employees and/or machinery such as job preparation, processing, packaging, and transportation of products as assembled according to customer specifications using a mutual time scale. The information collected on this chart shows challenges facing manufacturing SMEs in terms of quality, costs, reliability, safety, speedy delivery, and flexibility for improvement in their businesses.

Harmse, Engelbrecht, and Bekker (2016:7–9) contend that a multiple activity chart involves information regarding work activities carried out by employees and/or machinery such as job preparation, processing, packaging, and distribution of products in large batches (mass production) as per customer requirements using the a joint time scale. The information gathered on the chart identifies problem areas where manufacturing SMEs can reduce costs, build trust, encourage flexibility, ensure speedy delivery, and improve quality.

6.3.2.6 Flow Diagrams

Kaushik et al. (2012:7–8) show the flow diagram in the form of a SIPOC (supplier, input, process, output, customer) diagram, a diagram that is used in manufacturing SMEs for identifying the flow of material from supplier through manufacturing up to the customer. Burchart-Korol (2013:236–7) and

Alkaya and Demirer (2014:596–601) explain that a flow diagram depicts a flow of raw materials and energy as well as sustainable production technologies using process chart symbols such as operation, inspection, transport, delay, and storage for the workplace. The purpose of using these symbols is to identify problems facing manufacturing SMEs in their operation processes and to develop a proposed method for improving the manufacturing process by using energy efficiently, improved technology, and appropriate material to avoid unnecessary movements and delays so as to create cost savings in manufacturing SMEs.

As pointed out by Silva et al. (2015:176), flow process diagrams are common basic management tools, associated with flow process charts, that determine the sequence of events in a process in order to identify the unclear elements in the workplace in manufacturing SMEs. This diagram incorporates sequences of actions, inputs and outputs, decisions that should be made, and the people and time involved at each step and process measurements. As stated by Manne et al. (2016:33–9), a flow diagram involves the movement of operators and the flow of material using an indication of process chart symbols such as operation, monitoring, transport, delay, and storage for the workplace. The aim of using these process chart symbols is to identify problems in an existing flow diagram analyzing the problem to avoid unnecessary transport, delay, and storage in manufacturing SMEs. Storage in this case is avoided for flexibility and customer convenience.

6.3.2.7 Film Technique

Dai et al. (2012:53) identify the film technique for the manufacturing plan of manufacturing SMEs, which shows an image before and after problem solving intervention in the workplace. This picture is taken of activities in various manufacturing systems such as batch, mass, continuous, project, and assembly systems. Yuan, Zhai, and Dornfeld (2012:40–2) consider film technique as a technique used to take or acquire photos regarding input resource factors such as technology, material, and energy used in the manufacturing process. The aim of identifying these input resource factors is to improve the technological processes for reducing both material and energy consumptions in order to sustain the productivity of manufacturing SMEs. Furthermore, in terms of energy, in particular toxic substances, pictures assist management in manufacturing SMEs to ensure the health of employees in the working environment.

As is pointed out by Heidrich and Tiwary (2013:5884–91), the film technique uses photos to show how manufacturing SMEs are operating through resource planning and the adjustment of employees in relation to the environment. This helps management to comply with the environmental standard ISO 14001. Muruganantham et al. (2014:420–6) consider the film technique as the taking of photographs in order to identify problems in the manufacturing process and so reduce cycle time through eliminating waste and reducing employee fatigue by following proper operating procedures. Chay et al. (2015:1033–39) state that the film technique comprises photos taken or obtained to show whether employees lack skill and knowledge and need to be trained for improvement to avoid waste on the shop floor. Based on the literature studied, this achieved through business reengineering and just-in-time scheduling for continuous improvement in manufacturing SMEs.

According to Dora et al. (2016:6–8), the film technique involves photos indicating how the processes and operations are taking place before and after intervention for improvement in manufacturing SMEs. This process happens when the manufacturing SME has planned the number of employees to be assigned in specific work stations to avoid overstaffing as well as the working conditions required in terms of proper communication and good employee treatment in their sphere of operation.

6.3.2.8 Value Stream Mapping

In this section, value stream mapping (VSM) is reviewed and broken down to allow an understanding of the concept in manufacturing SMEs.

Chiarini (2012:681–8); Dora et al. (2013:125–37); and Matt (2014:334–43) explain that VSM includes waste elimination processes initiated by managers to meet set objectives and the same time continuously improving set goals in manufacturing SMEs. Furthermore, manufacturing capability and lean tools are used to ensure that value drivers are secured, enhancing productivity in manufacturing SMEs.

Dushyanth-Kumar-Kumar, Shivashankar, and Rajeshwar (2015:9–10) affirm that VSM maps out the current state of production with the intention of identifying bottlenecks and introducing measure for continuous improvement and the achievement of the proposed state. The VSM can be applied along the SIPOIC (supplier, input, process, output, customer) and maps the solution that will ensure productivity improvement in the manufacturing SMEs. Tools such as cause and effect are applied to allow the identification of problems and secure the sustainable solution required in manufacturing SMEs. Table 6.1 illustrates VSM symbols that can be applied through the SIPOC for productivity improvement in manufacturing SMEs.

Table 6.1 VSM Symbols

Symbol	*Description*
Customer/supplier	Used to show customers, suppliers, and outside manufacturing process
Dedicated process	This icon is a process, operation, machine, or department, through which material flows
Data box	Which carries customers, department, and manufacturing process
Inventory	This shows inventory between two process
Shipments	This icon represents movement of raw materials from suppliers to the receiving docks of the factory, or the customers
Push arrow	This icon represents the "pushing" of material from one process to the next process
Material pull	Pull of materials from supermarkets
External shipment	Shipments from suppliers to customers using external transport
Manual info	A straight, thin arrow shows general flow of information from memos, reports, or conversations
Electronic info	This wiggle arrow represents electronic flow such as electronic data interchange (EDI), the internet, intranets, LANs, WANs
Production kanban	This icon triggers production of a predefined number of parts. It signals a supplying process to provide parts to a downstream process
Withdrawal kanban	This icon represents a card or device that instructs a material handler to transfer parts from a supermarket to the receiving process
Signal kanban	It signals a changeover and production of a predetermined batch size of the part noted on the kanban
Go see	Gathering of information through visual means

(*Continued*)

Table 6.1 (Continued) VSM Symbols

Symbol	*Description*
Kaizen burst	These icons are used to highlight improvement needs and plan kaizen workshops at specific processes that are critical to achieving the VSM
Operator	This icon represents an operator. It shows the number of operators required to process the VSM family at a particular workstation
VA VA NVA NVA NVA Timeline	The timeline shows value added times (cycle times) and non-value added (wait) times. Use this to calculate lead time and total cycle time
Verbal information	This icon represents verbal or personal information flow
Sequenced pull	This icon represents a pull system that gives instruction to subassembly processes to produce a predetermined type and quantity of product
XOXO Load leveling	This icon is a tool to batch kanbans in order to level the production volume and mix over a period of time
MRP/ERP	Scheduling using MRP/ERP or other centralized systems

Source: Dushyanth-Kumar-Kumar, K.R. et al. *Industrial Engineering and Management*, 4(3):1–11, 2015.

Prasad, Khanduja, and Sharma (2016:408–13) state that VSM employs capabilities include product design, process design, service design, just-in-time delivery, layout strategies for safety and good housekeeping, location strategies for inventory space and control, maintenance systems, scheduling, material requirement planning and control, scheduling for setting up time, employee resources, and quality management for productivity improvement in manufacturing SMEs. The next sector focuses on the cause-and-effect diagram.

6.3.2.9 Cause-and-Effect Diagram

Gnanaraj et al. (2012:601–7); Silva ct al. (2013:174–86); Prashar (2014:111–2); and Gupta and Jain (2015:76) report that the cause-and-effect diagram (CAED) encompasses problem solving tools used in manufacturing process by management and employees through brainstorming sessions, considering input resource factors such as human, material, machine, technology, finance, location, layout, and management that are hindering productivity in the manufacturing process. The tool is normally used by production teams to identify problems in the manufacturing process and come up with solutions to the problem. This process of applying the CAED ends with an action plan and corrective measures to eliminate waste and defects and save costs for process improvement.

Srinivasan et al. (2016:810–14) explain that the CAED centers around input variables such as employees, material, machinery, standard operating procedures and environment, and key product and customer requirements. The CAED plays a significant role in the effective and efficient management of manufacturing SMEs. The tool can easily identify challenges such as poor maintenance systems, downtime, unavailability of machinery, and part, technical, and tooling problems. Important safety and health issues in manufacturing, such as dust, poor lighting, noise, and excessive vibration can be resolved, improving working conditions and thus productivity in manufacturing SMEs.

6.3.2.10 Business Process Reengineering

Poudelet et al. (2012:192–200) explain that business process reengineering (BPR) is a WS tool for process redesign, focusing on adding-value drivers such as cost, quality, service, and speed in order to improve productivity and competitiveness in manufacturing SMEs. This tool is applied in various manufacturing systems such as mass production and project systems. Various capabilities such as product, process and service design; quality management; JIT; SC; MRP; ERP; scheduling; maintenance; technology; human resources capability; location; and layout are also employed to ensure productivity improvement in manufacturing SMEs. For BPR to function properly, a number of different input resource factors need to be considered to ensure improvement in the manufacturing processes. These resources include employee health and capability, material safety and control, maintenance to achieve effective and efficient utilization of machinery, technology, and financial management. During the BPR process, manufacturing SMEs aim to produce a quality product that will meet customer demands.

Olhager (2013:6837–40) asserts that BPR is about making significant changes by reexamining the manufacturing process and optimizing existing conditions such as inventory control, time scheduling, material requirements planning, enterprise resource planning, supply chain management, shop-floor control, and production environment by focusing on quality, delivery, and flexibility.

BPR plays an important role in manufacturing systems such as mass and assembly systems.

BPR refers to the total transformation of a business: an unrestricted redesigning of all business processes, technologies, management systems (such as assembly line, batch, mass, project, and continuous production systems), and organizational structures by focusing on adding-value drivers such cost, quality, service, and speed in order to achieve a significant improvement in productivity (Böhme et al., 2014:6519–32).

As is pointed out by Low et al. (2015:703–19), BPR involves discovering how business processes operate in the present situation and how these processes can be redesigned to eliminate waste, improve efficiency, and implement process changes in order to exceed customer requirements for competitive advantage in manufacturing SMEs. Alaskari, Ahmad, and Pinedo-Cuenca (2016:68–7) state that BPR focuses on making meaningful amendments by reassessing the manufacturing process and designing a similar process in order to make improvements in terms of quality, delivery, and flexibility. Furthermore, BPR uses various manufacturing capabilities such as product design, process planning, facilities and layout, material purchasing, production planning and control, quality control, maintenance, employee resources, logistics and supply chain, new technology, and minimum cost production for productivity improvement in manufacturing SMEs.

6.4 Analysis of Current Applications of Method Study in South Africa

The focus of this chapter is the understanding of the method study in the workplace in manufacturing SMEs. This technique involve various method study techniques such as benchmarking, brainstorming, preliminary surveys, charts, diagrams, film techniques, standards, value stream mapping, CAED, and BPR. All these techniques are known to represent the best solutions for problem areas facing manufacturing industries, both large and small, worldwide. The challenge is how to create an understanding of these techniques to enhance the productivity of manufacturing SMEs for competitiveness in South Africa.

Even though Productivity SA is mandated by government to assist in improving the productivity of manufacturing SMEs in their businesses through application of method study as indicated in Chapter 1, WS specialists from industries and academics also need to collaborate through relevant advisory committees, conferences, and research workshops from various universities.

6.5 Summary

The purpose of this chapter is to provide a detailed background of various areas of WS relations, traits of WS specialists in manufacturing SMEs, expertise of WS specialists in manufacturing SMEs, and relations between WS experts and management, supervisors, workers, and trade unions in manufacturing SMEs. Furthermore, method study procedures are discussed, such as selection of the job to be studied; recording of the information regarding the task being carried out; critical examination of recorded information; development and selection of the alternative solution; definition of a new method; implementation of the new method; and maintenance of the new method. These procedures are followed by a definition of method study techniques such as benchmarking, brainstorming, preliminary surveys, charts, diagrams, film techniques, standards, value stream mapping, CAED, and BPR. These techniques are used to simplify the jobs in the workplace with the intention of boosting the productivity of manufacturing SMEs.

References

Alaskari, O., Ahmad, M.M., and Pinedo-Cuenca, R. 2016. Development of a methodology to assist manufacturing SMEs in the selection of appropriate lean tools. *International Journal of Lean Six Sigma*, 7(1): 62–84.

Alkaya, E. and Demirer, G.N. 2014. Sustainable textile production: A case study from a woven fabric manufacturing mill in Turkey. *Journal of Cleaner Production*, 65: 595–603.

Amin, M.A. and Karim, M.A. 2013. A time-based quantitative approach for selecting lean strategies for manufacturing organisations. *International Journal of Production Research*, 51(4): 1146–1167.

Bateman, N., Hines, P. and Davidson, P. 2014. Wider applications for lean: An examination of the fundamental principles within public sector organisations. *International Journal of Productivity and Performance Management*, 63(5): 550–568.

Battagello, F.M., Cricelli, L., and Grimaldi, M. 2016. Benchmarking strategic resources and business performance via an open framework. *International Journal of Productivity and Performance Management*, 65(3): 324–350.

Bechar, A. and Eben-Chaime, M. 2014. Hand-held computers to increase accuracy and productivity in agricultural work study. *International Journal of Productivity and Performance Management*, 63(2): 194–208.

Belekoukias, I., Garza-Reyes, J.A., and Kumar, V. 2014. The impact of lean methods and tools on the operational performance of manufacturing organisations. *International Journal of Production Research*, 52(18): 5346–5366.

Böhme, T., Deakins, E., Pepper, M., and Towill, D. 2014. Systems engineering effective supply chain innovations. *International Journal of Production Research*, 52(21): 6518–6537.

Burchart-Korol, D. 2013. Life cycle assessment of steel production in Poland: A case study. *Journal of Cleaner Production*, 54: 235–243.

Chay, T., Xu, Y., Tiwari, A., and Chay, F. 2015. Towards lean transformation: The analysis of lean implementation frameworks. *Journal of Manufacturing Technology Management*, 26(7): 1031–1052.

Chiarini, A. 2012. Lean production: Mistakes and limitations of accounting systems inside the SME sector. *Journal of Manufacturing Technology Management*, 23(5): 681–700.

Chompu-inwai, B., Jaimjit, B., and Premsuriyanunt, P. 2015. A combination of material flow cost accounting and design of experiments techniques in an SME: The case of a wood products manufacturing company in northern Thailand. *Journal of Cleaner Production*, 108: 1352–1364.

Coka, B. 2014. Annual Report 2013–2014. Midrand, Republic of South Africa: Productivity SA.

Dai, X. and Kuosmanen, T. 2014. Best practice benchmarking using clustering method: Application to energy regulation. *Omega*, 42: 179–188.

Dai, Q., Zhong, R., Huang, G.Q., Qu, T., Zhang, T., and Luo, T.Y. 2012. Radio frequency identification-enabled real-time manufacturing execution system: A case study in an automotive part manufacturer. *International Journal of Computer Integrated Manufacturing*, 25(1): 51–65.

Das, B., Ghosh, T., and Gangopadhyay, S, 2013. Child work in agriculture in West Bengal, India: Assessment of musculoskeletal disorders and occupational health problems. *Journal of Occupational Health*, 55(4): 244–258.

de Macedo Guimarães, L.B., Anzanello, M.J., Ribeiro, J.L.D., and Saurin, T.A. 2015. Participatory ergonomics intervention for improving human and production outcomes of a Brazilian furniture company. *International Journal of Industrial Ergonomics*, 49: 97–107.

Dinis-Carvalho, J., Moreira, F., Bragança, S., Costa, E., Alves, A., and Sousa, R. 2015. Waste identification diagrams. *Production Planning & Control*, 26(3): 235–247.

Dora, M., Kumar, M., and Gellynck, X. 2016. Determinants and barriers to lean implementation in food-processing SMEs-a multiple case analysis. *Production Planning and Control*, 27(1): 1–23.

Dora, M., Van Goubergen, D., Kumar, M., Molnar, A., and Gellynck, X. 2013. Application of lean practices in small and medium-sized food enterprises. *British Food Journal*, 116(1): 125–141.

Dushyanth-Kumar-Kumar, K.R., Shivashankar, G.S., and Rajeshwar, S.K. 2015. Application of value stream mapping in pump assembly process: A case study. *Industrial Engineering and Management*, 4(3): 1–11.

Elaswad, H., Islam, S., Tarmizi, S., Yassin, A., Lee, M.D., and Ting, C.H. 2015. Benchmarking of growth manufacturing SMEs: A review. *Science International*, 27(3): 2039–2048.

Esan, A.O., Khan, M.K., Qi, H.S., and Craig Naylor, C. 2013. Integrated manufacturing strategy for deployment of CADCAM methodology in a SMME. *Journal of Manufacturing Technology Management*, 24(2):257–273.

Ghosh, M. 2012. Lean manufacturing performance in Indian manufacturing plants. *Journal of Manufacturing Technology Management*, 24(1): 113–122.

Gnanaraj, S.M., Devadasan, S.R., Murugesh, R., and Sreenivasa, C.G. 2012. Sensitisation of SMEs towards the implementation of lean Six Sigma: An initialisation in a cylinder frames manufacturing Indian SME. *Production Planning and Control: The Management of Operations*, 23(8): 599–608.

Gould, O. and Colwill, J. 2015. A framework for material flow assessment in manufacturing systems. *Journal of Industrial and Production Engineering*, 32(1): 55–66.

Granly, B.M. and Welo, T. 2014. EMS and sustainability: Experiences with ISO 14001 and Eco-Lighthouse in Norwegian metal processing SMEs. *Journal of Cleaner Production*, 64: 194–204.

Gupta, S., Acharya, P., and Patwardhan, M. 2013. A strategic and operational approach to assess the lean performance in radial tyre manufacturing in India: A case based study. *International Journal of Productivity and Performance Management*, 62(6): 634–651.

Gupta, S. and Jain, S.K. 2014. The 5S and kaizen concept for overall improvement of the organisation: A case study. *International Journal of Lean Enterprise Research*, 1(1): 22–40.

Gupta, S. and Jain, K.S. 2015. An application of 5S concept to organize the workplace at a scientific instruments manufacturing company. *International Journal of Lean Six Sigma*, 6(1): 73–88.

Harmse, J.L., Engelbrecht, J.C., and Bekker, J.L. 2016. The impact of physical and ergonomic hazards on poultry abattoir processing workers: A review. *International Journal of Environmental Research and Public Health*, 13(197): 1–24.

Heidrich, O. and Tiwary, A. 2013. Environmental appraisal of green production systems: Challenges faced by small companies using life cycle assessment. *International Journal of Production Research*, 51(19): 5884–5896.

Herman, F. 2011. Textile disputes and two-level games: The case of China and South Africa. *Asian Politics and Policy*, 3(1): 115–130.

ILO. 2015. *World Employment and Social Outlook: Trends 2015*. International Labour Organization. Available from: www.ilo.org/wcmsp5/groups/public/---dgreports/---dcomm/---publ/documents/publication/wcms_337069.pdf.

Ingvaldsen, J.A., Holtskog, H., and Ringen, G. 2013. Unlocking work standards through systematic work observation: Implications for team supervision. *Team Performance Management: An International Journal*, 19(5/6): 279–291.

Jagoda, K., Lonseth, R., and Lonseth, A. 2013. A bottom-up approach for productivity measurement and improvement. *International Journal of Productivity and Performance Management*, 62(4): 387–406.

Jain, R., Gupta, S., Meena, M.L., and Dangayach, G.S. 2016. Optimisation of labour productivity using work measurement techniques. *International Journal of Productivity and Quality Management*, 19(4): 485–510.

Kafetzopoulos, D.M., Gotzamani, K.D., and Psomas, E.L. 2014. The impact of employees' attributes on the quality of food products. *International Journal of Quality and Reliability Management*, 31(5): 500–521.

Kaushik, P., Khanduja, D., Mittal, K., and Jaglan, P. 2012. A case study. *The TQM Journal*, 24(1): 4–16.

Kristianto, Y., Ajmal, M.M., and Sandhu, M. 2012. Adopting TQM approach to achieve customer satisfaction. *The TQM Journal*, 24(1): 29–46.

Kulkarni, P.P., Kshire, S.S., and Chandratre, K.V. 2014. Productivity improvement through lean deployment and WS methods. *International Journal of Research in Engineering and Technology*, 3(2): 429–434.

Kumar, S., Heustis, D., and Graham, J.M. 2015. The future of traceability within the U.S. food industry supply chain: A business case. *International Journal of Productivity and Performance Management*, 64(1): 129–146.

Kumar, E.S. and Tiwari, E.A. 2015. A work study on minimize the defect in aluminium casting. *International Journal of Emerging Technologies in Engineering Research*, 3(1): 32–38.

Lasrado, F., Arif, M., and Rizvi, A. 2015. Employee suggestion scheme sustainability excellence model and linking organizational learning: Cases in United Arab Emirates. *International Journal of Organizational Analysis*, 23(3): 425–455.

Low, S., Kamaruddin, S., and Azid, I.A. 2015. Improvement process selection framework for the formation of improvement solution alternatives. *International Journal of Productivity and Performance Management*, 64(5): 702–722.

Lyons, A.C., Vidamour, K., Jain, R., and Sutherland, M. 2013. Developing an understanding of lean thinking in process industries. *Production Planning and Control*, 24(6): 475–494.

Manne, P., Ahmad, S., and Waterman, J. 2016. Design and development of mine railcar components. *American Journal of Mechanical Engineering*, 4(1): 32–41.

Magu, P., Khanna, K., and Seetharaman, P. 2015. Path process chart: A technique for conducting time and motion study. *Procedia Manufacturing*, 3: 6475–6482.

Marodin, G.A. and Saurin, T.A. 2015. Managing barriers to lean production implementation: Context matters. *International Journal of Production Research*, 53(13): 3947–3962.

Maskaly, J. and Jennings, W. 2016. A question of style Replicating and extending Engel's supervisory styles with new agencies and new measures. *Policing: An International Journal of Police Strategies and Management*, 39(4): 620–634.

Mathur, A., Mittal, M.L., and Dangayach, G.S. 2012. Improving productivity in Indian SMEs. *Production Planning and Control: The Management of Operations*, 23(10–11): 754–768.

Matt, D.T. 2014. Adaptation of the value stream mapping approach to the design of lean engineer-to-order production systems. *Journal of Manufacturing Technology Management*, 25(3): 334–350.

Mehralian, G., Nazari, J.A., Nooriparto, G. and Rasekh, H.R. 2017. TQM and organizational performance using the balanced scorecard approach. *International Journal of Productivity and Performance Management*, 66(1): 111–125.

Muruganantham, V.R., Krishnan, P.N.. and Arun, K.K. 2013. Performance improvement and cost minimisation for manufacturing components in a fabrication plan by application of Lean with TRIZ principle. *International Journal of Productivity and Quality Management*, 12(4): 449–465.

Muruganantham, V.R., Krishnan, P.N., and Arun, K.K. 2014. Integrated application of TRIZ with lean in the manufacturing process in machine shop for productivity improvement. *International Journal of Productivity and Quality Management*, 13(4): 414–429.

Naber, A and Kolisch, R. 2014. MIP models for resource-constrained project scheduling with flexible resource profiles. *European Journal of Operational Research*, 239: 335–348.

Odesola, I.A., Okolie, K.C., and Nnametu, J.N. 2015. A comparative evaluation of labour productivity of wall plastering activity using work study. *PM World Journal a Comparative Evaluation of Labor Productivity*, IV(V): 1–10.

Oeij, P.R.A., De Looze, M.P., Have, K.T., Van Rhijn, J.W. and Kuijt-Evers, L.F.M. 2012. Developing the organisation's productivity strategy in various sectors of industry. *International Journal of Productivity and Performance Management*, 61(1): 93–109.

Ongkunaruk, P. and Wongsatit, W. 2014. An ECRS-based line balancing concept: A case study of a frozen chicken producer. *Business Process Management Journal*, 20(5): 678–692.

Ohu, I.P.N., Cho, S., Kim, D.H., and Lee, G.H. 2016. Ergonomic analysis of mobile cart–assisted stocking activities using electromyography. *Human Factors and Ergonomics in Manufacturing and Service Industries*, 26(1): 40–51.

Olhager, J. 2013. Evolution of operations planning and control: From production to supply chains. *International Journal of Production Research*, 51(23–24): 6836–6843.

Padhi, S.S., Wagner, S.M., Niranjan, T.T., and Aggarwal, V. 2013. A simulation-based methodology to analyse production line disruptions. *International Journal of Production Research*, 51(6): 1885–1897.

Panwar, A. Nepal, B., Jain, R., and Yadav, O.P. 2013. Implementation of benchmarking concepts in Indian automobile industry: An empirical study. *Benchmarking: An International Journal*, 20(6): 777–804.

Pandey, A., Singh, M., Soni, N., and Pachorkar, P. 2014. Process layout on advance CNG cylinder manufacturing. *International Journal of Application or Innovation in Engineering and Management (IJAIEM)*, 3(12): 113–116.

Poudelet, V., Chayer, J., Margni, M., Pellerin, R., and Samson, R. 2012. A process-based approach to operationalize life cycle assessment through the development of an eco-design decision-support system. *Journal of Cleaner Production*, 33: 192–201.

Prasad, S., Khanduja, D., and Sharma, S.K. 2016. An empirical study on applicability of lean and green practices in the foundry industry. *Journal of Manufacturing Technology Management*, 27(3): 408–426.

Prashar, A. 2014. Adoption of Six Sigma DMAIC to reduce cost of poor quality. *International Journal of Productivity and Performance Management*, 63(1): 103–126.

Psomas, E.L., Kafetzopoulos, D.P., and Fotopoulos, C.V. 2012. Developing and validating a measurement instrument of ISO 9001 effectiveness in food manufacturing SMEs. *Journal of Manufacturing Technology Management*, 24(1): 52–77.

Ramish, A. and Aslam, H. 2016. Measuring supply chain knowledge management (SCKM) performance based on double/triple loop learning principle. *International Journal of Productivity and Performance Management*, 65(5): 704–722.

Ray, S.J. and Teizer, J. 2012. Real-time construction worker posture analysis for ergonomics training. *Advanced Engineering Informatics*, 26: 439–455.

Sanjog, J., Patnaik, B., Patel, T., and Karmakar, S. 2016. Context-specific design interventions in blending workstation: An ergonomics perspective. *Journal of Industrial and Production Engineering*, 33(1): 32–50.

Schulze, M. Nehler, H., Ottosson, M., and Thollander, P. 2016. Energy management in industry: A systematic review of previous findings and an integrative conceptual framework. *Journal of Cleaner Production*, 112: 3692–3708.

Schwab, K. 2015. The Global Competitiveness Report 2014-2015: Insight report. World Economic Forum (WEF), Available from: www3.weforum.org/docs/WEF_GlobalCompetitivenessReport_2014-15.pdf.

Seth, D. and Gupta, V. 2016. Application of value stream mapping for lean operations and cycle time reduction: An Indian case study. *Production Planning and Control*, 16(1): 44–59.

Sharma, S. and Shah, B. 2016. Towards lean warehouse: Transformation and assessment using RTD and ANP. *International Journal of Productivity and Performance Management*, 65(4): 571–599.

Silva, D.A.S., Delai, I., de Castro, M.A.S., and Ometto, A.R. 2013. Quality tools applied to cleaner production programs: A first approach toward a new methodology. *Journal of Cleaner Production*, 47: 174–187.

Silva, D.A.S., Delai, I., de Castro, M.A.S., and Ometto, A.R. 2015. Quality tools applied to cleaner production programs: A first approach toward a new methodology. *Journal of Cleaner Production*, 47: 174–187.

Srinivasan, S., Ikuma, L.H., Shakouri, M., Nahmens, I., and Harvey, C. 2016. 5S impact on safety climate of manufacturing workers. *Journal of Manufacturing Technology Management*, 27(3): 364–378.

Srinivasan, K., Muthu, S., Devadasan, S.R., and Sugumaran, C. 2016. Enhancement of sigma level in the manufacturing of furnace nozzle through DMAIC approach of Six Sigma: A case study. *Production Planning and Control: The Management of Operations*, 27(10): 810–822.

Statistics SA 2015. Quarterly Labour Force Survey: Quarter 2: 2015. Available from: http://www.statssa.gov.za/publications/P0211/P02112ndQuarter2015.pdf.

Sylla, N., Bonnet. V., Colledani, F., and Fraisse, P. 2014. Ergonomic contribution of ABLE exoskeleton in automotive industry. *International Journal of Industrial Ergonomics*, 44: 475–481.

Tanvir, S.I. and Ahmed, S. 2013. Work study might be the paramount methodology to improve productivity in the apparel industry of Bangladesh. *Industrial Engineering Letters*, 3(7): 51–60.

Thomas A., Byard, P., Francis, M., Fisher, R., and White, G.R.T. 2016. Profiling the resiliency and sustainability of UK manufacturing companies. *Journal of Manufacturing Technology Management*, 27(1): 82–99.

Trianni, A., Cagno, E., and Farné, S. 2016. Barriers, drivers and decision-making process for industrial energy efficiency: A broad study among manufacturing small and medium-sized enterprises. *Applied Energy*, 162: 1537–1551.

Tyagi, S., Choudhary, A., Cai, X., and Yang, K. 2015. Value stream mapping to reduce the lead-time of a product development process. *International Journal of Production Economics*, 160: 202–212.

van Niekerk, W.P. 1986. *Productivity and Work Study*. 2nd edn. Durban: Butterworths Publishers.

Vijai, J.P., Somayaji, G.S.R., Swamy, R.J.R., and Aital, P. 2017. Taylor's principles to modern shop-floor practices: A benchmarking work study. *Benchmarking: An International Journal*, 24(2): 445–466.

Vimal, K.E.K. and Vinodh, S. 2013. Application of artificial neural network for fuzzy logic based leanness assessment. *Journal of Manufacturing Technology Management*, 24(2): 274–292.

Vinodh, S. and Joy, D. 2012. Structural equation modelling of lean manufacturing practices. *International Journal of Production Research*, 50(6): 1598–1607

Vinodh, S., Vasanth Kumar, S.V., and Vimal, K.E.K. 2014. Implementing lean sigma in an Indian rotary switches manufacturing organisation. *Production Planning and Control*, 25(4): 288–302.

Wajanawichakon, K. and Srimitee, C. 2012. ECRS's principles for a drinking water production plant. *IOSR Journal of Engineering*, 2(5): 956–960.

Wlazlak, P.G. and Johansson, G. 2014. R&D in Sweden and manufacturing in China: A study of communication challenges. *Journal of Manufacturing Technology Management*, 25(2): 258–278

Xia, P.J., Lopes, M., and Restivo, M.T. 2013. A review of virtual reality and haptics for product assembly (part 1): Rigid parts. *Assembly Automation*, 33(1): 68–77.

Yuan, C., Zhai, Q., and Dornfeld, D. 2012. A three dimensional system approach for environmentally sustainable manufacturing. *CIRP Annals: Manufacturing Technology*, 61: 39–42.

Chapter 7

Work Measurement (WM) Techniques

7.1 Introduction

In this chapter, work measurement (WM) as a concept and its techniques is examined and discussed on the basis of the literature sources with the aim of providing an in-depth understanding of these techniques in improving the productivity of manufacturing SMEs. WM refers to the utilization of techniques by the experienced worker to do detailed work in a given situation at a specified standard of performance (Wickramasinghe & Wickramasinghe, 2016:289–301). Work study characteristics are not indicated in this chapter since these characteristics are the same as in method study indicated in Chapter 6. Only WM procedures in small and medium-sized enterprises (SMEs) are addressed, which are different from those of method study and are presented as selection of the task to be studied, recording of the information regarding the task, analytical examination of the information collected, measurement of work, and compilation and definition of the method. Since preliminary survey studies have already been addressed in method study in Chapter 6 and could also be addressed similarly in work measurement in Chapter 7, when preparing a work study investigation in manufacturing SMEs, only time study, work sampling, predetermined motion time systems (PMTS), analytical and comparative estimation, synthesis, value stream mapping (VSM), statistical improvement techniques, and business process reengineering (BPR) are discussed in this chapter.

7.2 Work Measurement

WM addresses various quantitative techniques that involve direct observation such as time study and work (activity) sampling, and those that involve indirect observation with standard time already established, such as analytical estimating, comparative estimating (benchmarking), and PMTS as well as additional WM techniques such as control charts that involve Pareto analysis; graphs such as histogram, line, scattered, bar chart, run chart and statistical process control (SPC) graphs (refers to a diagram of a time study procedure in Garza-Reyes et al., 2012:181–5). All these techniques addressed earlier are utilized during the study investigation.

7.2.1 WM Procedure

WM procedure in manufacturing SMEs focuses on the following systematic procedures: select, record, examine, define, develop, install/implement, and maintain (SREDDIM). These procedures are used to firstly identify the problem experienced by manufacturing SMEs at the business level, in particular department, section, or unit, and to ensure the simplification of the job being carried out and its improvement for productivity enhancement in the business. The method *evaluation* is added if both method study and WM (i.e., work study) are applied in the study to improve productivity. When the method *evaluation* is added, the procedure for work study is abbreviated as SREDEDIM. The diagram for the WM procedure is highlighted and shown in Figure 7.1.

7.2.1.1 Select

Similar to method study, by selecting this means deciding on the task to be studied in manufacturing SMEs. Problems facing manufacturing SMEs are selected based on the following input resource factors: physical capital (human resources, material, machinery, layout, location, finance, technology); technological capital (tangible and intangible elements); and management. As indicated in Chapter 1, these elements need to be taken into account by manufacturing SMEs to ensure competitiveness in their market with the support of the government (ILO, 2015:61; Schwab, 2015:39).

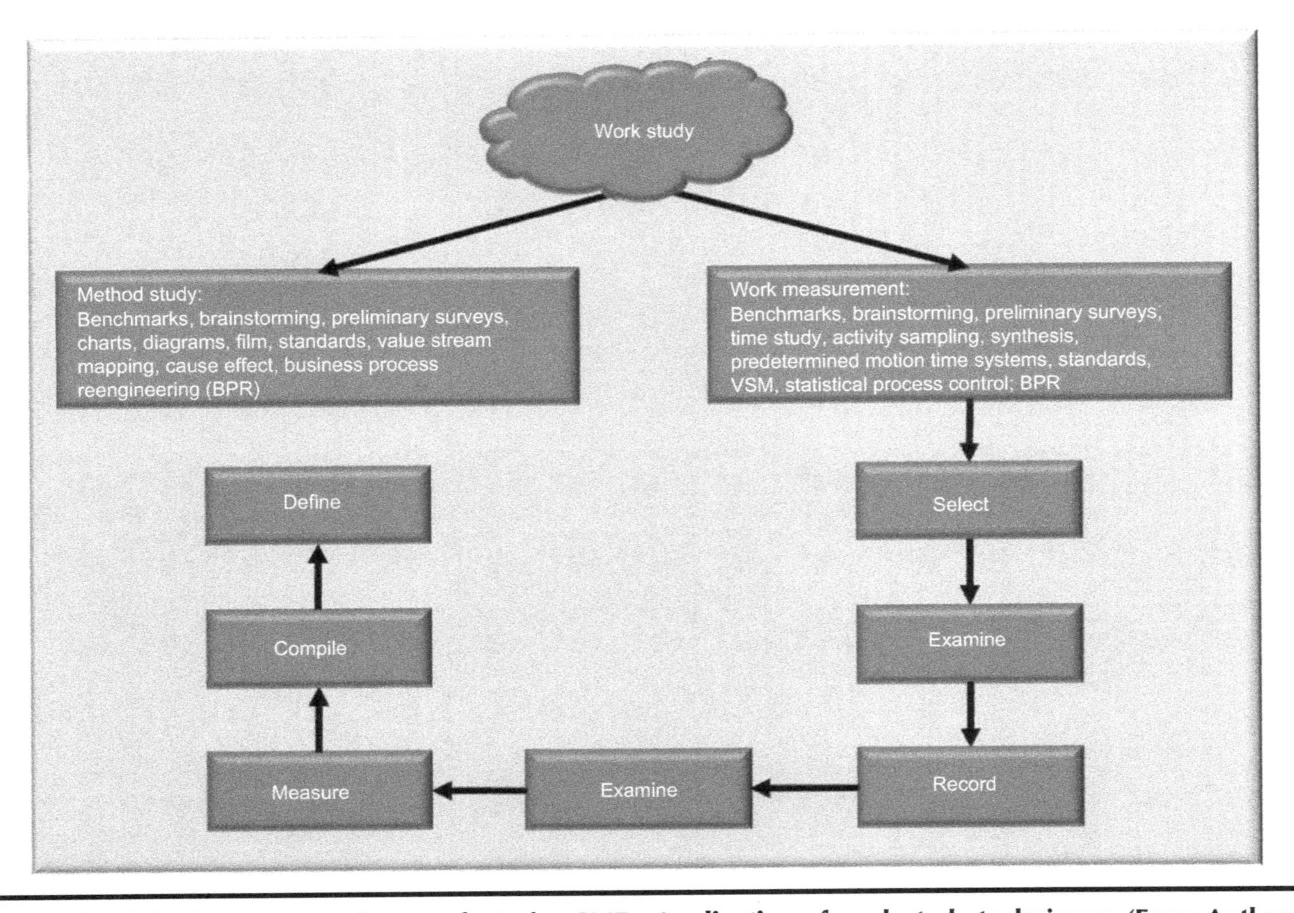

Figure 7.1 Productivity improvement in manufacturing SMEs: Application of work study techniques. (From Author, 2017.)

7.2.1.2 Recording

The next procedure to be followed is recording of the information regarding the task being carried out. Recording of the job in manufacturing SMEs is done by using various WM study techniques, the same as those used in method study. This recording procedure needs to have existing (current) and proposed (improved) information regarding the job being carried out. Even though the information is recorded for both current and improved information, improved information is exercised and the method is "developed."

7.2.1.3 Analytical Examination

Analytical examination of the information collected involves the current information collected during the method study investigation. The information is examined analytically by the work study expert using questions such as "What is being done?" "Who is doing it?" "Where is it done?" "When is it done?" "How is it done?" and "Why is done that way, by that person, in that place, and using that method?" The point of asking these questions is for the work study expert to get a clear idea of the problem being experienced by the manufacturing SME and to anticipate the improved method.

7.2.1.4 Measure

In this section, the development and selection of an alternative solution is the fourth step investigated and addressed, the aim being to provide the reader with an understanding of what "development" means when conducting a method study procedure. This method study procedure focuses only on proposed (improved) information regarding the job being carried in manufacturing SMEs with the aim of developing an improved method. These methods are time studies, work sampling, synthesis, PMTS, and analytical and comparative estimating as well as incentive schemes.

The information that has already being examined by the work study expert using questions such as "What is being done?" "Who is doing it?" "Where is it done?" "When is it done?" "How is it done?" and "Why is it being done that way, by that person, in that place, and using that method?" is supported by using the proposed questions to ensure productivity improvement in terms of the problems facing manufacturing SMEs. In this method, proposed questions are asked by the work study expert: "What else could be done?" "Who else could be doing it?" "Where else could it be done?" "When else could it be done?" "How else could it be done?" and "Why could it be done in that way, by that person, in that place, and using that method?"

7.2.1.5 Compile

The fifth step is the definition of the new method. By defining the new method, the work study expert introduces the proposed method to management for approval, so that, when approved, it can be implemented to improve the productivity of manufacturing SMEs.

7.2.1.6 Implementation

During the implementation stage, the qualified employee(s) is introduced to the proposed method to ensure the appropriate standard procedure for the job, focusing on various standards such as ISO 9001, ISO 14001, and ISO 18001.

7.2.1.7 Define

The last step in the method study procedure is to maintain the new method to avoid deviations from the proposed method.

7.2.2 WM Techniques (Quantitative)

7.2.2.1 Time Study

The diagram provided in Figure 7.2 is used to collect data based on time study investigation in manufacturing SMEs. The types of tools indicated are called time study equipment and include a stopwatch time study and time study clip board (van Niekerk, 1986:80).

Garza-Reyes et al. (2012:185) present a time study from the perspective of various scholars who define it in different ways. The first, Slack, Chambers, and Johnston (2009), cited in Garza-Reyes et al. (2012:185), refer to time study as a work measurement technique for recording the times and rate of working for the elements of a specified job, carried out under specified conditions, and for analyzing the data so as to obtain the time necessary for the carrying out of the job at a defined level of performance. Baines (1995), cited in Garza-Reyes et al. (2012:185), argues that time study, as well as being the oldest, is "the most flexible of the WM techniques as it can be applied to any type of work carried out in any environment."

Figure 7.2 Time study equipment. (Adapted from van Niekerk, W.P., *Productivity and Work Study*, Butterworths Publishers, Durban, 1986.)

According to Wajanawichakon and Srimitee (2012:956–9), time study in manufacturing SMEs involves recording the time taken by a skillful employee to carry out specific work activities under specified conditions, whereby data is analyzed and the necessary time (also referred to as *standard time*) is obtained for the employee to work at a defined rate of working.

Das, Ghosh, and Gangopadhyay (2013:247–9) report that time study in manufacturing SMEs focuses on writing information using time so as to analyze and obtain the necessary time for an employee who has acquired proper knowledge and has a relevant educational background to carry out a job under stipulated conditions at a standard performance. Time study in manufacturing SMEs is a WM technique that measures and controls the amount of time required by a well-trained, motivated, and qualified operator to do a specific task under normal conditions (Suwittayaruk, Van Goubergen, & Lockhart, 2014:725). According to Vinodh et al. (2015:380–91), time study in manufacturing SMEs involves recording the time required for a trained operator to perform a specific task using the proper standard operating procedure under the stated operating environment so as to obtain the information analyzed and time for the task being carried out a defined pace. Time study in manufacturing SMEs is a technique measuring how long it takes for a qualified employee to assemble a unit with all the parts on a continuous basis, using the stopwatch to determine cycle time at a predetermined event (Srinivasan et al., 2016:369).

An example of a time study is conducted at Company B, a business near Gauteng, South Africa, where caps are assembled. Two time study sheets are applicable to show how standard time is established for the efficient operation of the manufacturing process. These sheets involve a time study observation sheet and a study summary sheet.

The work activity undertaken is done with the application of time study by the work study intern at Company B, with the aim of determining the appropriate standard time to assemble the cap.

The time study observation sheet is used to calculate the individual basic time for the work activities being carried out under a specified work cycle. According to Stevenson (2012:298), a cycle in manufacturing SMEs is the series of work activities that is required to carry out work or a piece of a manufacturing operation. This sheet is shown in Table 7.2. There are various types of work activities carried out in the workplace in manufacturing SMEs. These activities include repetitive process activity, occasional process activity, manual work activity, machine work activity (Stevenson, 2012:238–305), constant element, variable element (McDermott & Prajogo, 2012:18–20), governing element (Adamson et al., 2017:371), and foreign element (Cochrana, Foley, & Bi, 2017:876).

According to Stevenson (2012:307–8), an activity that is also referred to as work activity is a well-defined portion of a particular work designated for accessibility of inspection or observation, measurement, and examination. Stevenson (2012:238–305) describes the various time study concepts as follows:

- Repetitive activity or process focuses on the work activity that takes place in every manufacturing run of an activity.
- Occasional element or nonrepetitive process involves a work activity that does not occur regularly in every work cycle of an activity.
- Manual work activity involves an activity carried out by hand by the employee in the manufacturing SME.
- Machine work activity is an element performed automatically—by machine, process, chemical, or otherwise—that, once started, cannot be influenced by a worker except to terminate it prematurely.

Table 7.1 Rest Allowance Guide Expressed as a Percentage of Basic Time

A. FATIGUE ALLOWANCE	%
1. Physical Exertion	
(a) Extremely light (2.5 kg)	0
(b) Light (2.5–5.0 kg)	1
(c) Medium (5.0–10.0 kg)	5
(d) Heavy (10.0–25.0 kg)	10
(e) Extremely heavy (25.0–50.0 kg)	15
2. Visual Strain	
(a) Normal attention to easy work	0
(b) Normal attention to complicated work	1
(c) Continuous attention to easy work	5
(d) Continuous attention to complicated work	10
3. Position	
(a) Alternate sitting and standing position for working	0
(b) Constant sitting position for working	1
(c) Constant standing position for working	5
(d) Body in squatting or bent position	10
(e) Hands above the shoulders	15
4. Working Conditions	
Atmospheric Conditions	
(a) Fresh air and sufficient ventilation	0
(b) Harmless but unpleasant odors	5
(c) Harmful dust and gases	10
Temperature	
(a) Cold (0°C–10°C)	1–5
(b) Normal (10°C– 27°C)	0
(c) Hot (>27°C)	1–20
Noise	
(a) Normal workshop noise	0
(b) Loud noise	5
(c) Extremely loud noise	10
General	
(a) Abnormal working conditions	1–5
(b) Wet floors	1–5
(c) Monotony	1–5
(d) Vibration	1–5
(e) Mental strain	1–10
(f) Protective clothing	1–20
B. PERSONAL ALLOWANCE	1–10

Source: Adapted from van Niekerk, W.P. 1986. *Productivity and Work Study*. 2nd edn. Durban: Butterworths Publishers.

According to McDermott and Prajogo (2012:18–20), variable work activity involves an activity for which basic time varies in relation to some characteristics of the product, equipment, or process, whereas constant work, also referred to as *stable work*, is considered as a work activity for which basic time remains the same, whenever the activity is carried out. As is explained by Adamson et al. (2017:371), governing work or process involves an employee activity that takes longer within the work cycle than any other activity that is carried out at the same time. A qualified worker is someone who has acquired the skill, knowledge, and other attributes to carry out work in hand to satisfactory standards of quantity, quality, and safety (Stevenson, 2012:304–5).

As pointed out by Cochrana et al. (2017:876), foreign work activity, also referred to as *indirect labor*, is an activity that does not form part of the operation being observed. Stevenson (2012:304–5) presents the following concepts with their formulae.

Firstly, *rating* is the assessment of the employees' measure of working relative to the observer's or work study practitioner concept of the rate corresponding to a standard pace. The second one is *basic*, also referred to as *normal* time, which refers to the time for carrying out an employee, machine, or process activity at a standard pace or rating. The formula for basic time is provided as follows:

$$\text{Basic time} = \frac{\text{Observed time} \times \text{observed rating}}{\text{Standard rating } (100)}$$

On the other hand, standard time is considered as the time in which work should be completed or finished at standard performance.

The formula for standard time is presented as follows:

$$\text{Standard time} = \text{Basic time} + \text{relaxation allowance} + \text{other allowances}$$

There are various types of allowance used in manufacturing, such as relaxation allowance, contingency allowance, policy allowance, machine allowance, and process allowance. The most commonly used allowance is relaxation or rest allowance, a supplement to basic or normal time aimed at providing an employee with an opportunity to recover from the physiological and psychological effects of carrying out a specified task under specified conditions and to allow attention to personal needs. Relaxation allowance consists of fixed and variable allowance. The second allowance is contingency allowance, which is included in the standard time to meet legitimate and expected items of work or delays.

A rest allowance guide is provided in Table 7.1.

Let us say that the rating for these activities is provided to follow the procedure of calculating the basic time. The aim is to determine the job carried out by the employee at standard performance in order to generate the required results by the company. After calculating basic time, the necessary allowances are included to calculate the standard time for the work activities being carried out. An example of a time study summary sheet used at Company B in Gauteng, South Africa, is presented in Tables 7.2 and 7.3.

The next section to be addressed is work or activity sampling.

7.2.2.2 Work (Activity) Sampling

Work sampling in manufacturing SMEs involves observations collected during the study for the assembly of parts completed in a period of time by using a workgroup, in several processes where machines are operated. The aim of this technique is to check the time when there is operation and idle time. Hasle et al. (2012:830–43) and Gupta, Acharya, and Patwardhan (2013:636–41) report that work sampling comprises observations made over a period of time of activities carried out by employees in the machine operation and manufacturing process. This is intended to help management to identify types of waste such as defects, inventory, waiting, overproduction, overprocessing, movement, and employee underutilization and to eliminate them so as to improve productivity.

Work sampling, also referred to as *activity sampling*, involves recording continuous observations of activities carried out by a group number of employees, machines, or processes in the assembly line. By recognizing what is taking place in the manufacturing environment, this technique helps to determine how the productive work is generating value for the customers and to discover non-value-adding activities with the purpose of eliminating non-productive work (Hasle, 2014:41–7).

Dinis-Carvalho et al. (2015:239) consider work sampling as a technique that focuses on how employees spend time working in various manufacturing processes. This technique helps management to identify non-value-added activities and costs incurred with the aim of eliminating waste in manufacturing SMEs. As reported by Czumanski and Lödding (2016:2934), work sampling comprises observations made of functions or departmental sections, groups of employees in engaging in their traditional work activities, and old-fashioned

Table 7.2 Time Study Summary Sheet Used at Company B in Gauteng, South Africa

Time Study Observation Sheet									
Name of the company		Company B			Division/Department/Section			Manufacturing	
Name of the work study specialist		Mr X			Name of the operator/ employee			Mr Y	
Study number		1			Sheet number			1	
Operator number		1			Job number			1	
Time start		7h00			Time finish			8h05	
Machine type		Knitting			Tool type			Scissors	
Product type		Cap			Date			16 June 2017	
Type	Worker		Material				Machine/ equipment		
Description of the task: Prepare caps for customers									
Step No.	Description of Work Activities	Observed Time	Observed Rating	Basic Time	Step No.	Description of Work Activities	Observed Time	Observed Rating	Basic Time
1	Ironing material	5.9	80	**4.7**					
2	Label pockets	3.1	80	**2.5**					

(*Continued*)

Table 7.2 (Continued) Time Study Summary Sheet Used at Company B in Gauteng, South Africa

3	Lining	2.8	100	**2.8**					
4	Joining	3.5	85	**3.0**					
5	Sewing	2.5	100	**2.5**					
6	Closing	5	85	**4.3**					
7	Joining lining	3.4	90	**3.1**					
8	Bending	3.3	100	**3.3**					
9	Lining	2.9	90	**2.6**					
10	Stitching	3.7	85	**3.1**					
11	Fabricking	3.8	85	**3.2**					
12	Closing	2.8	90	**2.5**					
13	Make stitches	2.7	100	**2.7**					
14	Lining	3.6	100	**3.6**					
15	Closing	5.7	85	**4.8**					
16	Eye leads	3.8	80	**3.0**					
17	Punching	3.4	100	**3.4**					

Table 7.3 Time Study Summary Sheet Used At Company B in Gauteng, South Africa

Time Study Observation Sheet						
Name of the company		Company B	Division/ Department/ Section		Manufacturing	
Name of the work study specialists		Mr X	Name of the operator/ employee		*Mr Y*	
Study number		1	Sheet number		*1*	
Operator number		1	Job number		*1*	
Time start		7h00	Time finish		*8h05*	
Machine type		Knitting	Tool type		*Scissors*	
Product type		Cap	Date		*16 June 2017*	
Type	Worker			Material	Machine/ equipment	
Description of the task: **Prepare caps for customers**						
Step No.	Description of Work Activities	BasicTime	Frequency	Selected Basic Time	Relaxation Allowances %	Actual Time
1	Ironing material	4.7	1/1	4.7	**12%**	5.29
2	Label pockets	2.5	1/1	2.5	**12%**	2.78
3	Lining	2.8	1/1	2.8	**10%**	3.08
4	Joining	3.0	1/1	3.0	**9%**	3.24

(Continued)

Table 7.3 (Continued) Time Study Summary Sheet Used At Company B in Gauteng, South Africa

5	Sewing	2.5	1/1	2.5	**12%**	2.80
6	Closing	4.3	1/1	4.3	**16%**	4.93
7	Joining lining	3.1	1/1	3.1	**13%**	3.46
8	Bending	3.3	1/1	3.3	**12%**	3.70
9	Lining	2.6	1/1	2.6	**12%**	2.92
10	Stitching	3.1	1/1	3.1	**10%**	3.46
11	Fabricking	3.2	1/1	3.2	**11%**	3.59
12	Closing	2.5	1/1	2.5	**11%**	2.80
13	Make stitches	2.7	1/1	2.7	**10%**	2.97
14	Lining	3.6	1/1	3.6	**12%**	4.03
15	Closing	4.8	1/1	4.8	**13%**	5.47
16	Eye leads	3.0	1/1	3.0	**13%**	3.44
17	Punching	3.4	1/1	3.4	**14%**	3.88
Total actual time						**61.82**
Contingency allowance (5%)						3.09
Standard time						**64.91**

machine operations in the manufacturing environment. The aim of the technique in manufacturing SMEs is to target detailed analyses of worker and machine operations in work systems of different production lines in order to improve manufacturing processes.

Based on the literature reviewed, with activity sampling, the study can be planned in such a way that the results fall within the desired accuracy limits for the selected confidence level. If management wishes to ensure the extent to which an employee or machine is being used in an 8-hour shift, the deduction could be made in such a way that the results show that an employee or machine is only engaged for a certain percentage of time and that the remaining time is idle time (Torrisi, 2014:772).

The formula for the number of observations is presented by Heizer and Render (2016:436) as follows:

$$N = \frac{\sigma^2 p\,(100 - p)}{h^2}$$

where:

N = required sample size
z or σ = number of standard deviations for the desired confidence level ($z=1$ for 68.26% confidence level, $z=2$ for 95.45% confidence level, and $z=99.73\%$ confidence level)
p = estimate of the sample proportion (of time an employee is recorded while busy working or idling)
h = acceptable error margin in the form of percentage
n = actual results acquired during the study justifying the error limit

Activity sampling as compared to time study is a less costly statistical method of continuous observation and is mainly based on probability (Abujiya, Lee, & Riazb, 2014:574).

As stated by Aven (2015:83–4), probability involves the degree to which it is possible for an occurrence to happen. During the carrying out of work activities, a curve of normal distribution may arise, as depicted in Figure 7.3.

This means that the study may show that 50% of work activities are certain to take place and 50% of other work activities are likely to happen. An example of proportion distribution of “head” and “tails” is shown in Table 7.4 and the curve of normal distribution in Figure 7.4.

Various elements have to be followed when conducting activity sampling in manufacturing SMEs. These elements are representative period, accuracy

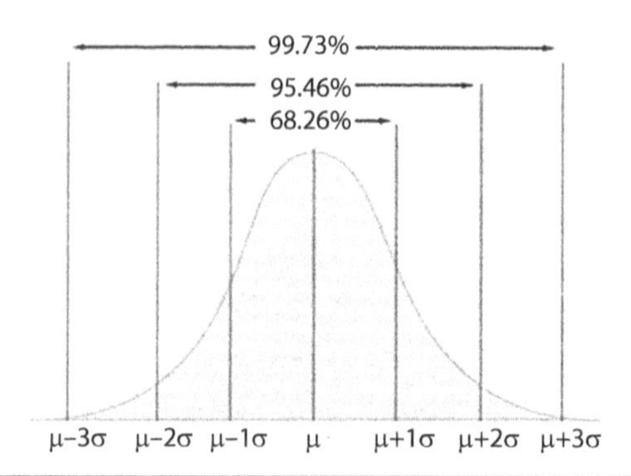

Figure 7.3 Bell-shaped normal distribution. (Adapted from Frankfort-Nachmias, F., and Leon-Guerrero, A., *Social Statistics for a Diverse Society,* SAGE Publications, London, 2011.)

(Abujiya et al., 2014:570–88), stratification (Jayaram, Dixit, & Motwan, 2014:474), work activities, confidence level, number of observations, and observation time (Torrisi, 2014:761–81). These elements are exercised during the study when data is collected on the work activities that are carried out in the workplace in manufacturing SMEs. An example in Company A is when a steel rod is prepared for customers.

These activities are done in manufacturing SMEs using WM applications, such as activity sampling. For example, an activity sampling is conducted in Company A, whereby activity sampling procedures are followed to determine the proportion of time being assigned to each activity. This technique is used by the work study specialist to show the proportion of time spent on productive work and on nonproductive work. The aim of

Table 7.4 Proportion Distribution of "Head" and "Tails" (100 Tosses of Five Coins at a Time)

Activity Number	*Heads*	*Tails*	*Number of Observations*
1	5	0	4
2	4	1	18
3	3	2	28
4	2	3	28
5	1	4	18
6	0	5	4

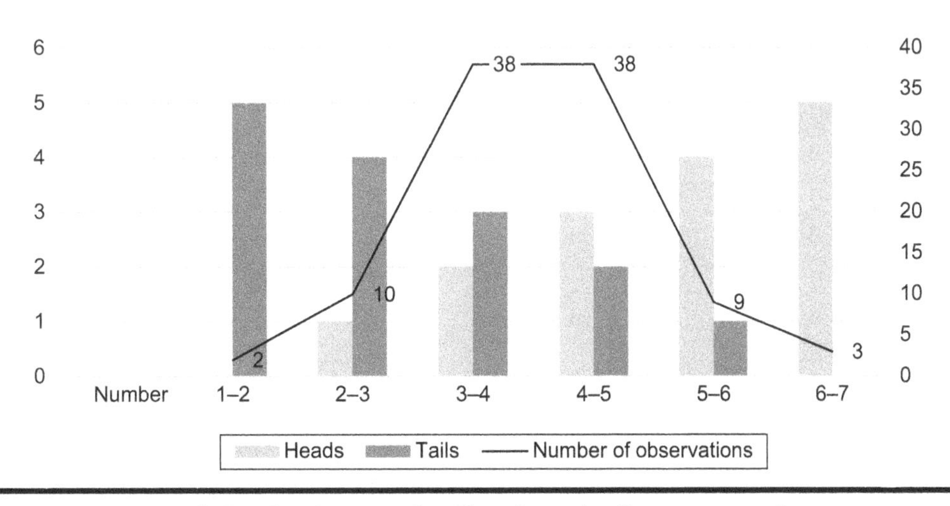

Figure 7.4 Normal distribution graph of heads and tails. (From Author, 2017.)

using activity sampling is to eliminate nonproductive work by ensuring that the workers' ability is improved through management commitment and support in areas such as employee training, motivation, exposure to appropriate working conditions, job enrichment, and other areas that can equip employees.

A sequence of activity sampling procedures is followed at Company A. The first procedure is identifying and defining the problem in the company. The purpose of the study is to assist management in ensuring that employees are efficiently and effectively utilized to ensure productivity improvement.

The work activities to be studied should be clearly understood and grouped in order to ensure their differences. The aim of the study is to find a suitable way to apply an activity sampling technique in order to facilitate the process of completing the project of work activities being carried out in Company A. These activities are either productive or nonproductive. An example of observations regarding work activities being carried out with a code for each activity is presented below.

A pilot of the following work activities carried out in Company A is presented in Table 7.5.

The second procedure is designing an activity sampling observation sheet, with headings provided under which the random observations can be classified. An example of an activity sampling observation sheet used for a pilot study is depicted in Table 7.6, showing the various activities of an 8-hour shift in Company A.

An example is the time allotted for an activity taking place from 07:00 to 16:00 with a subtraction for a tea break of 15 minutes and lunch of 45 minutes.

Table 7.5 Description of Tasks: Prepare Rod for Customers

Activity Number	*Code*	*Activities*
1	CSA	Cutting steel
2	MSA	CNC milling
3	TSA	Herbst threading
4	SSA	Slotting milling
5	DSA	Dispatch
6	IDA	Idle

The remaining time is time spent by an employee, which is 8 hours per day and is converted into 480 minutes.

The third step is planning the investigation and setting of the observation for Company A. In this case, work study specialists use random numbers by selecting as many as there are number of observations for the day. Ten-minute intervals between each activity recorded during the study are determined by the work study specialists as shown in Table 7.7, from 1 to 48 rounds of observations. The recording of numbers is then allocated in order with the aim of determining the observation times.

If a random number table is not provided, there are other alternatives for randomization of observation, which in the case of Company A, is based on 10-minute intervals.

An example of a random number table and the pilot study for the following work activities is presented in Table 7.8.

A random number table is used to do a pilot study of six observations and to determine the number of observations, the error limit, the percentage of each work activity, and standard time activity, including the relaxation and contingency allowances, in Company A. Table 7.9 is an example of activities for preparting a rod for customers, which is provided below.

The fourth step is communication between the work study specialists and the supervisors and employees in Company A. The aim of the meeting between work study specialists and supervisors, including employees, is to ensure that employees know about the investigation to avoid making them suspicious about what the specialists are doing.

In the fifth step, the work study specialist conducts a pilot study in Company A to acquire an approximate idea of the percentage occurrence of the different

Table 7.6 Activity Sampling Observation Sheet Used at Company A in Gauteng, South Africa

<table>
<tr><td colspan="8">Activity Sampling Observation Sheet</td></tr>
<tr><td colspan="2">In</td><td colspan="2">Company A</td><td colspan="3">Division/Department/Section</td><td>Manufacturing</td></tr>
<tr><td colspan="2">Name of the work study specialist</td><td colspan="2">Mr X</td><td colspan="3">Name of the operator/employee</td><td>Mr Y</td></tr>
<tr><td colspan="2">Study number</td><td colspan="2">1</td><td colspan="3">Sheet number</td><td>1</td></tr>
<tr><td colspan="2">Operator number</td><td colspan="2">1</td><td colspan="3">Job number</td><td>1</td></tr>
<tr><td colspan="2">Time start</td><td colspan="2">7h00</td><td colspan="3">Time finish</td><td>16h00</td></tr>
<tr><td colspan="2">Machine type</td><td colspan="2">Cutting</td><td colspan="3">Tool type</td><td>Cutting tool</td></tr>
<tr><td colspan="2">Product type</td><td colspan="2">Rods</td><td colspan="3">Date</td><td>16 June 2017</td></tr>
<tr><td colspan="8"></td></tr>
<tr><td>Type</td><td>Worker</td><td></td><td colspan="2">Material</td><td></td><td>Machine/equipment</td><td></td></tr>
<tr><td colspan="8">Description of the task: Preparing rods for customers</td></tr>
<tr><td>Observation time</td><td>Code</td><td>Observation time</td><td>Code</td><td>Observation time</td><td>Code</td><td></td><td></td></tr>
<tr><td>08:00</td><td>CSA</td><td>11:10</td><td>SSA</td><td>14:20</td><td>CSA</td><td></td><td></td></tr>
<tr><td>08:10</td><td>CSA</td><td>11:20</td><td>SSA</td><td>14:30</td><td>MSA</td><td></td><td></td></tr>
<tr><td>08:20</td><td>CSA</td><td>11:30</td><td>DSA</td><td>14:40</td><td>MSA</td><td></td><td></td></tr>
<tr><td>08:30</td><td>MSA</td><td>11:40</td><td>DSA</td><td>14:50</td><td>TSA</td><td></td><td></td></tr>
<tr><td>08:40</td><td>MSA</td><td>11:50</td><td>CSA</td><td>15:00</td><td>TSA</td><td></td><td></td></tr>
</table>

(Continued)

Table 7.6 (Continued) Activity Sampling Observation Sheet Used at Company A in Gauteng, South Africa

08:50	MSA	12:00	CSA	15:10	IDA		
09:00	TSA	12:10	SSA	15:20	TSA		
09:10	TSA	12:20	SSA	15:30	SSA		
09:20	TSA	12:30	DSA	15:40	DSA		
09:30	SSA	12:40	IDA	15:50	DSA		
09:40	DSA	12:50	CSA	16:00			
09:50	DSA	13:00	CSA				
10:00	CSA	13:10	MSA				
10:10	CSA	13:20	TSA				
10:20	MSA	13:30	TSA				
10:30	MSA	13:40	TSA				
10:40	TSA	13:50	SSA				
10:50	TSA	14:00	IDA				
11:00	TSA	14:10	DSA				

Table 7.7 Order Numbers and Observations Times Used at Company A in Gauteng, South Africa

Order Numbers	*Observation Time*
06	08:00
11	08:50
14	09:20
18	10:00
21	10:30
27	11:30
31	12:10
41	13:50
43	14:10
48	15:00

activities taking place. An example of the task based on the preparation of rods by Company A for its customers is shown in Table 7.9.

The sixth step is determining a number of observations for Company A. For most studies using activity sampling, a suitable confidence level for an absolute accuracy limit is 95%. The more observations are recorded, the more

Table 7.8 Random Number Table

10	92	62	82	94	39
9	6	3	49	80	115
65	19	55	60	92	75
80	65	41	39	69	83
8	72	39	36	38	47
36	5	14	72	67	39
82	65	45	8	48	52
16	105	65	80	77	84
51	64	12	47	49	17
7	45	38	67	100	19
47	49	39	51	92	68

Table 7.9 Prepare Rods for Customers

Activity Number	*Activities*	*Incidence*	*Percentage Observations*
1	Cutting steel	10	21
2	CNC milling	8	17
3	Herbst threading	12	25
4	Slotting milling	7	15
5	Dispatch	8	17
6	Idle	3	5
Total		**48**	**100**

accurate the results of the study will be. A confidence level of 95% means that the actual results found should be correct for 95 out of 100 events or incidents in manufacturing SMEs. The sigma (σ) for a 95% confidence level is 1.96 at a 5% margin of error. The highest percentage for p-value needs to be selected as representative of the study as a percentage number.

The formula for the number of observations is presented as follows:

$$N = \frac{\sigma^2 p\left(100 - p\right)}{h^2}$$

where:

N = required sample size
p = 2 for 95.45% confidence level
p = estimates of idle proportion = 25% or 0.25
h = acceptable error margin of ±5% (any selected error margin between 1% and 5%)

$$n = \frac{\sigma^2 p\left(100 - p\right)}{h^2}$$

$$= \frac{3.84 \times 25\left(100 - 25\right)}{5^2}$$

$$= \frac{96 \times 75}{25}$$

$$= \frac{7200}{25}$$

= 288 observations

Therefore,

$$=288-48$$

= 240 more observations required to have the accuracy of the actual results

The formula subsequent to the formula for the number of observations is error margin, and this is presented as follows:

$$h=\frac{\sqrt{\sigma^2 p\ (100-p)}}{n}$$

$$=\frac{\sqrt{3.84\times 25\ (100-25)}}{48}$$

$$=\frac{\sqrt{96\times 75}}{48}$$

$$=\frac{\sqrt{7200}}{48}$$

$$=\sqrt{150}$$

$$\mathbf{=12.25\%}$$

> 4% carry out more observations

The seventh step focuses on the actual study that needs to be conducted, which is directed by the number of observations that needs to be recorded to acquire accurate results for Company A. These observations are guided by the pilot study that has already been conducted for Company A as indicated in Table 7.6.

Let us say that the actual study to cater for 288 observations is done over 6 days, including the 48 observations done during the pilot study in order to reach the accurate results. An example of observations is presented in Table 7.10.

The eighth step focuses on determining the standard time for the activity that has been carried out in a day. Let us say the standard time has been counted for the pilot study. An example of observations is presented for Company A in Table 7.11, as indicated in Table 7.9.

Table 7.10 Prepare Rods for Customers for Six Days

Activity Number	*Activities*	*Incidence*	*Percentage Observations*
1	Cutting steel	60	20
2	CNC milling	48	17
3	Herbst threading	72	24
4	Slotting milling	42	16
5	Dispatch	48	15
6	Idle	18	8
Total		**288**	**100**

In activity sampling, there are also various procedures that manufacturing SMEs have to follow when conducting an investigation. An example of activity sampling is conducted at Company A, using a control chart in order to see how Company A is affected by late delivery. Let us suppose that Company A is being investigated for 10 days, observing the number of frequencies the company is experiencing late delivery, using the control chart in Table 7.12. Let us suppose that 100 observations have been collected over the 10 days of the study. Late delivery has been normal for only 8 out of 10 days.

The formula to determine control limits is presented as follows:

$$p = p \pm 3\frac{\sqrt{p(1-p)}}{n}$$

Table 7.11 Determine Standard Time for Preparing a Rod for Customers (from Pilot Study)

	CSA	*MSA*	*TSA*	*SSA*	*DSA*	*IDA*	*Total*
% Activity	21	17	25	15	17	5	100
Basic time	100.80	81.60	120.00	72.00	81.60	24.00	480
RA% (10%)	10.08	8.16	12.00	7.20	8.16	2.4	
Actual time	110.88	89.76	132.00	79.20	89.76	26.40	
CA% (5%)	5.54	4.49	6.60	3.96	4.49	1.32	
Standard time	116.42	94.25	138.60	83.16	94.25	27.72	

Table 7.12 Control Chart for Late Delivery at Company A in Gauteng, South Africa

Date	*Total Studies*	*Occurrences of Late Delivery*	*Percentages of Late Delivery*
Day 1	100	8	8
Day 2	100	12	12
Day 3	100	9	9
Day 4	100	7	7
Day 5	100	6	6
Day 6	100	35	35
Day 7	100	36	36
Day 8	100	9	9
Day 9	100	13	13
Day 10	100	15	15
Total	**1000**	**150**	

Source: Adapted from Stevenson, W.J., *Operations Management: Theory and Practice*, McGraw-Hill Publishers, London, 2012.

Before calculating the control limits, the percentage period for the number of late deliveries and the total number days for the studies being conducted need to be determined in Company A, where percentage period is p and the total number of days for the studies is n.

Therefore,

$$p = \frac{n}{N}$$

$$= \frac{150}{1000}$$

$$= 0.20 \text{ or } 20\%$$

and

$$n = \frac{N}{d}$$

$$= \frac{1000}{10}$$

$$= 100$$

$$\text{Control limits for p} = p \pm 3\frac{\sqrt{p(1-p)}}{n}$$

$$\text{Control limits for p} = 0.20 \pm 3\frac{\sqrt{0.20(1-0.20)}}{100}$$

$$\text{Control limits for p} = 0.20 \pm 3(0.04)$$

$$\text{Control limits for p} = 0.20 \pm 0.12$$

Therefore,

$$\text{Control limits for p} = 0.32 \text{ or } 0.08$$

An example of observations for late delivery in Company A is presented in Figure 7.5.

In a similar situation, a work study investigation using activity sampling study through the application of the control chart is able to plot the daily or cumulative results of the sampling study regarding late deliveries by Company A. If the graph plotted falls outside the control limits, it is likely that there is an unusual condition caused by inexperienced senior personnel in high positions forecasting the progress of the company. The next concept to be defined is PMTS.

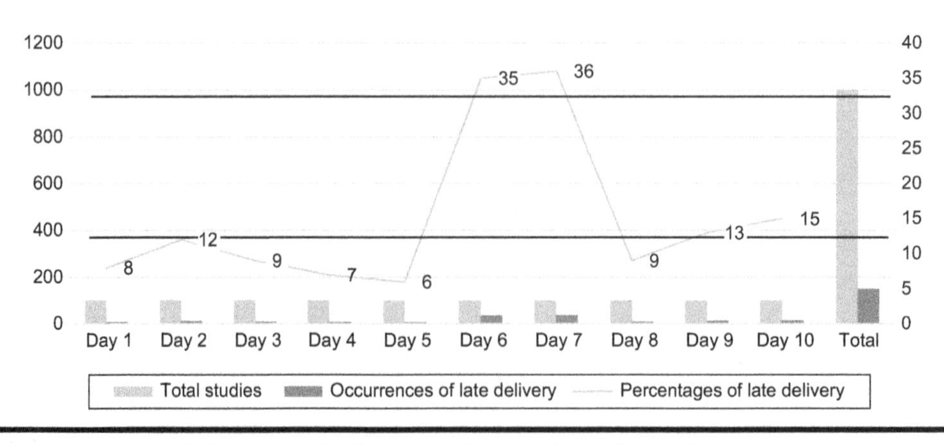

Figure 7.5 Control chart graph showing observation of late delivery in Company A. (From Author, 2017.)

7.2.3 Predetermined Motion Time Systems

According to Kuo and Wang (2012:6519–22), PMTS in manufacturing SMEs involve motions such as get, place, operate (handle tool), hand move (reach), body move (body motion), visual check (eye-motion), and process time used on an assembly line. This technique is aimed at improving standard time for productivity improvement, taking into consideration employee safety, health, and well-being, and this is done by simulating and assessing manual tasks in advance and checking the manufacturing system is designed for a new process.

Larsson (2013:891) comment that PMTS is time estimated using the knowledge practical experience of work study specialists or industrial engineers, whereby the time estimated for the motions in an operation is highly standardized without considering the thinking process required when providing a specialized service in manufacturing SMEs. This technique takes the form of established computer-based manufacturing system for the time used to make products and provide services to customers.

According to Suwittayaruk et al. (2014:727), PMTS involves using a stopwatch or a video at 100% pace to record the time for work activities in motion, such as reaching, grasping, moving, positioning, and releasing. As is reported by Battini et al. (2015:825–7), PMTS in manufacturing SMEs focuses on the times of basic human movements used to build up the time for a job with related normal time values, together with a set of procedures for applying the data to analyze manual tasks and establish normal and standard times for the analyzed tasks.

Wu et al. (2016:292) emphasize that PMTS in manufacturing SMEs encompasses determining WM and production standards in the assembly using integration of information, standard operating procedures, techniques, and motion times to determine accuracy of time and thereby improve productivity.

7.2.4 Analytical Estimating

According to Mathur, Mittal, and Dangayach (2012:761), analytical estimating is an assumption made by work study experts based on their skills, knowledge, and experience from previous methods and times studies conducted in manufacturing SMEs. These assumptions are made to simplify the job using time constraints of well-trained and qualified employees on quality, delivery, flexibility, and cost of the product produced in improving productivity.

As is pointed out by Sharma and Sharma (2014:751–3), analytical estimating is time used by the employee to carry out a task, starting with preparation, task inspection, task adjustment, and operation, based on their skills, knowledge, training, and experience of the manufacturing process. Padhi et al. (2013:1885–94) explain that analytical estimating in manufacturing SMEs is an assumption of time recorded by work study experts on the activities done by employees guided by the machine pace for completion of a task in the manufacturing process line layouts based on the skill acquired. As is pointed out by Sharma and Sharma (2014:751–3), analytical estimating is a tool used to determine the time taken by employees to carry out a task, starting with preparation, task inspection, task adjustment, and operation based on their skills, knowledge, training, and experience of the manufacturing process.

Colledania, Rattia, and Senanayake (2015:357–61) report that analytical estimating is regarded as a tool for measuring employees' tasks based on their knowledge, using time and speed to analyze work activities: the loading process; welding, hemming, and curing on the assembly line; and the unloading of parts after operation. Analytical estimating is time required to carry out work activities at a distinct pace of working as determined by the employees' skills, knowledge, training, and experience of the manufacturing process. These employees carry out their repetitive work activities daily in the manufacturing environment of manufacturing SMEs. These activities occur in assembly at work stations (Małachows & Korytkowski, 2016:166–70).

7.2.5 Comparative Estimating (Benchmarking)

This is a WM technique in which the time for a task is evaluated by comparing it with the work in a series of similar tasks (benchmarks) whose work contents have been measured. The arrangement of tasks into broad bands of time is referred to as *slotting*.

Comparative estimating is a WM technique for an employee to carry out the a task at a specific time whereby this task is evaluated by comparing it with the work in a series of similar tasks in manufacturing.

Comparative estimating is a WM technique for measuring time, whereby the current workload distribution and the process flow performance of the manufacturing process is compared with the traditional manufacturing process. By comparing tasks, this type of estimating assists in the assembly line and continuous process to reduce throughput times in order to improve quality, minimize operational costs, deliver on time, adapt to situations; depending on each other in every situation of the supplier, manufacturing process, and customer. All these value-adding forces contribute to various manufacturing capabilities such as JIT; employee resource and job design; product, process and service design; TQM; scheduling; SCM; technology; maintenance; location; and layout (Garza-Reyes et al., 2012:178–94).

As reported by Boulter, Bendell, and Dahlgaard (2013:201–3), comparative estimating, also referred to as *benchmarking* of tasks, involves comparing the time taken in measuring employees' actual work performance with the work in a series of similar tasks whose work contents have been measured in manufacturing SMEs. By comparing the time of the tasks, manufacturing capabilities such as TQM; employee resources management; improved product and process design; supply chain management; customer satisfaction; technology; JIT; and time scheduling are considered in manufacturing SMEs. The type of manufacturing system used in manufacturing SMEs by comparing a series of similar task is continuous process.

Mandal (2014:1–5) considers comparative estimating as a technique involving the application of tool condition monitoring methods used in conventional manufacturing process, compared with similar ones proposed for the manufacturing process in manufacturing SMEs. This manufacturing process occurs in batch and continuous processes to achieve cost reduction and speed enhancement. Cost reduction and speed enhancement in manufacturing SMEs depend on various manufacturing capabilities such as product and process design, employee resources, scheduling, maintenance, location, and layout strategies.

Wang et al. (2015:221–26) explain that comparative estimating refers to a technique involving the manner in which the speed of employee performance in a typical manufacturing process is compared with the similar speed projected from the previous study in order to monitor the rate of productivity in the manufacturing SMEs. The type of system used for this kind of estimating tool in manufacturing SMEs is the assembly line.

As is stated by Kate et al. (2016:361–76), comparative estimating as a technique is considered as involving the utilization of machine learning techniques by the employee, which can vary widely depending upon the task being carried out which relies on the employee work force. The type of work being carried out is based on process chart symbols such as operation and movement for energy cost improvement and, in turn, productivity improvement in manufacturing SMEs.

7.2.6 Synthesis

Santosa, Vidal, and Moreira (2012:1656–61) refer to synthesis, also referred to as *standard approach*, as a technique for building up cycle time for a task carried out by a qualified employee in a project or continuous manufacturing process, whether in the form of an individual employee or a team. The aim of this technique in manufacturing SMEs is to establish the correct or appropriate standard time for the job taking place in manufacturing settings, taking into consideration safety, fatigue recovery, and cost reduction.

As explained by Doltsinis, Ratchev, and Lohse (2013:85–90), synthesis is a technique for developing time for a task performed by a trained employee in a manufacturing process, taking into account cost, time, flexibility, and quality, using the normal time for the job depending on the experience of shop-floor foremen or supervisors, who use their knowledge to guide the decision-making process. This manufacturing process happens in assembly and project manufacturing systems in manufacturing SMEs.

Beyer (2014:1–6) reports that synthesis in manufacturing SMEs involves time built up parts of the task or the task carried out by a trained employee in an assembly and mass manufacturing process in manufacturing SMEs for the appropriate standard time for the job. This manufacturing process is measured based on cost and time savings in manufacturing SMEs.

According to Glock and Grosse (2015:6637–45), synthesis in manufacturing is a technique for building up cycle time capacity on a job carried out by a qualified employee in order to use the correct standard time depending on the process quality to produce a quality product for the customer starting from the supplier. The characteristics of a qualified employee include knowledge, skills, relevant educational background, and experience acquired to perform the task, and these characteristics are used in mass, batch, continuous, and assembly manufacturing systems in manufacturing SMEs.

Pedersen and Slepniov (2016:44–56) report that synthesis in manufacturing comprises the manufacturing cycle for building up speed in carrying out the job by an employee with training, knowledge, and experience at standard performance. The types of systems where synthesis takes place are mass, continuous, and assembly manufacturing systems. The application of synthesis helps experts to determine the cost implications affecting manufacturing SMEs and for manufacturing SMEs to come up with improvements to reduce costs, so that these SMEs can improve the productivity of their businesses.

7.2.7 Value Stream Mapping (Quantitative)

As emphasized by Chiarini (2012:683–7), VSM is the mapping of processes to target waste for elimination.

In terms of quantitative techniques, as reported by Dushyanth, Shivashankar, and Rajeshwar (2015:1–9), the concept of VSM involves presenting work activities where value is added to the product for improvement of productivity using control charts such as line graphs and bar charts. The aim of using these graphs and bar charts is to compare the current results to past ones that show poor productivity results, which are referred to as *non-value activities* (NVA); these are such things as employees' lack of knowledge, poor process, and delay in material flow. Value-added activities (VA) include work in process reduction (WIP), quality improvement, enhanced flexibility, reduced transactions, simplified scheduling, improved communications, reduced costs, better on-time deliveries, increased sales, and improved space utilization. The literature studied provides a comparison of NVA and VA on how manufacturing SMEs can be improved using VSM. For example, the results indicate that after the application of VSM to the assembly line of manufacturing SMEs, the NVA time of the total processing time is reduced from 300 hours to 160 hours (53.3%), and the VA is improved from 100 to 140, representing 46.7%. This VSM process improved the productivity of manufacturing SMEs through inspection by the work study expert of the assembly line for quality of the assembly to short cycle time. Thus, the focal point in VSM to reduce NVA is 20.1% and to improve VA it is 40.1%.

Calculations of NVA and VA before and after improvement are indicated in the following table.

Dushyanth et al. (2015:10) report that VSM is considered as identifying work activities where value is added to the product for improvement of productivity in manufacturing SMEs. VSM maps out the current state of a production line in order to ensure improvement of the mapping design for the future of the operational process in manufacturing SMEs. This mapping reduces work-in-progress (WIP) inventory, improves quality, reduces space utilization, and designs a more appropriate workplace. For manufacturing SMEs to ensure progress in the manufacturing process through VSM, current flow process charts are used to record information for the work activities carried out and for material and information flow along the SIPOC (supplier, input, process, output, customer) system for the assembly line. During the completion of the flow process chart by work study officials, an analysis of those work activities and material and information flow is done critically through the use of a cause-and-effect diagram. For example, problem areas identified through cause-and-effect diagrams are employees, machinery, material, method, and environment. Employee effort considers issues such as poor planning, lack of training, and poor communication. Machinery used considers poor maintenance and lack

Activities	*Before*	*After*
NVA	Percentage of NVA Time $= \frac{\text{Results after}}{\text{Results before } 1}$ $= \frac{53.3}{66.7} \times \frac{100}{1}$ $= \mathbf{79.9\%}$ Therefore, if it is less than 100, minus the small figure from 100. $=100-79.9$ **=20.1% NVA reduction**	Percentage of NVA Time $= \frac{\text{Total NVA time in days} \times 100}{\text{Total processing time in days } 1}$ $= \frac{46.7}{33.3} \times \frac{100}{1}$ $= \mathbf{140.1\%}$ Therefore, if the value of the results is greater than 100, minus the 100 from that figure. $= 140.1-79.9$ **= 40.1% VA increase**

Activities	*Before*	*After*
NVA	Percentage of NVA Time $= \frac{\text{Total NVA time in days} \times 100}{\text{Total processing time in days } 1}$ $= \frac{200}{300} \times \frac{100}{1}$ $= \mathbf{66.7\%}$	Percentage of NVA Time $= \frac{\text{Total NVA time in days} \times 100}{\text{Total processing time in days } 1}$ $= \frac{160}{300} \times \frac{100}{1}$ $= \mathbf{53.3\%}$
VA	Percentage of VA Time $= \frac{\text{Total NVA time in days} \times 100}{\text{Total processing time in days } 1}$ $= \frac{100}{300} \times \frac{100}{1}$ $= \mathbf{33.3\%}$	Percentage of VA Time $= \frac{\text{Total NVA time in days} \times 100}{\text{Total processing time in days } 1}$ $= \frac{140}{300} \times \frac{100}{1}$ $= \mathbf{46.7\%}$

of space for spares. Material considers incorrect material and poor inventory control, and, lastly, the method used indicates improper standard operating procedures and the use of the wrong tools. When work study experts finish using the cause-and-effect diagram, the assembly line manufacturing system is then redesigned, aiming at integrating appropriate shop-floor practices to improve human and machine productivity.

7.2.8 Statistical Improvement Techniques

In this section, statistical techniques are studied and presented with the aim of providing insight into the composition of techniques in manufacturing SMEs.

Statistical techniques are also referred to as WM under the umbrella of work study. These techniques were presented earlier in the introduction to method study. SPC is a tool that uses control charts and process capability studies to control variability in the manufacturing process in manufacturing SMEs. A control chart is built by plotting the results of a process on a graph, providing a visual presentation of variation in the process and allowing simple discovery of changes over time within the process. Process capability, meanwhile, determines the ability of a manufacturing process to meet manufacturing specifications. Both control charts and process capability require a full understanding of the manufacturing process for adjustments within the specification. In addition, SPC uses Pareto analysis in the form of a diagram reporting 80% of the problems caused by 20% of the errors done during the operation in manufacturing SMEs. This tool is applicable to various manufacturing systems such as the mass, batch, assembly, project, and continuous process systems used in various manufacturing SMEs (Mathur et al., 2012:756–67).

SPC uses the control chart as an effective tool for monitoring the manufacturing process variation in industrial applications among manufacturing SMEs. The control chart is used to monitor the manufacturing process in an assembly line and continuous process. The aim of using this tool is to reduce costs on processes and improve the quality of the manufacturing process in order for manufacturing SMEs to be competitive in the market place. The ability to monitor and reduce process variation for cost reduction, reliability, speed, flexibility, and quality improvement in industrial processes plays a critical role in the success of an enterprise in today's globally competitive market place (Du & Lv, 2013:377–86).

As found by Mishra and Sharma (2014:524–41), SPC is a tool that involves determining the process capability by means of a control chart (moving average) and a range chart capturing individual variation in manufacturing SMEs. Statistical process enforces the quality practice in production process, detecting the failure before it happens and applying corrective action, resulting in minimizing cost and product variability and inconsistencies, and thus improving productivity in manufacturing SMEs. This tool uses a histogram, Pareto diagram, and run chart to measure percentage defects occurring in the manufacturing process, resulting in quality failure of their businesses. The aim of using these measures in terms of SPC is to ensure that there is improvement in quality and, in turn, productivity progress in manufacturing SMEs. The types of systems that use SPC in manufacturing SMEs include mass volumes produced in the form of batches and continuous projects focusing on costs, quality, delivery, flexibility, and dependability so that the business can remain competitive in the market.

Ahmed, Ramadan, and Saghbini (2015:26) state that SPC entails identifying necessary design and process modifications in the form of projects for achieving customer satisfaction and productivity progress. This achievement can only be realized through process monitoring using Pareto and control charts to ensure quality, cost, reliability, and delivery with the intention of sustaining and improving productivity in manufacturing SMEs.

SPC is to a tool that detects out-of-control occurrences in manufacturing operation work stations in order to maintain the quality level of manufacturing SMEs. These work stations show how raw materials pass through various operations to form a complete product and are inspected through SIPOC for quality, cost, and time to ensure waste reduction and productivity progress. This tool is applied in manufacturing systems such as mass, assembly, and continuous process systems (Zhu, Zhang, & Deng, 2016:1804–6).

7.2.9 Business Process Reengineering

As indicated in Chapter 6, business process reengineering (BPR) is also used by work study experts through the application of WM, is a similar concept to that explained in Section 6.3.2 involving all work study tools, in particular quantitative tools, in order to measure the manufacturing process in manufacturing SMEs to improve their productivity level.

7.3 Analysis of Current Applications of Work Measurement in South Africa

The focus of this chapter is on understanding WM in the workplace in manufacturing SMEs. WM involves various techniques such as benchmarking, brainstorming, preliminary surveys, time study, activity sampling, synthesis, PMTS, standards, VSM, SPC, and BPR.

All these techniques are known to represent the best solution for problem areas facing manufacturing industries, both large and small, worldwide. The challenge is how to develop understanding of these techniques to enhance the productivity of manufacturing SMEs for competitiveness in South Africa. Even though Productivity SA is mandated by government to assist in improving the productivity of manufacturing SMEs in their businesses through application of WM, as indicated in Chapter 1, work study specialists from industries and academics also need to collaborate through relevant advisory committees, conferences, and research workshops from various universities.

7.4 Summary

This chapter addresses the same work study relations used in method study and is also focused on WM procedures such as selection of the task to be studied, recording of the information regarding the task, analytical examination of the information collected, measurement of work, compilation, and definition of the method. The next section to be discussed is the understanding of techniques forming WM in manufacturing SMES. These techniques include benchmarking, brainstorming, preliminary surveys, time study, activity sampling, synthesis, PMTS, standards, VSM, SPC, and BPR. These techniques are used to measure work and provide the appropriate standard time for the job being carried out with the aim of enhancing the productivity of manufacturing SMEs. Chapters 8 through 10 examine the impact of work study on physical capital, technological capital, and management for productivity improvement in manufacturing SMEs.

References

Abujiya, M.R., Lee, M.H., and Riazb, M. 2014. Improving the performance of exponentially weighted moving average control charts. *Quality and Reliability Engineering International*, 30: 571–590.

Adamson, G., Wang, L., Holm, M., and Moore, P. 2017. Cloud manufacturing: A critical review of recent development and future trends. *International Journal of Computer Integrated Manufacturing*, 30(4–5): 347–380.

Ahmed, A.M.M.B., Ramadan, M.Z., and Saghbini, H.A. 2015. Sustainable improvement for United Arab Emirates' SMEs: A proposed approach. *International Journal of Customer Relationship Marketing and Management*, 6(3): 25–32.

Aven, T. 2015. Implications of black swans to the foundations and practice of risk assessment and management. *Reliability Engineering and System Safety*, 134: 83–91.

Baines, A. 1995. Work measurement: The basic principles revisited. *Work Study*, 44(7): 10–14.

Battini, D., Delorme, X., Dolgui, A., Persona, A., and Sgarbossa, F. 2015. Ergonomics in assembly line balancing based on energy expenditure: A multi-objective model. *International Journal of Production Research*, 54(3): 824–845.

Beyer, C. 2014. Strategic implications of current trends in additive manufacturing. *Journal of Manufacturing Science and Engineering*, 136: 1–8.

Boulter, L., Bendell, T., and Dahlgaard, J. 2013. Total quality beyond North America. *International Journal of Operations & Production Management*, 33(2): 197–215.

Chiarini, A. 2012. Lean production: Mistakes and limitations of accounting systems inside the SME sector. *Journal of Manufacturing Technology Management*, 23(5): 681–700.

Cochrana, D.S., Foley, J.T., and Bi, Z. 2017. Use of the manufacturing system design decomposition for comparative analysis and effective design of production systems. *International Journal of Production Research*, 55(3): 870–890.

Colledania, M., Rattia, A., and Senanayake, C. 2015. An approximate analytical method to evaluate the performance of multi-product assembly manufacturing systems. *Procedia CIRP*, 33: 357–363.

Czumanski, T. and Lödding, H. 2016. State-based analysis of labour productivity. *International Journal of Production Research*, 54(10): 2934–2950.

Das, B., Ghosh, T., and Gangopadhyay, S. 2013. Child work in agriculture in West Bengal, India: Assessment of musculoskeletal disorders and occupational health problems. *Journal of Occupational Health*, 55(4): 244–258.

Dinis-Carvalho, J., Moreira, F., Bragança, S., Costa, E., Alves, A., and Sousa, R. 2015. Waste identification diagrams. *Production Planning and Control*, 26(3): 235–247.

Dushyanth-Kumar, K.K.R., Shivashankar, G.S., and Rajeshwar S.K. 2015. Application of value stream mapping in pump assembly process: A case study. *Industrial Engineering and Management*, 4(3): 1–11.

Doltsinis, S.C., Ratchev, S., and Lohse, N. 2013. A framework for performance measurement during production ramp-up of assembly stations. *European Journal of Operational Research*, 229: 85–94.

Du, S. and Lv, J. 2013. Minimal Euclidean distance chart based on support vector regression for monitoring mean shifts of auto-correlated processes. *International Journal of Production Economics*, 141: 377–387.

Frankfort-Nachmias, F. and Leon-Guerrero, A. 2011. *Social Statistics for a Diverse Society*. 6th edn. London: SAGE Publications.

Garza-Reyes, J.A., Oraifige, L., Soriano-Meier, H., Forrester, P.L., and Harmanto, D. 2012. The development of a lean park homes production process using process flow and simulation methods. *Journal of Manufacturing Technology Management*, 23(2): 178–197.

Glock, C.H. and Grosse, E.H. 2015. Decision support models for production ramp-up: A systematic literature review. *International Journal of Production Research*, 53(21): 6637–6651.

Gupta, S., Acharya, P., and Patwardhan, M. 2013. A strategic and operational approach to assess the lean performance in radial tyre manufacturing in India: A case-based study. *International Journal of Productivity and Performance Management*, 62(6): 634–651.

Hasle, P. 2014. Lean production: An evaluation of the possibilities for an employee supportive lean practice. *Human Factors and Ergonomics in Manufacturing & Service Industries*, 24(1): 40–53.

Hasle, P., Bojesen, A., Jensen, P.L., and Bramming, P. 2012. Lean and the working environment: A review of the literature. *International Journal of Operations & Production Management*, 32(7): 829–849.

Heizer, J. and Render, B. 2016. *Operations Management: Sustainability and Supply Chain Management.* 11th edn. New York: Pearson Publishing.

ILO. 2015. *World Employment and Social Outlook: Trends 2015.* International Labour Organization. Available from: www.ilo.org/wcmsp5/groups/public/---dgreports/---dcomm/---publ/documents/publication/wcms_337069.pdf.

Jayaram, J., Dixit, M., and Motwan, J. 2014. Supply chain management capability of small and medium sized family businesses in India: A multiple case study approach. *International Journal of Production Economics*, 147: 472–485.

Kate, R.J., Swartz, A.M., Welch, W.A., and Strath, S.J. 2016. Comparative evaluation of features and techniques for identifying activity type and estimating energy cost from accelerometer data. *Physiological Measurement*, 37: 360–379.

Kuo, C. and Wang, M.J. 2012. Motion generation and virtual simulation in a digital environment. *International Journal of Production Research*, 50(22): 6519–6529.

Larsson, A. 2013. The accuracy of surgery time estimations. *Production Planning & Control*, 24(10–11): 891–902.

Małachows, B. and Korytkowski, P. 2016. Competence-based performance model of multi-skilled workers. *Computers & Industrial Engineering*, 91: 165–177.

Mandal, S. 2014. Applicability of tool condition monitoring methods used for conventional milling in micromilling: A comparative review. *Journal of Industrial Engineering*, 1–8.

Mathur, A., Mittal, M.L., and Dangayach, G.S. 2012. Improving productivity in Indian SMEs. *Production Planning & Control: The Management of Operations*, 23(10–11): 754–768.

McDermott, C.M. and Prajogo, D.I. 2012. Service innovation and performance in SMEs. *International Journal of Operations & Production Management*, 32(2): 216–237.

Mishra, P. and Sharma, P.K. 2014. A hybrid framework based on SIPOC and Six Sigma DMAIC for improving process dimensions in supply chain network. *International Journal of Quality & Reliability Management*, 31(5): 522–546.

Padhi, S.S., Wagner, S.M, Niranjan, T.T., and Aggarwal, V. 2013. A simulation-based methodology to analyse production line disruptions. *International Journal of Production Research*, 51(6): 1885–1897.

Pedersen, P. and Slepniov, D. 2016. Management of the learning curve: a case of overseas production capacity expansion. *International Journal of Operations & Production Management*, 36(1): 42–60.

Santosa, M.S., Vidal, M.C.R., and Moreira, S.B. 2012. The RFad Method: A new fatigue recovery time assessment for industrial activities. *Work*, 41: 1656–1663.

Schwab, K. 2015. *The Global Competitiveness Report 2014-2015: Insight Report.* World Economic Forum (WEF). Available from: www3.weforum.org/docs/WEF_GlobalCompetitivenessReport_2014-15.pdf.

Sharma, R.K. and Sharma, R.G. 2014. Integrating six sigma culture and TPM framework to improve manufacturing performance in SMEs. *Quality and Reliability Engineering International*, 30: 745–765.

Slack, N., Chambers, S., and Johnston, R. 2009. *Operations Management*. 6th edn. London: FT/Prentice-Hall.

Srinivasan, S., Ikuma, L.H., Shakouri, M., Nahmens, I., and Harvey, C. 2016. 5S impact on safety climate of manufacturing workers. *Journal of Manufacturing Technology Management*, 27(3): 364–378.

Stevenson, W.J. 2012. *Operations Management: Theory and Practice*. 11th edn. London: McGraw-Hill Publishers.

Suwittayaruk, P., Van Goubergen, D., and Lockhart, T.E. 2014. A preliminary study on pace rating using video technology. *Human Factors and Ergonomics in Manufacturing & Service Industries*, 24(6): 725–738.

Torrisi, B. 2014. A multidimensional approach to academic productivity. *Scientometrics*, 99: 755–783.

van Niekerk, W.P. 1986. *Productivity and Work Study*. 2nd edn. Durban: Butterworths Publishers.

Vinodh, S., Kumar, S.V., and Vimal, K.E.K. 2014. Implementing lean sigma in an Indian rotary switches manufacturing organisation. *Production Planning & Control*, 25(4): 288–302.

Vinodh, S., Selvaraj, T., Chintha, S.K., and Vimal, K.E.K. 2015. Development of value stream map for an Indian automotive components manufacturing organization. *Journal of Engineering, Design and Technology*, 13(3): 380–399.

Wajanawichakon, K. and Srimitee, C. 2012. ECRS's principles for a drinking water production plant. *International Organization of Scientific Research (IOSR) Journal of Engineering*, 2(5): 956–960.

Wang, Q., Huang, P., Li, J., and Ke, Y. 2015. Uncertainty evaluation and optimization of INS installation measurement using Monte Carlo Method. *Assembly Automation*, 35(3): 221–233.

Wickramasinghe, V. and Wickramasinghe, G.L.D. 2016. Variable pay and job performance of shop-floor workers in lean production. *Journal of Manufacturing Technology Management*, 27(2): 287–311.

Wu, S., Wang, Y., BolaBola, J.Z., Qin, H., Ding, W., Wen, W., and Niu, J. 2016. Incorporating motion analysis technology into modular arrangement of predetermined time standard (MODAPTS). *International Journal of Industrial Ergonomics*, 53: 291–298.

Zhu, H., Zhang, C., and Deng, Y. 2016. Optimisation design of attribute control charts for multi-station manufacturing system subjected to quality shifts. *International Journal of Production Research*, 54(6): 1804–1821.

Chapter 8

Impact of Work Study on Physical Capital for Productivity in Manufacturing SMEs

8.1 Introduction

Chapters 8 through 10 consider the impact of work study on physical capital, technological capital, and management, respectively, for productivity improvement in manufacturing SMEs. The diagram for these three chapters is presented in Figure 8.1, which is similar to Figure 1.1 in Chapter 1. The reason for presenting this diagram is because these three chapters unpack the structure of the diagram in detail. According to Kaydos (1999:3), if the job cannot be measured it cannot be controlled. So for manufacturing small and medium enterprises (SMEs) to survive in terms of production planning and control for productivity improvement, the performance of these SMEs in terms of all input factors needs to measured. The key technique used in this book to measure these input factors is work study. The reason for using this technique, as indicated earlier, is that limited scientific research has been conducted on how work study impacts on physical capital, technological capital, and management in relation to the productivity of manufacturing SMEs. The diagram for productivity improvement in manufacturing SMEs through the application of work study techniques to physical capital factors is addressed in Figure 8.1.

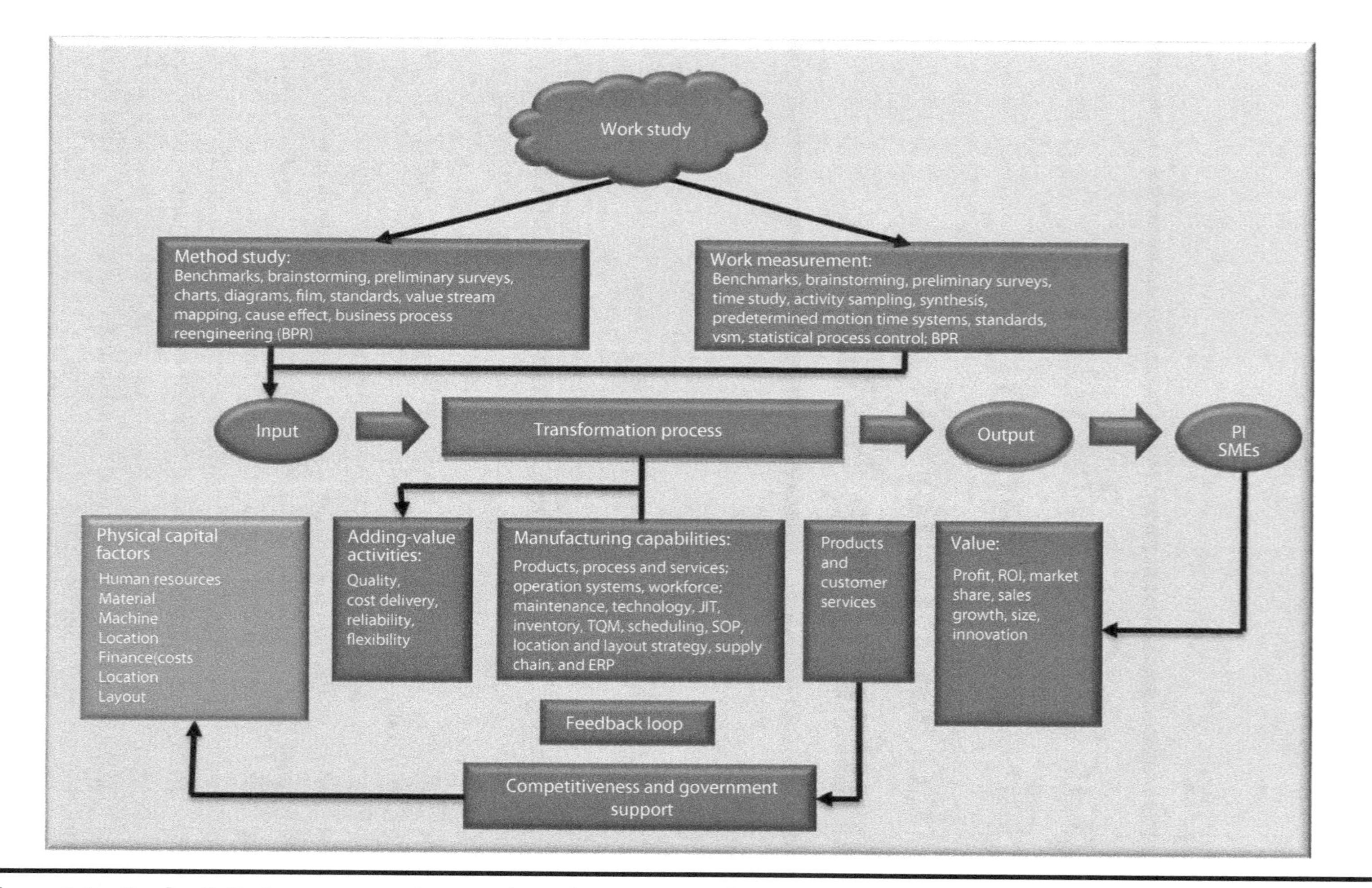

Figure 8.1 Productivity improvement in manufacturing SMEs: Application of work study techniques. (From Author, 2017.)

Chapter 8 focuses on the impact of work study on physical capital for productivity in manufacturing SMEs. The structure of work study is broken down into method study and work measurement, and this is followed by a discussion of how these sub-tools impact on physical capital and thus the productivity of manufacturing SMEs.

8.2 The Impact of Method Study Tools on Manufacturing SMEs

In this section, research on the impact of method study on physical capital for productivity in manufacturing SMEs is studied and discussed. The aim is to determine whether there is a relationship between method study and physical capital in the productivity of manufacturing SMEs. The input resource factors measured using method study in this chapter are human resources, material, machinery, finance, location, and layout.

8.2.1 Preliminary Surveys Used in Method Study for Human Resources

Preliminary surveys used by work study experts through the application of method study for human resources are studied and discussed to provide a detailed look at how interviews, observation, and questionnaires are used to measure the impact of human resources on the productivity of manufacturing SMEs. Manufacturing SMEs use any or all of these preliminary surveys, depending on the situation they are facing.

Gupta, Acharya, and Patwardhan (2013:635–45) investigate how interviews are conducted by work study experts with employees to determine the productivity of manufacturing SMEs. The outcome of the literature review shows that when interviews are conducted to determine the background to what employees are doing in their workplace, the interviews enable experts to identify where employees are going wrong and to show employees how to simplify their work activities with the aim of improving productivity. These work study experts bring attention to various manufacturing process issues, such as the setting of operational improvement targets through waste elimination. When employees are informed about waste elimination, these employees ensure that there is flexibility, cost-effectiveness, product quality, and on-time delivery in the operational process for productivity improvement.

Pandey et al. (2014:113–4) examine the use of observation by work study experts of employee performance in manufacturing SMEs. The findings of the literature studied show that observations made using a process chart, such as an outline process chart, enable work study experts to provide an overall picture of the information recorded in sequences using only operations and inspections. By making notes from observations, work study practitioners help management to see the current position of their businesses, using process chart symbols such as operation, inspection, transport, delay, and storage in the process chart and in the flow diagram to show where these activities are inappropriate and to improve them for productivity progress. On the contrary, when notes are not recorded by work study experts concerning the employees' working background, manufacturing SMEs will not be able to determine progress in employee performance. The lack of written notes may negatively impact on or worsen the productivity of manufacturing SMEs.

Marodin and Saurin (2015:69–71) identify the role of questionnaires used by work study experts to determine employee performance in manufacturing SMEs. Findings from the results of the literature indicate that information from the questionnaires, as noted down by work study experts, shows management the risks facing employees in manufacturing SMEs. When the risk factors are not inspected to ensure improvement in operational processes and to ensure an increase in competitiveness in the market for manufacturing SMEs, the productivity of these SMEs decreases.

8.2.2 Brainstorming Used in Method Study for Human Resources

Mathur, Mittal, and Dangayach (2012:755–7) examine brainstorming among employees to identify problem areas in the workplace and come up with improved solutions. Based on the findings of the literature reported, the research indicates that the use of the cause-and-effect diagram in brainstorming aids in measuring employees' creativity in terms of factors such as methods of working, material specification, machinery capacity, methods used, procedures followed, and the environment where the work activities are taking place to ensure productivity progress in manufacturing SMEs.

Gupta et al. (2013:635–45) investigate the use of brainstorming to examine employee behavior in relation to the productivity of manufacturing SMEs. Based on the results of literature studied, during the brainstorming session, the supervisor and employees use their ability, generating innovative ideas and making possible decisions on job cards and proper operational tools for carrying out work activities in manufacturing SMEs. The aim of using the guide and proper operational tools is to ensure that waste is eliminated. The types of waste expected in the manufacturing process as perceived by employees and their superiors are defects/scrap, inventory, waiting, overprocessing, and underutilization of employee efforts. When waste is eliminated, the productivity of manufacturing SMEs tends to increase.

According to Gupta and Jain (2014:23–27), brainstorming plays an important role in employee performance in relation to productivity. Based on the literature, brainstorming by supervisors and employees is being done to identify the various problems facing manufacturing SMEs in their businesses. These problem areas include absence of appropriate methodology to assure quality, a disorganized workplace, a lack of training, and poor quality of raw material. During brainstorming sessions, suggestions are made to deal with the problems identified in the manufacturing process. When brainstorming sessions are attended by supervisors and employees to identify problems facing manufacturing SMEs in their working environment, precautionary measures are dealt with and suggestions made to eliminate waste and save costs. By eliminating waste and saving costs, the productivity of manufacturing SMEs improves.

Chompu-inwai, Jaimjit, and Premsuriyanunt (2015:1359) examine brainstorming among supervisors and employees in relation to the productivity of manufacturing SMEs. The outcome of the results indicates that brainstorming sessions enable employees to come up with ideas by sharing a range of information regarding the cause of problems in the manufacturing process. These problems occur in areas such as the standard procedures used by employees, material specification, technical machinery, specification, and work methods. If brainstorming sessions are not conducted, there will be a lack of information about these factors, resulting in a lack of improvement in productivity.

Sharma and Shah (2016:582–99) assess brainstorming into employee attitudes in relation to productivity in manufacturing SMEs. Findings from the literature reviewed, report that brainstorming in manufacturing SMEs takes place during meetings or seminars on problem issues requiring clarification and knowledge sharing discussing problems on employees abilities on ideas coming from employees, supervisors, and management from the top working as a team. Training is assured and provided to employees by managers after the brainstorming session to improve coordination, motivation, and trust, with the intention of encouraging employees to be dedicated to accomplishing results that will enhance productivity.

8.2.3 Method Study Process Charts and Time Scale, Diagrams, and Films for Human Resources

Method study process charts consist of charts such as process flow, outline, and two-handed process charts as well as process charts based on time scales, such as multiple activity charts.

Wajanawichakon and Srimitee (2012:956–7) investigate the use of flow process charts to look at the behavior of employees on jobs being carried out in the workplace in manufacturing SMEs. Based on the findings of the literature reviewed, the flow process chart (man type) plays an important role in showing management how employees experience problem areas faced in the manufacturing process so that manufacturing SMEs can deal with these problems. When a problem is identified, manufacturing SMEs have the opportunity to deal with the problem immediately to ensure progress in the business. The manufacturing process can only be justified by bringing forth the proposed flow process chart (man type) for improvement of manufacturing process in the manufacturing SMEs. Finally, the improved flow process chart (man type) is compared with the current one through the use of process chart symbols such as operation inspection, transport, delay, and storage for improvement of productivity. The aim of comparing the improved chart with the existing chart is that management can improve productivity by eliminating waste, rearranging, combining (synthesizing) work, and simplifying work activities.

According to Ray and Teizer (2012:439–55); Xia, Lopes, and Restivo (2013:69–74); Sylla et al. (2014:475–7); de Macedo Guimarães et al. (2015:105); and Ohu et al. (2016:48–9), the two-handed process chart plays a pivotal role in the productivity of manufacturing SMEs. This chart includes process chart symbols such as operation, inspection, transport, and delay to indicate the movement of employees' hands in the workplace. The aim is to determine how the use of employees' hands increases or decreases productivity in the manufacturing SME. The results found show that using an existing two-handed process chart helps to identify the ineffective use of employees' hands and enables work study experts to introduce a proposed two-handed process chart to close gaps or bottlenecks by improving employee performance through efficient and effective use of their hands. Bottlenecks result from employee fatigue, injury, and an unsafe working environment due to discomfort in the hands when handling material and carrying out activities. By exposing their hands to discomfort, resulting in fatigue, employees do not engage their efforts in improving manufacturing processes, and this employee behavior negatively affects performance, leading to a decrease in productivity. On the contrary, if employees can use their hands efficiently, then the productivity of manufacturing SMEs improves.

Sanjog et al. (2016:33–45) explore flow process charts in terms of the movement and activities of employees in the operations of manufacturing. Based on the literature investigated, it can be seen that when the work of the manufacturing process is visualized in terms of the employees' behavior and the background of the job, a flow process chart (man type) makes the process simple and effective by scrutinizing a particular operation. The productivity of manufacturing SMEs then improves when information about employee activities is reflected in the existing flow process chart (man type) for bottlenecks affecting the manufacturing process, with the intention of introducing and preparing a proposed chart in order to eliminate unnecessary transport and delay, which result in waste at workstations.

Lyons et al. (2013:476); Pandey et al. (2014:113); Gould and Colwill (2015:55); and Manne, Ahmad, and Waterman (2016:40) explore the use of the outline process chart to examine employees' activities in the workplace for the productivity of manufacturing SMEs. These work activities are monitored in different types of manufacturing systems such as batch (Lyons et al., 2013:477–80; Pandey et al., 2014:113–4; Gould & Colwill, 2015:61); mass (Lyons et al., 2013:477–80; Gould & Colwill, 2015:64); continuous (Lyons et al., 2013:477–80; Gould & Colwill, 2015:61); project (Lyons et al., 2013:477–80; Manne et al., 2016:40); and assembly systems (Lyons et al., 2013:477–80; Gould & Colwill, 2015:63) through the use of process chart symbols like operations and inspections. Findings from the literature indicate that when the manufacturing process is inspected in terms of employee behavior and the background of the job, the outline process chart has an impact on manufacturing operations. The problem areas identified through the use of an outline process chart are poor performance by unskilled employees, a poor working environment, inconsistent machinery maintenance, poor scheduling, and a lack of standard operating procedures (SOPs); these issues need serious attention in manufacturing SMEs.

Based on the literature studied, during the designed duration of the employees' work in the workplace, productivity is seen to decline when the percentage of the employees' time spent on productive work is less. However, for manufacturing SMEs to resolve this problem and improve productivity, work study is the key tool to give direction to management through a proposed outline process chart; based on the existing outline process chart, it identifies areas of improvement where the employee can eliminate less productive work; this can be achieved through employee training, good housekeeping, scheduled machinery maintenance, and appropriate SOPs.

Based on the literature review studied, we explore how multiple activity charts impact on different activities taking place simultaneously in the workplace in manufacturing SMEs. These work activities take place in different manufacturing systems such as batch (Dai et al., 2012:51–62); mass (Harmse, Engelbrecht, & Bekker, 2016:7–9); continuous (Dai et al., 2012:51–62); project (Padhi et al., 2013:1885–7; Dai et al., 2012:51–62); and assembly systems (Padhi et al., 2013:1886–7; Dinis-Carvalho et al., 2015:235–44) and are applied in the manufacturing processes in manufacturing SMEs. This type of chart can be categorized into an existing or a proposed method for the improvement of manufacturing processes. Thus, the outcome of the results from the literature review studied report that by generating information on work activities carried out by the employees and/or machinery, such as job preparation, operation, and inspection of products, a multiple activity chart enables work study experts to identify gaps made up of nonproductive work and to improve manufacturing operations. When manufacturing operations are improved, the productivity of manufacturing SMEs improves (Dai et al., 2012:53–61; Padhi et al., 2013:1885–7; Dinis-Carvalho et al., 2015:241–4; Harmse et al., 2016:2–16).

8.2.4 Flow Diagram for the Environment

Kaushik et al. (2012:7–8) and Manne et al. (2016:33–9) investigate the use of a flow diagram to identify the flow of material through the use of operator and machine in the manufacturing process from the warehouse to the customer at the dispatch area in the manufacturing SMEs. Based on the literature study reviewed, the flow diagrams are associated with flow process charts that determine the sequence of events used in process chart symbols such as operation, monitoring (Kaushik et al., 2012:7–8; Manne et al., 2016:33–9), transport, delay, and storage. The outcome of the results of the literature review show that the flow diagram, in the form of a SIPOC (supplier, input, process, output, customer) diagram, identifies gaps where there are problems facing manufacturing SMEs, and, through the proposed flow diagram, work study experts can provide guidance for the improvement of productivity of SMEs.

As is investigated by Burchart-Korol (2013:236-7); Alkaya and Demirer (2014:596–601); and Silva et al. (2013:176), flow diagrams play an important role in ensuring progress in terms of productivity in manufacturing SMEs.

Process chart symbols such as operation, inspection, transport, delay, and storage are used through the application of flow diagrams for sustainable manufacturing operations in manufacturing SMEs. Findings from the literature review report that when there is a flow of raw materials from the starting point in the storeroom to the transformation process where this material is transformed into a product, problem areas can be identified using a current flow diagram. These problem areas cause symptoms of productivity decline in manufacturing SMEs, which spur them to resolve the problems. The application of a proposed flow diagram is introduced to improve the layout where work activities are taking place. By attending to this problem and ensuring compliance with appropriate SOPs, the workplace is improved, thus enhancing productivity.

8.2.5 Flow Process Chart for Material

Wajanawichakon and Srimitee (2012:956–8) investigate the use of flow process charts to look at the flow of material on the job taking place on the shop floor in manufacturing SMEs. Based on the findings of the literature studied, the flow process chart (material type) is critical to problem areas faced in the manufacturing process in terms of material flow. When a problem is recognized, manufacturing SMEs can present it immediately with the intention of improving material flow. This existing process can only be validated by introducing a proposed flow process chart (material type) for enhancement of the manufacturing process. As a final point, the proposed flow process chart (material type) is then linked with the present chart through the use of process chart symbols like operation, inspection, transport, delay, and storage for the advancement of productivity. The aim of comparing the proposed chart with the existing chart is to allow management to eliminate waste, rearrange and collaborate on work, and streamline work activities in their businesses.

Schmidt and Nakajima (2013:361–4) study current flow process charts (material type) to examine the material used in the sequence of manufacturing processes for product generation. The results of the literature studied indicate that material is shown from the time it flows from preparation through the manufacturing process to the end product with the intent of identifying unnecessary elements such as excessive transport and delays. Excessive transport and delays then persuade operations managers to involve a WS specialist to continue with the proposed method to resolve the problem. Thus, the application of a proposed flow process chart is introduced by the WS specialist for the improvement of material flow and cost savings to enhance the productivity of manufacturing SMEs.

As found by Chauhan and Agrawal (2014:408–20) and Jasch (2015:1376), a current flow process material chart helps ensure that management knows where the problems are in the current manufacturing processes in terms of the material work flow. Based on the literature studied, the research reports that work study specialists introduced the current flow process material chart to assist management in monitoring defects in the products made from the material to make corrections in terms of material preparation, work in process, and finished products. The results show that when an inspection is made using the current chart, work study specialists introduce the proposed chart with the intention of assisting management to reduce the rejection of products based on quality and in turn improve productivity of manufacturing SMEs.

Sanjog et al. (2016:37–47) explore the use of flow process charts to investigate the flow of material on the job taking place in various workstations in manufacturing SMEs. This material is conveyed into the machinery for operational transformation into products.

Findings from the literature studied report that when a problem is identified, the use of a present flow process chart allows manufacturing SMEs to resolve the problem, guided by work study specialists, in order to improve material flow. One such problem, in terms of material, is continuous manual handling of material, resulting in discomfort and delays in material flow. The current process of manual continuous handling of material at various workstations can only be supported by initiating the projected flow process material type chart in order to improve the manufacturing process through the sequence of workstations. When the material flow is improved through the use of machine rather than manual handling, the chance that productivity will improve in manufacturing SMEs increases.

8.2.6 Process Flow Chart (Machinery/Equipment Type) for Machinery or Equipment

Wajanawichakon and Srimitee (2012:956–7) examine the use of flow process charts to study the utilization of machinery in the workplace in manufacturing SMEs. The outcome of the results in the literature studied show that the flow process chart (machinery type) impacts on the utilization of machinery in terms of delays in the manufacturing process. In terms of machinery, the results are found to be positive for both the existing and the proposed method. The machinery used is not considered a challenge, since the process of utilizing the machinery is appropriate in the manufacturing process, and the productivity of manufacturing SMEs appears to be improving continuously due to the consistent and effective use of machinery.

Silva et al. (2013:176–86) and Prashar (2014:112) inspect the use of flow process charts (machinery type) to study the utilization of machinery in the workplace for productivity in manufacturing SMEs. On the basis of the literature reviewed, it can be seen that a current flow process chart (machinery type) impacts on the utilization of machinery in terms of maintenance, availability, and breakdowns in the manufacturing process; manufacturing SMEs can refer to the flow process machinery chart to replace or fix machines. By replacing or fixing machines on the basis of a proposed flow process machinery chart, manufacturing SMEs aim to optimize machine maintenance procedures for the improvement of productivity.

8.2.7 Film Technique for Employee, Material, Machine, Finance, Location, and Environment

Dai et al. (2012:53) examine the application of film techniques for the improvement of productivity in manufacturing SMEs. These films are generated in the form of durable images revealing resource input factors such as employee images, material specification and flow, product specification, machinery, and workstation processes where work activities are taking place. The techniques involve images of the workplace in various manufacturing systems such as batch, mass, continuous, project, and assembly systems so that supervisors can see and reflect on the current situation facing the business as guided by management. The aim is to identify areas that need improvement, such as poor working environment, material damage, poor maintenance, and high inventory. Findings from the literature investigated show that film techniques positively contribute to the productivity of manufacturing SMEs. When images of various resource input factors, as indicated earlier, are generated, the situation in the workplace drives management to act on the problem immediately. Supervisors are driven to make decisions through authority from top management and advice from work study experts to improve the working environment in terms of employee safety, waste material elimination, scheduled maintenance, and minimization of inventory. Thus, when the environment is improved, the productivity of manufacturing SMEs improves. As investigated by Yuan, Zhai, and Dornfeld (2012:40–2), film techniques play a vital role in the productivity of manufacturing SMEs. Management in manufacturing SMEs requires photos to identify input resource factors such as employees, material, and the environment to ensure that employees are not exposed to ill-health, material is not damaged, and the area where activities are taking place is not dirty. The outcome of the results in the literature reviewed indicates that when an image of the environment is generated through the resource input factors addressed earlier, management is able to act in a timely way, through the guidance of the work study expert, by comparing the proposed image with the current image for improvement of the working environment for employees and material. As a result, the productivity of the manufacturing SME improves.

As discovered by Heidrich and Tiwary (2013:5884–91) and Muruganantham, Krishnan and Arun (2014:420–6), film techniques are critical to the manufacturing environment of manufacturing SMEs. As reviewed by the literature, when images of photos are generated and standards of procedure are indicated in the form of the manual for compliance in manufacturing SMEs, management and employees are motivated to adhere to standards in the workplace. Management also monitors operators through the planning of standards and adjustment of employee behaviors toward respect for the environment. These standards are ISO 9001 in terms of quality product for the customers and ISO 14001 certification in terms of an enabling environment for employees (Heidrich & Tiwary, 2013:5884–91). The literature reviewed infers that when these standards are adopted, waste and costs are reduced and the environment is taken care of in terms of cleanliness, making it easier for manufacturing SMEs to improve the productivity of their businesses.

Chay et al. (2015:1033–39) state that film techniques involve photos taken or obtained to show employees that they lack skill and knowledge and need to be trained to avoid waste on the shop floor. Based on the literature studied, this process is done through business reengineering and just-in-time (JIT) scheduling for continuous improvement. According to Dora, Kumar, and Gellynck (2016:6–8), film techniques involve photos indicating how processes and operations are taking place before and after improvement. This process happens when the manufacturing SME has planned the number of employees to be assigned to specific work stations to avoid overstaffing as well as appropriate working conditions in terms of proper communication and good employee treatment in their sphere of operation.

8.2.8 Value Stream Mapping (Qualitative)

Chiarini (2012:681–90) investigates value stream mapping (VSM) in physical capital factors for productivity in manufacturing SMEs by analyzing manufacturing capabilities such as employees' abilities under the supervision of the first-line manager and middle managers, material flow, machine maintenance, layout, and costs involved, in order to ensure that the manufacturing process is operating effectively. VSM not only focuses on capabilities but also on using adding-value activities that strengthen these capabilities to ensure that productivity increases without waste being incurred. These adding-value activities include quality, cost, speed, flexibility, and delivery. Using VSM in the absence of appropriate use of capabilities and adding-value activities results in waste in manufacturing SMEs. As a result, productivity declines.

As found by Dora et al. (2013:126–36) and Matt (2014:333–45), VSM plays an important role in physical capital factors in relation to the productivity of manufacturing SMEs. Findings from the results of the literature studied show that absence of VSM for the identification of inventories and waste along the supply chain with the aim of reducing costs and increasing customer satisfaction impacts on the productivity of manufacturing SMEs. Furthermore, lack of visualization of additional capabilities such as JIT, total productivity maintenance (TPM), human resource training and employee work involvement, supply chain, quality management, location, working environment, scheduling, and poor working environment with the absence of management commitment results in failure to identify waste in manufacturing SMEs. Lack of ensuring adding-value drivers such as cost, improved quality, delivery, dependability, flexibility, and reliability also lead manufacturing SMEs to fail to track areas of waste within the manufacturing process, which represents a risk to productivity.

Based on the results of the literature reviewed, the application of VSM to physical capital factors contributes to the inspection of waste, focusing on various capabilities such as employee abilities, material flow, costs involved, location, layout, inventory control, supply chain control, quality improvement, scheduling, time management, production planning and control, and product, process and service design in order to ensure that manufacturing SMEs do not incur waste, which may negatively impact on the productivity of their businesses. Adding-value activities are also taken into consideration since they impact on the effective use of capabilities applied by manufacturing SMEs in their businesses. These activities are quality, cost, delivery, reliability, dependability, and flexibility (Tyagi et al., 2015:202–13).

Dushyanth, Shivashankar, and Rajeshwar (2015:2–11) examine the application of VSM to physical capital factors to identify work activities influencing productivity in manufacturing SMEs. Based on the findings of the literature evaluated, employee abilities, product design, process design, service design, inventory management, quality management, layout, location, supply chain management, planning, TPM, scheduling, JIT, and efficient transportation through SIPOC have a positive impact on the productivity of manufacturing SMEs. So, failure to consider the impact of adding-value activities such as quality, cost, speed, delivery, and flexibility on the above manufacturing capabilities results in waste and high costs, and the productivity of manufacturing SMEs declines.

As explored by Prasad, Khanduja, and Sharma (2016:409–424), failure to use VSM on physical capital factors in order to identify waste and product defects means that the probability that the manufacturing SMEs can improve productivity is low. The only way for manufacturing SMEs to ensure productivity progress in their businesses is to use VSM to identify waste and defects in manufacturing capabilities such as JIT, production systems, workplace housekeeping, total productive maintenance (TPM), and total quality management (TQM). Adding-value activities such as quality, cost, speed, dependability, and delivery can also be used to ensure productivity improvement in manufacturing SMEs.

The next section to be discussed is the impact of cause-and-effect diagrams on productivity in manufacturing SMEs.

8.2.9 Cause-and-Effect Diagram for Environment

As is reported by Gnanaraj et al. (2012:603–5), cause-and-effect diagrams play a vital role in ensuring productivity progress in manufacturing SMEs. The cause-and-effect diagram is used to identify resource input factors such as human resources, material, machinery, or equipment, along with environment and procedures, to assist management in taking corrective measures by eliminating waste, rearranging, combining (synthesizing) work, and simplifying work activities for productivity improvement. So, if corrective measures in manufacturing SMEs are not pinpointed, employee inability (due to poor management skills, lack of planning, inadequate training, limited availability of funds, and fewer human resources); material defects; machine misalignment; lack of machine maintenance; and power shut downs result in a decline in productivity. Furthermore, a failure to take into consideration employee safety, health, and satisfaction also has a negative impact on the productivity of manufacturing SMEs.

Silva et al. (2013:175–85) investigate the use of cause-and-effect diagrams on physical capital factors in the manufacturing process. This process is done by detecting possible causes, paying attention to the effect of machinery, method, measurement, people, materials, and the environment on productivity. Findings from the literature reviewed report that a cause-and-effect diagram can indicate the commitment of management to employee training, improved machinery, proper operational documentation, and improved quality of material, as well as improvements in workers' health, safety, and satisfaction and lower absenteeism, thereby improving the productivity of manufacturing SMEs.

Prashar (2014:112) provides a survey of various input factors in a cause-and-effect diagram with the aim of identifying problem areas affecting the productivity of manufacturing SMEs. These factors involve employee errors and wrong fitment, wrong specification and noncompliance to SOPs, poor machine failure, and damage to machine tools due to lack of maintenance, as well as poor environment, such as a lack of cleanliness and poor air quality. Thus, the shortcomings of management on these factors cause productivity decline.

As found by Gupta and Jain (2015:76), cause-and-effect diagrams contribute to the productivity of manufacturing SMEs. The outcomes of the results in the literature studied indicate that problem areas indicated in the diagram, such as lack of employee SOPs, unorganized material and shop floor, and lost or misplaced material or machinery, including tools, results in a decrease in productivity in manufacturing SMEs.

As reported by Srinivasan et al. (2016:8–10), cause-and-effect diagrams impact on the operation of manufacturing SMEs. By determining key process input variables, such as employees, material, machinery, method, and environment; and key process output variables, such as product and customer requirements, the productivity of manufacturing SMEs can be improved. Insufficient training, poor attitude, poor motivation, and inadequate supervision result in a decline in productivity in manufacturing SMEs. In addition, when poor material specification, poor maintenance, improper speed, poor tools, the wrong gauges, and environmental factors such as dust, noise, temperature, poor lighting, and humidity are experienced, the productivity of manufacturing SMEs decreases. The next tools to be discussed are work measurement tools.

8.3 Work Measurement Techniques

In this section, research is studied and discussed on the impact of work measurement on physical capital for productivity in manufacturing SMEs. The aim is to determine whether there is a relationship between work measurement and physical capital in terms of the productivity of manufacturing SMEs. Similar to the method study addressed in Section 8.1, the input resource factors measured using work measurement study in this chapter are human resources, material, machinery, finance, location, and layout. The tools used, as indicated in Chapter 7, are time study, work (activity) sampling, predetermined motion time systems (PMTS), analytical estimating, comparative estimating (benchmarking), synthesis, VSM, statistical process control, and business process reengineering.

8.3.1 Preliminary Surveys Used in Method Study for Human Resources

Preliminary surveys used by work study experts through the application of work measurement are also similar to the surveys used in Section 8.1.1, utilizing tools such as interviews, observation, and questionnaires, which are used to measure employee resources for productivity in manufacturing SMEs (Gupta et al., 2013:635–45; Pandey et al., 2014:113–4; Marodin & Saurin, 2015:69–71). The difference between method study and work measurement is that method study tools are shown in Section 8.1, whereas work measurement focuses on tools outlined in Section 8.2.

Preliminary surveys used by work study (WS) experts through the application of method study for human resources are studied and discussed to provide a detailed look at how interviews, observation, and questionnaires are used to measure how human resources impact on productivity in manufacturing SMEs. Manufacturing SMEs use any of or all these preliminary surveys depending on the situation they are facing.

Gupta et al. (2013:635–41) investigate how interviews and questionnaires are conducted by WS specialists to guide management on employee performance in terms of productivity. The outcomes of the literature review show that when interviews are conducted and questionnaires provided to employee members with the support of operations managers in order to determine the standard time for the job being carried out, the interviews and questionnaires enable these experts to identify where employees are going wrong in terms of their standard of performance. As a result, WS specialists help to identify ineffective time spent and to improve the appropriate standard time for the job, which will contribute to the improvement of productivity in manufacturing SMEs. The appropriate standard time brings attention to the setting of operational improvement targets through waste elimination.

Pandey et al. (2014:113–4) examine the use of observation by WS specialists in guiding operations managers in terms of establishing standard time for employee performance in their workplace. The outcome of the results in the literature shows that observations made using standard time on the job being carried out by employees in terms of manufacturing and transport enable the WS experts to provide an overall picture of the information recorded, focusing on the standard of employee performance. By making notes based on observations, WS specialists help operations managers to see the current position of their businesses by indicating the standard time for the job for manufacturing SMEs, showing where times are inappropriate and improving those work activities for productivity progress. On the contrary, when notes are not recorded by the WS expert about the employees' working environment, manufacturing SMEs will not be able to determine the correct standard time for the performance of employees. The lack of written notes may negatively impact on or decrease productivity in manufacturing SMEs.

Marodin and Saurin (2015:69–71) investigate preliminary surveys used by WS specialists to enlighten operations managers before work activities can be carried out with the aim of achieving the results required by manufacturing SMEs. Based on the results of the literature reviewed, the research reports that information on the manufacturing process from the interviews conducted, questionnaires used, and notes recorded during observation by WS specialists on the time study conducted, aids management to be aware of the problem areas affecting the workplace in manufacturing SMEs. This information is also gathered by using the results collected from previous studies done through analytical estimating regarding work activities carried out in the manufacturing process. By ensuring that these preliminary surveys are undertaken by WS specialists with the support of management, the productivity of manufacturing SMEs can be projected and improved. On the contrary, when operations managers do not disclose information on their shortcomings, such as lack of knowledge, technical skills to guide employees, and management support on all resources in carrying out operational activities in the workplace, the probability of a decline in productivity in the manufacturing SME will be high.

Hooi and Leong (2017:2–13) examine the use of observation and questionnaires by WS specialists on the information gathered from operations managers with regard to the set-up time of employees' work activities in their workplace within the manufacturing SME. The outcome of the results in the literature review show that the lack of record keeping through observations and questionnaires means that there is no correct information on the required standard time being taken for various job. Consequently, a lack of guidance by WS specialists in this regard delays the communication between the parties involved from top to bottom, which, in turn, means that manufacturing SMEs cannot reach their goals in terms of productivity.

8.3.2 Time Study for Human Resources, Material, and Machinery

As reported by Garza-Reyes et al. (2012:181–5) and Vinodh et al. (2015:380–91), time study is one of the main flexible work measurement techniques employed to analyze any work activity performed by employees or machinery in any environment, as well as the material and tools used in any location, so as to influence the productivity of manufacturing SMEs. All these input resource factors are applicable in the manufacturing process of manufacturing SMEs. Manufacturing capabilities such as product, process, and service design; JIT; supply chain management; scheduling; TPM; inventory management; production planning and control; location; and layout are implemented by employees through the support of adding-value activities such as quality, cost, speed, reliability, flexibility, dependability, and delivery. So, if a time study is applied in the manufacturing process, measuring work activities focusing on both manufacturing capabilities and adding-value activities, this technique as used by a WS specialist can help manufacturing SMEs to identify bottlenecks for waste reduction, product variation, and elimination of defects taking place in the current process flow and encourage the implementation of a proposed process flow in order to improve productivity. All these activities take place in mass, batch, continuous, assembly, and project operational systems in manufacturing SMEs.

Das, Ghosh, and Gangopadhyay (2013:247–9) examine the impact of time study on the manufacturing process, focusing on input resource factors such as employee abilities, material, machinery, layout, and location in relation to the productivity of manufacturing SMEs. The outcome of the results in the literature is that if time study is not conducted in the manufacturing process where employees' activities are carried out, management will not detect the appropriate standard time for the job in sequence, with employees' greater workload and delays incurred during the operation. So, lack of standard time results in delays and a poor method of doing the job resulting in poor performance. Secondly, employees' workload contributes to employee fatigue, which may lead to accidents. So, when manufacturing SMEs are not supported by a WS specialist with the application of time study to determine the correct standard time for the job, poor performance, employee fatigue, and delays are not realized, which negatively impacts on the productivity of manufacturing SMEs.

The use of time study as indicated by Suwittayaruk, Van Goubergen, and Lockhart (2014:725) is critical to the productivity of manufacturing SMEs in their manufacturing process. Findings from the literature reviewed show that if the employee is not well trained, motivated, and qualified to carry out operational tasks using a particular machine in a specified environment, the time study will not be carried out properly, and the symptoms of productivity decline will prevail in the manufacturing SME.

Based on the literature discussed, time study in manufacturing SMEs is regarded as the driver of productivity in the manufacturing process. The outcome of the literature examined shows that if cycle times are measured using a stopwatch by the WS specialist on the work activities being carried out by employees in the workplace, SOP should be considered. In cases where SOPs are not followed during the current work flow, the time study will identify ineffective time spent and delays resulting in waste. Such waste causes the productivity of manufacturing SMEs to suffer, and therefore the WS specialist will propose a projected work flow that will assist manufacturing SMEs to improve productivity. However, even though time study improves the productivity of manufacturing SMEs, manufacturing capabilities are crucial to the manufacturing process and need support from adding-value activities such as quality, cost, reliability, and delivery, which need to be considered to ensure improved productivity (Srinivasan et al., 2016:364–76).

8.3.3 Work (Activity) Sampling for Human Resources, Material, and Machinery

As reported by Garza-Reyes et al. (2012:186–88), work sampling is also one of the main measurement techniques utilized to analyze any work activity performed in any environment as well as the material and tools used in any location. The type of activities that are measured using work sampling in manufacturing SMEs involve a successive number of observations done over a period time, focusing on employees, machines, or processes. An observation of individual operations or a group of operations done by employees, machines, or processes records what is transpiring at any given moment, whereby idle time ranges from a low to a high percentage. A high percentage indicates that work activities are not equally distributed among operations or a group of operations done by employees, machines, or processes in the work place. When the work activities are not equally distributed, the workload in workstations builds up and the output of the manufacturing SMEs becomes delayed, leading to excessive waste and high costs. As a result, productivity is negatively affected.

Gupta et al. (2013:636–41) investigate the impact of work sampling on the productivity of manufacturing SMEs. Findings from the literature reported indicate that WS specialists use work sampling to identify various types of wastes such as defects, inventory, waiting, overproduction, overprocessing, movement, and employee underutilization in work activities carried out by employees, by machinery, or on processes. When these types of waste are eliminated, the productivity of manufacturing SMEs is enhanced.

Based on the results of the literature reviewed, work sampling plays a pivotal role in discovering non-value-adding activities caused by employees or processes in manufacturing SMEs. By recognizing what is taking place in the manufacturing environment, this technique then contributes to productive work, generating value for the manufacturing SMEs in terms of productivity improvement in their businesses and, in turn, good customer service (Hasle, 2014:41–7; Dinis-Carvalho et al., 2015:235–9).

Czumanski and Lödding (2016:2946) examine the impact of work sampling on the productivity of manufacturing SMEs. Observations are made on work activities taking place in manufacturing functions or departmental sections by groups of employees in engaging in their work activities, as well as via machine operations in the manufacturing environment. The aim of the technique used is to target detailed analyses of workers, machine operations, and process in work systems of different production lines for the improvement of manufacturing processes, the ultimate aim being improved productivity.

8.3.4 Predetermined Motion Time Systems for Employees

According to Kuo and Wang (2012:6519–22), PMTS contribute to the productivity of manufacturing SMEs. Findings from the literature studied show that PMTS, by considering motion, improves the standard time of the work activities being carried out in manufacturing SMEs. These motions need to be done by taking into consideration standards in place such as employee posture in terms of safety, health, and well-being. When these standards are adhered to, the productivity of manufacturing SMEs improves.

Larsson (2013:891) explores the impact of PMTS on the productivity of manufacturing SMEs. When WS specialists or industrial engineers are knowledgeable and experienced in using this tool, these specialists can estimate the time required for the various motions in an operation. As a result, these specialists have the opportunity to ensure that employees comply with the required standards in order to ensure productivity progress in manufacturing SMEs.

Suwittayaruk et al. (2014:725) emphasize the importance of PMTS in the manufacturing process in relation to the productivity of manufacturing SMEs. Based on the results of the literature examined, if the employee operating the machine is not well trained, motivated, and qualified to carry out work activities under specified conditions, the application of time study by WS experts will not provide adequate results, and there will be indications of productivity deterioration.

As reported by Battini et al. (2015:825--7) and Wu et al. (2016:292), PMTS plays a fundamental role in ensuring that basic human movement contributes to the manufacturing process of manufacturing SMEs. So, when employees follow the SOP in their workplace when doing work activities involving motion, these employees are able to adhere to safety measures put in place for the proper execution of work activities. Consequently, through compliance with safety measures in the workplace, employee abilities enrich the productivity of manufacturing SMEs.

8.3.5 Analytical Estimating for Human Resources, Material, and Machinery

As indicated by Mathur et al. (2012:761); and Sharma and Sharma (2014:751–65), the literature studied explores the impact of analytical estimating on the activities done by previous methods and time studies in relation to the productivity of manufacturing SMEs. The research finds that when WS experts are skilled, knowledgeable, and experienced in these studies, these experts are able to assist employees who are qualified in their sphere of operation to simplify their job within a limited time on the basis of quality, delivery, flexibility, and cost of the product produced. Thus, considering these adding-value activities by employees through the guidance of WS experts promotes the productivity of manufacturing SMEs.

Padhi et al. (2013:1885–94); and Colledania, Rattia, and Senanayake (2015:357–61) investigate analytical estimating in the manufacturing process line layout in relation to the productivity of manufacturing SMEs. This tool is applied by WS experts to ensure that there is adequate time for the chain or sequence of manufacturing processes; inspection of the operational frequency and skill of the employees; and changing of tools during the operation and machine operation in manufacturing SMEs. By focusing on the time used in the sequence of manufacturing processes, inspection of the operational frequency and skill of the employees, and changing of tools during the operation and machine operation in the workplace, these experts guide employees to enable manufacturing SMEs to improve productivity.

Findings from the literature show that analytical estimating is one of the driving factors of the work measurement technique influencing the productivity of manufacturing SMEs. When WS experts use these tools, these experts use their knowledge to encourage management to ensure that employees not only use their competencies such as skills, knowledge, training, and experience but also have the power to make their own decisions to cope with complex situations they face in the manufacturing process, especially in cases when there is no one superior in attendance while they are working. By making these decisions, employees become more confident, which enables them to engage in improving the manufacturing processes of manufacturing SMEs. These improved processes lead to improved productivity (Małachows & Korytkowski, 2016:166–70).

8.3.6 Comparative Estimating (Benchmarking)

Findings from the literature explored report that comparative estimating plays an important role in the manufacturing operation of the working environment in manufacturing SMEs. With comparative estimating, the employees' current work activities are compared with the historical data of the previous time study. When comparing the current results with previous data of the time study being carried, manufacturing SMEs are able to evaluate the trend of the employees' performance and anticipate the productivity progress of these SMEs in their businesses (Matawale, Datta, & Mahapatra, 2014:154–62; Moon et al., 2016:85)

8.3.7 Synthesis

Santosa, Vidal, and Moreira (2012:1656–61); and Doltsinis, Ratchev, and Lohse (2013:85–90) investigate synthesis as an indirect work measurement technique for building up cycle time for a task carried out by a qualified employee in relation to the productivity of manufacturing SMEs. Findings from the literature reviewed show that when the correct time is established for the job taking place in manufacturing settings, taking into consideration safety, fatigue recovery, and cost, employees are able to perform well in contributing to the continuous improvement of manufacturing processes, be it in the form of an individual employee or a team. Thus, the productivity of manufacturing SMEs continues to increase.

Beyer (2014:1–6); Glock and Grosse (2015:6637–45); and Pedersen and Slepniov (2016:44–56) report that the application of synthesis plays a vital role in ensuring that productivity of manufacturing SMEs improves. This technique becomes efficient if employees have such things as knowledge, skills, a relevant educational background, and experience acquired to perform the task in manufacturing SMEs.

8.3.8 Value Stream Mapping (Quantitative)

Chiarini (2012:681–90) investigates VSM in physical capital factors for productivity in manufacturing SMEs by analyzing manufacturing capabilities such as employee costs and cycle time. Findings from the literature indicate that a reduction in employee costs and cycle time leads manufacturing SMEs to eliminate waste and save costs. Consequently, waste elimination and cost saving results in improved productivity.

As found by Dora et al. (2013:126–36); Matt (2014:333–45); Dushyanth et al. (2015:2–11); and Tyagi et al. (2015:202–13), VSM contributes to the productivity of manufacturing SMEs. The outcome of the results in the literature reviewed not only focus on cycle time and cost but also on challenging areas such scrap and delivery time profitability, sales, customers, employees, product, and quality in terms of price. These areas, when reduced and improved, positively contribute to the productivity of manufacturing SMEs.

Prasad et al. (2016:409–424) contributed further by looking at VSM considering statistical process control in terms of defects and waste to ensure productivity progress in manufacturing SMEs. Similar to other research authors discussed earlier, the focus was also on challenging areas such as profitability, sales, customers, employees, product, and quality in terms of price in manufacturing SMEs.

8.4 Case Study Results and Analysis for South African SMEs

This section examines the results found in various case studies focusing on Companies A, B, C, and D in Gauteng, South Africa. The purpose of the research conducted was to identify productivity problems facing these companies that were attributed to manufacturing capabilities and adding-value activities exercised in physical capital factors. The interventions necessary to resolve these problems are then introduced with the intention of improving productivity in the companies addressed in the case studies.

8.4.1 Company A (Small Company)

The research conducted commenced with Company A, in which services were provided by employees to customers involved in mining, gas, and electrical businesses as well as households. Even though injuries and damages were stringently avoided over and above inappropriate suppliers' delivery, customer returns, and elimination of complaints, Company A did not comply with ISO 14001 in terms of the environment or ISO 18001 with regard to occupational safety and health. Furthermore, employees were provided with limited training. Employees were faced with challenges in terms of management, whereby these employees differed with management in terms of decision making and were resistant to change. This was due to the management's uncertainty and inability to motivate the employees. Incentives such as formal training and benefits were ignored by management, and this led to employees not making an effort to perform well in carrying out their work activities.

In addition, stock accuracy was also a serious situation facing Company A. During the manufacturing process, employees converted material through cutting, computer numeric control, drilling, and tapping processes.

The challenge faced by Company A was poor inventory control, delays, and waste incurred during the operations. Company A was focusing on the business using old machinery for manufacturing. There was insufficient space for employees to work and for the storage of inventory delivered by suppliers. Employees failed to comply with safety measures, and negligence impacted on the costs incurred by the company. There were also no safety demarcations marking out a walking route for employees to focus on so as to avoid accidents.

8.4.2 Company B (Medium Company)

The second research conducted was at Company B, where a variety of headwear is produced by employees for officers in the police, armed forces, and security firms in Southern Africa and other Asian countries, including islands off the east coast of South Africa. The type of material used by Company B to produce headwear includes wool, fabric, and wire. The suppliers of Company B were located both within the country and abroad. Employees were empowered with on-the-job training, which is training that is provided while carrying out the job in the workplace, rather than a career path that will develop the employee to feel motivated. This career path involves a certain level of educational background that should be provided to the employees in the form of reward by manufacturing SMEs. Supervisors improved the job carried out by employees through the use of job descriptions for every task, and standards were put in place using a time study carried out by a WS intern. In addition, filming techniques were also applied in the form of pictures of employees and of the environment where employees carry out activities. Other tools used, as advised by Productivity SA, included cause-and-effect analysis to identify areas for improvement in the company.

The assembly process at Company B consists of stores, sample department, cutting room, preparation, embroidery, quality check, and, finally, packaging. The average standard time for the whole process is 64.17 minutes. The company produced 1000 caps per day. The challenge of the human element was that most employees in this company are aging and face a challenge when carrying heavy boxes, which is demotivating and results in poor employee performance. Incentives such as benefits and medical aids were not provided to employees. Communication between management and supervisors and employees was also a major problem in terms of group decision making.

Company B was faced with poor inventory control, delays, and waste. The company had insufficient space for working and the storage of material and products to be delivered. Excessive sitting resulted in chronic ill-health among employees.

8.4.3 Company C (Small Company)

The third company focused on packing colored cocky pens via parallel work stations. Employees in Company C focused on the packaging process, starting with the packaging of the pens and finishing with the wrapping of boxes packed with pens with plastic, incorporating corners to protect the finished products from damage. Employees packaged 2250 cocky pens per hour. Various challenges were experienced by employees in Company C, including inappropriate SOPs and job rotation, resulting in employee redundancy. The conveyor belt was operated by employees to transport completed products to pallets for final wrapping. In terms of the location, Company C uses the same door for both receiving material and issuing packaged products. Employees were resistant to change due to lack of incentives, which resulted in many employees leaving the company. Old machinery was being utilized, and the flow of material and waste were major concerns within Company C.

Haphazard areas were identified such as insufficient space for working and storage of material and products to be delivered to customers. Poor safety control and poor SOPs were experienced, and there were no safety demarcations in place.

8.4.4 Company D (Medium Company)

The final research was conducted at Company D, which produces components in the form of batches. Company D supplies a wide range of radio communication systems for mobile communications based on sites and at control centers both locally and internationally. Employees in Company D generated 2500 components per hour with 102 employees hired. Employees were motivated by receiving extra payment from overtime and bonuses when they exceeded the company's expectations. All these incentives were paid according to what is planned by the manufacturing department at Company D.

Delivery of items produced by Company D was done using the company's transport or the use of couriers to reduce costs. Cycle time variation impacted on the efficiency of the manufacturing process in Company D. Waste and damages were experienced by employees due to poor quality systems. Employees' material handling was very poor, whereby employees had to carry some of the material physically, resulting in severe ill-health. Even though machine maintenance was done on a monthly basis, there were no safety demarcations to guide employees.

8.5 Comparative Analysis and Discussion

Based on the case study results discussed and analysis made in terms of physical capital factors used in South African manufacturing SMEs, such as Companies A, B, C, and D, a comparative analysis and discussion of challenges facing these companies is addressed in Table 8.1 in order to highlight discrepancies that exist between these companies. Furthermore, a comparative analysis of the application of WS is also made based on these companies to determine differences in the manner in which they contribute to productivity in their businesses. See the outline in Table 8.1.

Based on the background of these companies, Companies A, B, and C are exposed to a number of WS tools to ensure productivity progress in their companies. These tools are exercised through method study and involve benchmarking, brainstorming, work flow charts, flow diagrams, cause-and-effect diagrams, and filming techniques. With work measurement, however, these companies focus on measuring time. The difference between Companies A, B, and C is that Companies A and C have supervisors conduct their own time measurement, whereas Company B makes use of the WS intern to conduct a time study for the job being carried out by employees. The WS, as indicated earlier around method study and work measurement in Companies A, B, C, and D, is exercised on physical capital factors indicated in Table 8.1 for productivity improvement in the businesses.

8.6 Summary

This chapter has focused on the impact of WS on physical capital for productivity improvement in manufacturing SMEs as well as addressing productivity challenges facing companies in Gauteng, South Africa.

The types of WS techniques used to measure physical capital factors for productivity performance in manufacturing SMEs were method study and work measurement, respectively. Method study used sub-techniques such as preliminary surveys, brainstorming, method study process charts and time scales, flow diagrams, filming techniques, and VSM along with cause-and-effect diagrams. Work measurement utilized preliminary surveys, brainstorming, and benchmarking just as the method study did, differing only in terms of the techniques used, such as time study, work or activity sampling, PMTS, analytical estimating, comparative estimating and synthesis, and VSM, accompanied by the statistical process control cause-and-effect diagram.

The physical capital factors measured involved human resources, material, machinery, location, environment, and finance, and these factors were exercised in the manufacturing process of manufacturing SMEs. When WS is applied to physical capital factors implemented in the manufacturing processes—that is, exposed to management manufacturing capabilities such as

Table 8.1 Comparative Analysis and Discussion of Challenges Facing Companies A, B, C, and D

Background of manufacturing	*Company A*	*Company B*	*Company C*	*Company D*
Challenges facing manufacturing capabilities based on Input resource factors utilized in companies	**Human element** Employee resistance to change, uncertainty, lack of incentives No compliance with standard operating procedures in terms of ISO 14001 and ISO 18001 **Material** Poor inventory control, delays, and waste **Machine** Aging machines **Location** Haphazard areas; insufficient space for working and storage of material and products to be delivered **Environment/layout** Safety clothing but no compliance with safety measures No safety demarcations	**Human element** Mostly aging employees, maternity in the case of most young women, heavy box handling, no incentives, and poor communication by management **Material** Poor inventory control, delays, and waste **Machine** Aging machines **Location** Insufficient space for working and storage of material and products to be delivered **Environment/layout** Excessive sitting of employees while working may lead to life-long illness No safety demarcations	**Human element** Employee resistance, no incentives, and high employee turnover **Material** Too much line-based material flow and waste The issues affecting material include bad material, cracks, and hardness **Machine** Tooling problems; old machines are utilized, and as a result delays are encountered by Company C **Location** Haphazard areas; insufficient space for working and storage of material and products to be delivered **Environment/layout** Poor safety controls, poor standard operating procedures, no standards in place No safety demarcations	**Human element** Severe ill-health, absenteeism **Material** Waste and damages are experienced due to poor quality systems and poor material handling **Machine** Maintenance on monthly basis **Location** Haphazard areas; insufficient space for working and storage of material and products to be delivered **Environment/layout** Safety measures complied with No safety demarcations

(*Continued*)

Table 8.1 (Continued) Comparative Analysis and Discussion of Challenges Facing Companies A, B, C, and D

Background of manufacturing	*Company A*	*Company B*	*Company C*	*Company D*
Work study tools available in these companies	Standards such as Quality ISO 9001 certified	Quality plan SABS certified standards	Quality plan neither SABS nor ISO 9001 certified standards	Quality ISO 9001 certified standards
Work study: Method study	Benchmark with productivity SA, brainstorming, work flow chart, flow diagram, cause-and-effect chart, filming techniques	Benchmark with productivity SA, brainstorming, work flow chart, flow diagram, cause-and-effect chart, filming techniques	Benchmark with other companies, work flow chart	Benchmark with other companies, brainstorming, work flow chart, flow diagram, cause-and-effect chart, filming techniques
Work study: Work measurement	Measure time from start to end for cycle time	Use work study expert to measure time using time study	N/A	Measuring of time using computer recording

product, process and service design; JIT; supply chain management; scheduling; TPM; inventory management; production planning and control; location; and layout—the job is easily carried out, resulting in improved employee health and safety, eliminated waste, cost savings, proper standards, and low absenteeism as well as quality products and services.

Even though these capabilities are driven by WS for the improvement of manufacturing processes, these capabilities need to be integrated along with value-adding activities exercised on physical capital factors for the continuous improvement of productivity in manufacturing SMEs. However, in the case of companies in Gauteng, South Africa, various productivity challenges were addressed. These challenges were identified on the basis of input resource factors such as employee abilities and involvement, material, machinery, location, and environment. The challenges being faced differed from one company to another. These challenges contributed to productivity delays, negatively impacting on these companies. The impact of WS into technological capital for productivity in manufacturing SMEs is addressed in the next chapter.

References

Alaskari, O., Ahmad, M.M., and Pinedo-Cuenca, R. 2016. Development of a methodology to assist manufacturing SMEs in the selection of appropriate lean tools. *International Journal of Lean Six Sigma*, 7(1): 62–84.

Alkaya, E. and Demirer, G.N. 2014. Sustainable textile production: A case study from a woven fabric manufacturing mill in Turkey. *Journal of Cleaner Production*, 65: 595–603.

Al-Najem, M., Dhakal, H., Labib, A., and Bennett, N. 2013. Lean readiness level within Kuwaiti manufacturing industries. *International Journal of Lean Six Sigma*, 4(3): 280–320.

Amin, M.A. and Karim, M.A. 2013. A time-based quantitative approach for selecting lean strategies for manufacturing organisations. *International Journal of Production Research*, 51(4): 1146–1167.

Aschehoug, S.H., Boks, C., and Støren, S. 2012. Environmental information from stakeholders supporting product development. *Journal of Cleaner Production*, 31: 1–13.

Battagello, F.M., Cricelli, L., and Grimaldi, M. 2016. Benchmarking strategic resources and business performance via an open framework. *International Journal of Productivity and Performance Management*, 65(3): 324–350.

Battini, D., Delorme, X., Dolgui, A., Persona, A., and Sgarbossa, F. 2015. Ergonomics in assembly line balancing based on energy expenditure: A multi-objective model. *International Journal of Production Research*, 54(3): 824–845.

Belekoukias, I., Garza-Reyes, J.A., and Kumar, V. 2014. The impact of Lean methods and tools on the operational performance of manufacturing organisations. *International Journal of Production Research*, 52(18): 5346–5366.

Beyer, C. 2014. Strategic implications of current trends in additive manufacturing. *Journal of Manufacturing Science and Engineering*, 136: 1–8.

Böhme, T., Deakins, E., Pepper, M., and Towill, D. 2014. Systems engineering effective supply chain innovations. *International Journal of Production Research*, 52(21): 6518–6537.

Burchart-Korol, D. 2013. Life cycle assessment of steel production in Poland: A case study. *Journal of Cleaner Production*, 54: 235–243.

Chauhan, P.S. and Agrawal, C.M. 2014. Identification of manufacturing defects leading to rejection during manufacturing of stabiliser bar. *International Journal of Productivity and Quality Management*, 14(4): 408–422.

Chay, T., Xu, Y., Tiwari, A., and Chay, F. 2015. Towards lean transformation: The analysis of lean implementation frameworks. *Journal of Manufacturing Technology Management*, 26(7): 1031–1052.

Chiarini, A. 2012. Lean production: Mistakes and limitations of accounting systems inside the SME sector. *Journal of Manufacturing Technology Management*, 23(5): 681–700.

Chompu-inwai, B., Jaimjit, B., and Premsuriyanunt, P. 2015. A combination of material flow cost accounting and design of experiments techniques in an SME: The case of a wood products manufacturing company in northern Thailand. *Journal of Cleaner Production*, 108: 1352–1364.

Colledania, M., Rattia, A., and Senanayake, C. 2015. An approximate analytical method to evaluate the performance of multi-product assembly manufacturing systems. *Procedia CIRP*, 33: 357–363.

Czumanski, T. and Lödding, H. 2016. State-based analysis of labour productivity. *International Journal of Production Research*, 54(10): 2934–2950.

Dai, Q., Zhong, R., Huang, G.Q., Qu, T., Zhang,T., and Luo, T.Y. 2012. Radio frequency identification-enabled real-time manufacturing execution system: A case study in an automotive part manufacturer. *International Journal of Computer Integrated Manufacturing*, 25(1): 51–65.

Das, B., Ghosh, T., and Gangopadhyay, S. 2013. Child work in agriculture in West Bengal, India: Assessment of musculoskeletal disorders and occupational health problems. *Journal of Occupational Health*, 55(4): 244–258.

de Macedo Guimarães, L.B., Anzanello, M.J., Ribeiro, J.L.D., and Saurin, T.A. 2015. Participatory ergonomics intervention for improving human and production outcomes of a Brazilian furniture company. *International Journal of Industrial Ergonomics*, 49: 97–107.

Dinis-Carvalho, J., Moreira, F., Bragança, S., Costa, E., Alves, A., and Sousa, R. 2015. Waste identification diagrams. *Production Planning & Control*, 26(3): 235–247.

Doltsinis, S.C., Ratchev, S., and Lohse, N. 2013. A framework for performance measurement during production ramp-up of assembly stations. *European Journal of Operational Research*, 229: 85–94.

Dora, M., Kumar, M., and Gellynck, X. 2016. Determinants and barriers to lean implementation in food-processing SMEs-a multiple case analysis. *Production Planning and Control*, 27(1): 1–23.

Dora, M., Van Goubergen, D., Kumar, M., Molnar, A., and Gellynck, X. 2013. Application of lean practices in small and medium-sized food enterprises. *British Food Journal*, 116(1): 125–141.

Dushyanth, K.K.R., Shivashankar, G.S., and Rajeshwar, S.K. 2015. Application of value stream mapping in pump assembly process: A case study. *Industrial Engineering & Management*, 4(3): 1–11.

Elaswad, H., Islam, S., Tarmizi, S., Yassin, A., Lee, M.D., and Ting, C.H. 2015. Benchmarking of growth in manufacturing SMEs: A review. *Science International*, 27(3): 2039–2048.

Ene, S.A., Teodosiu, C., Robu, B., and Volf, I. 2013. Water footprint assessment in the winemaking industry: A case study for a Romanian medium size production plant. *Journal of Cleaner Production*, 43: 122–135.

Forno, A.J.J.D.D., Forcellini, F.A., Kipper, L.M., and Pereira, F.A. 2016. Method for evaluation via benchmarking of the lean product development process: Multiple case studies at Brazilian companies. *Benchmarking: An International Journal*, 23(4): 1–14.

Garza-Reyes, J.A., Oraifige, L., Soriano-Meier, H., Forrester, P.L., and Harmanto, D. 2012. The development of a lean park homes production process using process flow and simulation methods. *Journal of Manufacturing Technology Management*, 23(2): 178–197.

Ghosh, M. 2012. Lean manufacturing performance in Indian manufacturing plants. *Journal of Manufacturing Technology Management*, 24(1): 113–122.

Glock, C.H. and Grosse, E.H. 2015. Decision support models for production ramp-up: A systematic literature review. *International Journal of Production Research*, 53(21): 6637–6651.

Gnanaraj, S.M., Devadasan, S.R., Murugesh, R., and Sreenivasa, C.G. 2012. Sensitisation of SMEs towards the implementation of lean six sigma: An initialisation in a cylinder frames manufacturing Indian SME. *Production Planning & Control: The Management of Operations*, 23(8): 599–608.

Gould, O. and Colwill, J. 2015. A framework for material flow assessment in manufacturing systems. *Journal of Industrial and Production Engineering*, 32(1): 55–66.

Gupta, S., Acharya, P., and Patwardhan, M. 2013. A strategic and operational approach to assess the lean performance in radial tyre manufacturing in India: A case based study. *International Journal of Productivity and Performance Management*, 62(6): 634–651.

Gupta, S. and Jain, S.K. 2014. The 5S and kaizen concept for overall improvement of the organisation: A case study. *International Journal of Lean Enterprise Research*, 1(1): 22–40.

Gupta, S. and Jain, K.S. 2015. An application of 5S concept to organize the workplace at a scientific instruments manufacturing company. *International Journal of Lean Six Sigma*, 6(1): 73–88.

Harmse, J.L., Engelbrecht, J.C., and Bekker, J.L. 2016. The impact of physical and ergonomic hazards on poultry abattoir processing workers: A review. *International Journal of Environmental Research and Public Health*, 13(197): 1–24.

Hasle, P. 2014. Lean production an evaluation of the possibilities for an employee supportive lean practice. *Human Factors and Ergonomics in Manufacturing and Service Industries*, 24(1): 40–53.

Heidrich, O. and Tiwary, A. 2013. Environmental appraisal of green production systems: Challenges faced by small companies using life cycle assessment. *International Journal of Production Research*, 51(19): 5884–5896.

Jasch, C. 2015. Governmental initiatives: The UNIDO (United Nations Industrial Development Organization) TEST approach. *Journal of Cleaner Production*, 108: 1375–1377.

Jin, S. and Mirka, G.A. 2015. A systems-level perspective of the biomechanics of the trunk flexion-extension movement: Part 1—Normal low back condition. *International Journal of Industrial Ergonomics*, 46: 7–11.

Kaushik, P., Khanduja, D., Mittal, K., and Jaglan, P. 2012. A case study. *The TQM Journal*, 24(1): 4–16.

Kaydos, W. 1999. *Operational Performance Measurement: Increasing Total Productivity*. London: St. Lucie Press.

Khrais, S., Omar Al-Araidah, O., Aweisi, A.M., Fadia Elias, F., and Al-Ayyoub, E. 2013. Safety practices in Jordanian manufacturing enterprises within industrial estates. *International Journal of Injury Control and Safety Promotion*, 20(3): 227–238.

Klewitz, J. and Hansen, E.G. 2014. Sustainability-oriented innovation of SMEs: A systematic review. *Journal of Cleaner Production*, 65: 57–75.

Kumar, E.S. and Tiwari, E.A. 2015. A work study on minimize the defect in aluminium casting. *International Journal of Emerging Technologies in Engineering Research*, 3(1): 32–38.

Kumar, S., Heustis, D., and Graham, J.M. 2015. The future of traceability within the U.S. food industry supply chain: A business case. *International Journal of Productivity and Performance Management*, 64(1): 129–146.

Kuo, C. and Wang, M.J. 2012. Motion generation and virtual simulation in a digital environment. *International Journal of Production Research*, 50(22): 6519–6529.

Larsson, A. 2013. The accuracy of surgery time estimations. *Production Planning and Control*, 24(10–11): 891–902.

Low, S., Kamaruddin, S. and Azid, I.A. 2015. Improvement process selection framework for the formation of improvement solution alternatives. *International Journal of Productivity and Performance Management*, 64(5): 702–722.

Lyons, A.C., Vidamour, K., Jain, R., and Sutherland, M. 2013. Developing an understanding of lean thinking in process industries. *Production Planning & Control*, 24(6): 475–494.

Małachows, B. and Korytkowski, P. 2016. Competence-based performance model of multi-skilled workers. *Computers and Industrial Engineering*, 91: 165–177.

Manne, P., Ahmad, S. and Waterman, J. 2016. Design and development of mine railcar components. *American Journal of Mechanical Engineering*, 4(1): 32–41.

Marodin, G.A. and Saurin, T.A. 2015. Managing barriers to lean production implementation: Context matters. *International Journal of Production Research*, 53(13): 3947–3962.

Matawale, C.R., Datta, S., and Mahapatra, S.S. 2014. Leanness estimation procedural hierarchy using interval-valued fuzzy sets (IVFS). *Benchmarking: An International Journal*, 21(2): 150–183.

Mathur, A., Mittal, M.L., and Dangayach, G.S. 2012. Improving productivity in Indian SMEs. *Production Planning and Control: The Management of Operations*, 23(10–11): 754–768.

Matt, D.T. 2014. Adaptation of the value stream mapping approach to the design of lean engineer-to-order production systems. *Journal of Manufacturing Technology Management*, 25(3): 334–350.

Moon, T., Hur, W., Ko, S., Kim, J., and Yoo, D. 2016. Positive work-related identity as a mediator of the relationship between compassion at work and employee outcomes. *Human Factors and Ergonomics in Manufacturing & Service Industries*, 26(1): 84–94.

Muruganantham, V.R., Krishnan, P.N., and Arun, K.K. 2014. Integrated application of TRIZ with lean in the manufacturing process in machine shop for productivity improvement. *International Journal of Productivity and Quality Management*, 13(4): 414–429.

Naber, A. and Kolisch, R. 2014. MIP models for resource-constrained project scheduling with flexible resource profiles. *European Journal of Operational Research*, 239: 335–348.

Ohu, I.P.N., Cho, S., Kim, D.H., and Lee, G.H. 2016. Ergonomic analysis of mobile cart–assisted stocking activities using electromyography. *Human Factors and Ergonomics in Manufacturing & Service Industries*, 26(1): 40–51.

Olhager, J. 2013. Evolution of operations planning and control: From production to supply chains. *International Journal of Production Research*, 51(23–24): 6836–6843.

O'Neill, P., Sohal, A., and Teng, C.W. 2016. Quality management approaches and their impact on firms' financial performance: An Australian study. *International Journal of Production Economics*, 171: 381–393.

Ortolano, L., Sanchez-Triana, E., Afzal, J., Ali, C.L., and Rebellón, S.A. 2014. Cleaner production in Pakistan's leather and textile sectors. *Journal of Cleaner Production*, 68: 121–129.

Padhi, S.S., Wagner, S.M., Niranjan, T.T., and Aggarwal, V. 2013. A simulation-based methodology to analyse production line disruptions. *International Journal of Production Research*, 51(6): 1885–1897.

Pandey, A., Singh, M., Soni, N., and Pachorkar, P. 2014. Process layout on advance CNG cylinder manufacturing. *International Journal of Application or Innovation in Engineering & Management (IJAIEM)*, 3(12): 113–116.

Panwar, A., Nepal, B., Jain, R., and Yadav, O.P. 2013. Implementation of benchmarking concepts in Indian automobile industry: An empirical study. *Benchmarking: An International Journal*, 20(6): 777–804.

Parthanadee, P. and Buddhakulsomsiri, J. 2014. Production efficiency improvement in batch production system using value stream mapping and simulation: A case study of the roasted and ground coffee industry. *Production Planning & Control*, 25(5): 425–446.

Pedersen, P. and Slepniov, D. 2016. Management of the learning curve: A case of overseas production capacity expansion. *International Journal of Operations and Production Management*, 36(1): 42–60.

Petek, J., Glavič, P., and Kostevšek, A. 2016. Comprehensive approach to increase energy efficiency based on versatile industrial practices. *Journal of Cleaner Production*, 112: 2813–2821.

Poudelet, V., Chayer, J., Margni, M., Pellerin, R., and Samson, R. 2012. A process-based approach to operationalize life cycle assessment through the development of an eco-design decision-support system. *Journal of Cleaner Production*, 33: 192–201.

Prasad, K.D., Jha, S.K., and Prakash, A. 2015. Quality, productivity and business performance in home based brassware manufacturing units. *International Journal of Productivity and Performance Management*, 64(2): 270–287.

Prasad, S., Khanduja, D., and Sharma, S.K. 2016. An empirical study on applicability of lean and green practices in the foundry industry. *Journal of Manufacturing Technology Management*, 27(3): 408–426.

Prasanna, M. and Vinodh, S. 2013. Lean six sigma in SMEs: An exploration through literature review. *Journal of Engineering, Design and Technology*, 11(3): 224–250.

Prashar, A. 2014. Adoption of six sigma DMAIC to reduce cost of poor quality. *International Journal of Productivity and Performance Management*, 63(1): 103–126.

Psomas, E.L., Kafetzopoulos, D.P., and Fotopoulos, C.V. 2012. Developing and validating a measurement instrument of ISO 9001 effectiveness in food manufacturing SMEs. *Journal of Manufacturing Technology Management*, 24(1): 52–77.

Quintana, R. and Leung, M.T. 2012. A case study of Bayesian belief networks in industrial work process design based on utility expectation and operational performance. *International Journal of Productivity and Performance Management*, 61(7): 765–777.

Ray, S.J. and Teizer, J. 2012. Real-time construction worker posture analysis for ergonomics training. *Advanced Engineering Informatics*, 26: 439–455.

Sangwan, K.S., Bhamu, J., and Mehta, D. 2014. Development of lean manufacturing implementation drivers for Indian ceramic industry. *International Journal of Productivity and Performance Management*, 63(5): 569–587.

Sanjog, J., Patnaik, B., Patel, T., and Karmakar, S. 2016. Context-specific design interventions in blending workstation: An ergonomics perspective. *Journal of Industrial and Production Engineering*, 33(1): 32–50.

Santosa, M.S., Vidal, M.C.R., and Moreira, S.B. 2012. The RFad method: A new fatigue recovery time assessment for industrial activities. *Work*, 41: 1656–1663.

Schmidt, M. and Nakajima, M. 2013. Material flow cost accounting as an approach to improve resource efficiency in manufacturing companies. *Resources*, 2: 358–369.

Schulze, M., Nehler, H., Ottosson, M., and Thollander, P. 2016. Energy management in industry: A systematic review of previous findings and an integrative conceptual framework. *Journal of Cleaner Production*, 112: 3692–3708.

Seth, D. and Gupta, V. 2016. Application of value stream mapping for lean operations and cycle time reduction: An Indian case study. *Production Planning & Control*, 16(1): 44–59.

Sharma, R.K. and Sharma, R.G. 2014. Integrating six-sigma culture and TPM framework to improve manufacturing performance in SMEs. *Quality and Reliability Engineering International Journal*, 30: 745–765.

Sharma, S. and Shah, B. 2016. Towards lean warehouse: Transformation and assessment using RTD and ANP. *International Journal of Productivity and Performance Management*, 65(4): 571–599.

Silva, D.A.L., Delai, I., de Castro, M.A.S., and Ometto, A.R. 2013. Quality tools applied to cleaner production programs: A first approach toward a new methodology. *Journal of Cleaner Production*, 47: 174–187.

Srinivasan, S., Ikuma, L.H., Shakouri, M., Nahmens, I., and Harvey, C. 2016. 5S impact on safety climate of manufacturing workers. *Journal of Manufacturing Technology Management*, 27(3): 364–378.

Srinivasan, K., Muthu, S., Devadasan, S.R., and Sugumaran, C. 2016. Enhancement of sigma level in the manufacturing of furnace nozzle through DMAIC approach of Six Sigma: A case study. *Production Planning & Control: The Management of Operations*, 810–822.

Suárez-Barraza, M.F. and Ramis-Pujol, J. 2012. An exploratory study of 5S: A multiple case study of multinational organizations in Mexico. *Asian Journal on Quality*, 13(1): 77–99.

Suwittayaruk, P., Van Goubergen, D., and Lockhart, T.E. 2014. A preliminary study on pace rating using video technology. *Human Factors and Ergonomics in Manufacturing and Service Industries*, 24(6): 725–738.

Sylla, N., Bonnet, V., Colledani, F., and Fraisse, P. 2014. Ergonomic contribution of ABLE exoskeleton in automotive industry. *International Journal of Industrial Ergonomics*, 44: 475–481.

Thurner, W. and Roud, V. 2016. Greening strategies in Russia's manufacturing: From compliance to opportunity. *Journal of Cleaner Production*, 112: 2851–2860.

Tortorella, G., Viana, S., and Fettermann, D. 2015. Learning cycles and focus groups: A complementary approach to the A3 thinking methodology. *The Learning Organization*, 22(4): 229–240.

Trianni, A., Cagno, E., and Farné, S. 2016. Barriers, drivers and decision-making process for industrial energy efficiency: A broad study among manufacturing small and medium-sized enterprises. *Applied Energy*, 162: 1537–1551.

Tyagi, S., Choudhary, A., Cai, X., and Yang, K. 2015. Value stream mapping to reduce the lead-time of a product development process. *International Journal of Production Economics*, 160: 202–212.

Vidyadhar, R., Kumar, R. S., Vinodh, S., and Antony, J. 2016. Application of fuzzy logic for leanness assessment in SMEs: A case study. *Journal of Engineering, Design and Technology*, 14(1): 78–103.

Vimal, K.E.K. and Vinodh, S. 2013. Application of artificial neural network for fuzzy logic based leanness assessment. *Journal of Manufacturing Technology Management*, 24(2): 274–292.

Vinodh, S. and Joy, D. 2012. Structural equation modelling of lean manufacturing practices. *International Journal of Production Research*, 50(6): 1598–1607.

Vinodh, S., Selvaraj, T., Chintha, S.K., and Vimal, K.E.K. 2015. Development of value stream map for an Indian automotive components manufacturing organization. *Journal of Engineering, Design and Technology*, 13(3): 380–399.

Vinodh, S., Vasanth Kumar, S.V., and Vimal, K.E.K. 2014. Implementing lean sigma in an Indian rotary switches manufacturing organisation. *Production Planning & Control*, 25(4): 288–302.

Wajanawichakon, K. and Srimitee, C. 2012. ECRS's principles for a drinking water production plant. *IOSR Journal of Engineering*, 2(5): 956–960.

Wlazlak, P.G. and Johansson, G. 2014. R&D in Sweden and manufacturing in China: A study of communication challenges. *Journal of Manufacturing Technology Management*, 25(2): 258–278.

Wu, S., Wang, Y., BolaBola, J.Z., Qin, H., Ding, W., Wen, W., and Niu, J. 2016. Incorporating motion analysis technology into modular arrangement of predetermined time standard (MODAPTS). *International Journal of Industrial Ergonomics*, 53: 291–298.

Xia, P.J., Lopes, M., and Restivo, M.T. 2013. A review of virtual reality and haptics for product assembly (part1): Rigid parts. *Assembly Automation*, 33(1): 68–77.

Yuan, C. Zhai, Q., and Dornfeld, D. 2012. A three dimensional system approach for environmentally sustainable manufacturing. *CIRP Annals: Manufacturing Technology*, 61: 39–42.

Chapter 9

Impact of Work Study on Technological Capital for Productivity in Manufacturing SMEs

9.1 Introduction

Chapter 9 focuses on the impact of work study (WS) on technological capital for productivity improvement in manufacturing small and medium enterprises (SMEs). A diagram for productivity improvement in manufacturing SMEs through the use of WS techniques is shown in Figure 9.1.

The structure of technological capital is split into tangible assets and intangible assets, which are measured through WS to improve productivity in manufacturing SMEs. The aim is to determine whether there is a relationship between WS and technological capital for the productivity of manufacturing SMEs. Tangible assets include equipment such as hardware, trackers, and scanners, whereas intangible assets comprise software, technological tools, systems, innovation, networking, and technological skills.

9.2 Work Study and Tangible Assets in Relation to the Productivity of Manufacturing SMEs

This research focuses on WS and tangible assets in relation to the productivity of manufacturing SMEs. As pointed out by Panizzolo et al. (2012:776),

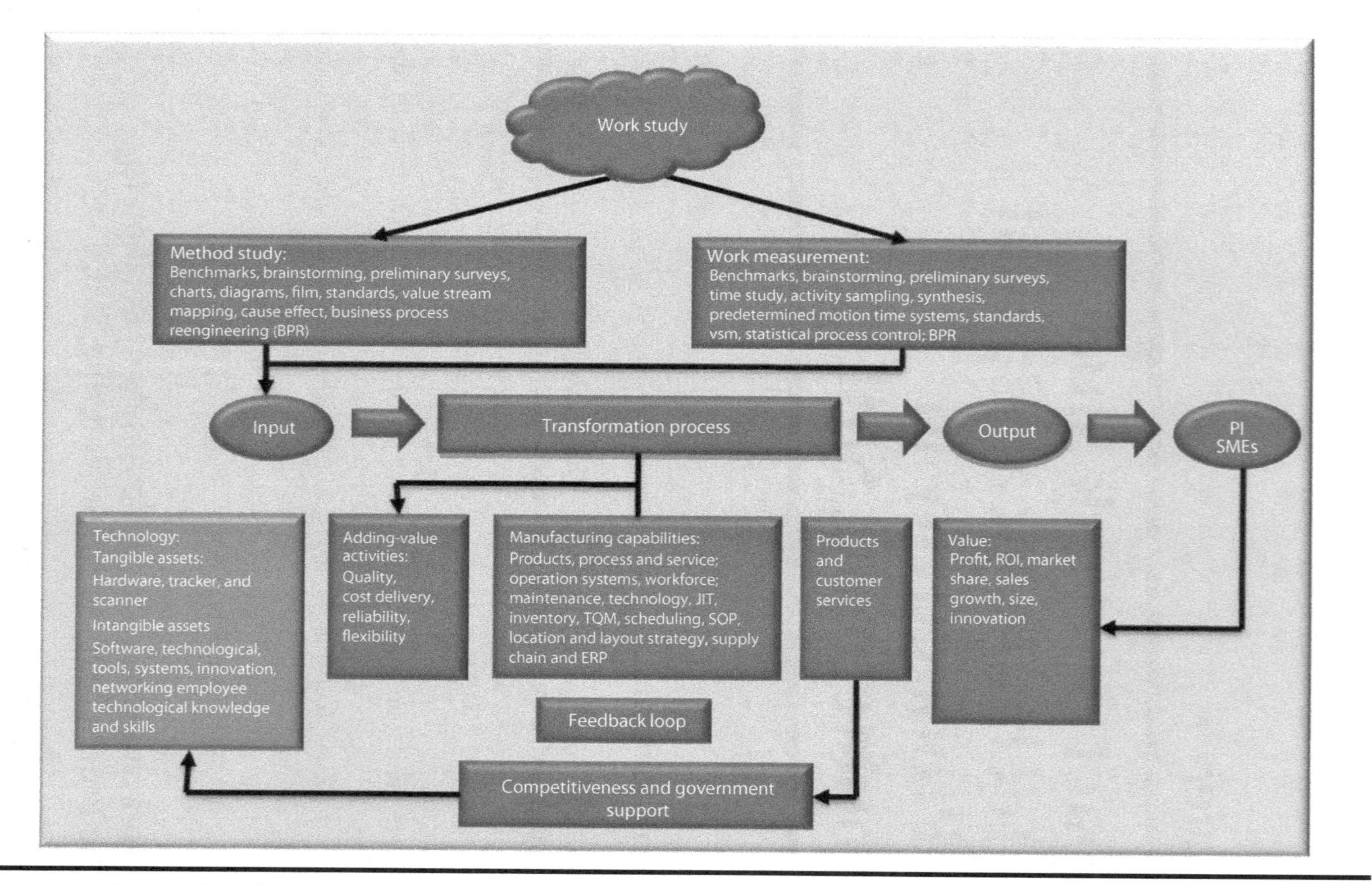

Figure 9.1 Productivity improvement in manufacturing SMEs: Application of work study techniques. (From Author, 2017.)

electronic equipment such as barcodes plays an important role before, during, and after the manufacturing process for the purpose of monitoring and control. Monitoring and control of the manufacturing process through barcodes results in productivity improvement in manufacturing SMEs.

Gupta, Acharya, and Patwardhan (2013:638–44) investigate how WS is exercised in terms of technology to ensure productivity improvement in manufacturing SMEs. Findings from the literature show that the simplification of the job being carried out in terms of the flow of information can be facilitated by ensuring that tangible assets such as computers are in place and intangible assets such as software, systems, and employee innovation, skills, and expertise are used adequately. By improving the effective use of tangible assets and intangible assets, the productivity of manufacturing SMEs is accelerated. As found by Al-Somali, Gholami, and Clegg (2015:10), face-to-face and telephone interviews are found to be a useful tool in research for customer care, customer loyalty, information exchange, and servicing a niche market. By following such an interview procedure, manufacturing SMEs can increase their productivity and become competitive in the global marketplace.

Wu, Yu, and Wu (2012:164) and Esan et al. (2013:265–70) explore hardware in relation to the productivity of manufacturing SMEs. Based on the literature reviewed, the research reports that computer integration with software facilitates the control of communication between management, employees, and customers. As a result, economic efficiency is encouraged and productivity levels are raised. Despite the extensive consumption of energy among manufacturing SMEs, the literature studied reports that the recording of energy bills using energy-saving equipment, investment in new energy equipment, energy tracking through meters (equipment), energy tracking for plant, and energy tracking for equipment results in improvements in efficiency. By improving efficiency through these energy-saving technologies, productivity progress prevails in the business of manufacturing SMEs (Kostka, Moslener, & Andreas, 2013:59–61; Chaplin, Heap, & O'Rourke, 2016:130).

9.3 Work Study and Intangible Assets in Relation to the Productivity of Manufacturing SMEs

This research focuses on WS and intangible assets in relation to the productivity of manufacturing SMEs.

9.3.1 Software

As described by Wu et al. (2012:164); Gupta et al. (2013:638–44); Guo and Wang (2014:258); and José Rodríguez-Gutiérrez, Moreno, and Tejada (2015:204), technological capital in manufacturing SMEs, such as software, facilitates information flow between management and employees for customer satisfaction. As a result, easy control of information improves business performance and, in turn, improves productivity within manufacturing SMEs. As stated by Heidrich and Tiwary (2013:5884–6) and Cuerva, Triguero-Cano, and Córcoles (2014:105–8), technologies such as material recycling play an important role in influencing the productivity of manufacturing SMEs. Based on the literature studied, recycling of material as a tangible element reduces environmental impact and maximizes resource efficiency for cost and waste reduction in manufacturing SMEs.

9.3.2 Technological Tools

Cantamessa, Montagna, and Neirotti (2012:209–10) report that a tool such as information and communications technologies (ICT) ensures that the product lifecycle management system supports the management of information for new product design and development, which influences the productivity of manufacturing SMEs. Findings from the literature studied show that the absence of ICT in most manufacturing SMEs hinders their productivity. As a result, these SMEs are not able to compete in the market either locally or worldwide.

May, Stahl, and Taisch (2016:633) advise that the use of ICT ensures monitoring and energy consumption patterns that support operational management in making informed decisions on the use per product. This monitoring and identification of energy consumption patterns through the use of sufficient information reduces costs and contributes to the productivity of manufacturing SMEs.

9.3.3 Technological Systems

As noted by Panizzolo et al. (2012:769–76), associating technological capital, involving tools such as just-in-time (JIT) and total quality control (TQC), with manufacturing SMEs' processes influences material flow and boost productivity. As indicated in the literature and addressed by Gupta et al. (2013:638–44), WS contributes to the effective use of both office and manufacturing systems in terms of JIT, product and service design, maintenance, supply chain, planning and scheduling, inventory management, workplace and layout, and employees' ability to accelerate the productivity of manufacturing SMEs.

Poba-Nzaou, Raymond, and Fabi (2014:495) advise that the use of automated administrative systems and the standards compliance of these processes eliminate errors in manufacturing SMEs. Elimination of errors using automated administrative processes and adhering to standards can only work if barcodes are used for tracking. As result, tracking drives the productivity of manufacturing SMEs. Despite the fact that the productivity of manufacturing SMEs is improved through sales, turnover, and employment by using input manufacturing factors, if equipment is not used efficiently, serious backlogs result in decreased productivity (Guo & Wang, 2014:258; José Rodríguez-Gutiérrez et al., 2015:204).

As detected by Poba-Nzaou, Raymond, and Fabi (2014:495–7) and Huang and Handfield (2015:3), technological capital, involving aligning a tool such as enterprise resource planning (ERP) with manufacturing SMEs' processes, influences operational costs and in turn the productivity of manufacturing SMEs. Based on the literature studied, the research reports that the absence of ERP systems delays material flow, cost estimation, unawareness of material shortage, and lack of integration on sales; manufacturing; financing; quality; procurement and logistics; and engineering and information communication technology for material to be ordered for the customer. As a result, productivity of manufacturing SMEs is delayed.

Kumar, Heustis, and Graham (2015:131–3) advise that technological capital systems such as the radio frequency identification (RFID) technological system reduce stock-outs by tracking, managing inventory, avoiding shrinkage, correcting in-stock, identifying stock, and authentication through identification of the person responsible for the stock level in the warehouse and the process for the final product. Furthermore, this system also deals with traceability on processes, packaging, and stock kept in the warehouse through the use of barcode. This system aims to control product that has a short life span in order to avoid waste and damages that are costly for the manufacturing SMEs. Therefore, using RFID systems to avoid waste and damage positively contributes to the productivity of manufacturing SMEs.

Ball (2015:414–9) declares that, by adding value to technology, manufacturing SMEs follow the waste management hierarchy for categorizing waste-management strategies, which contain a number of steps in sequence such as prevent, reduce, reuse, recycle, recover (energy), and dispose. This categorization is referred to as material, product, and packing in preparation for the customer. On the other hand, waste-management hierarchy systems in terms of energy include prevent, reduce, reuse, and dispose steps. This hierarchy can only function if a JIT tool is used to reduce costs, improve the quality of the product, and ensure quick delivery resulting in productivity improvement of manufacturing SMEs.

9.3.4 Innovation

Gao et al. (2011:436–8) investigate how technological capital such as innovation influences productivity on the assembly line in manufacturing SMEs. The result of the literature studied shows that when there is a lack of innovation, such as the use of an unknowledgeable and unskilled cheap workforce as well as the unavailability of networks, high costs are incurred and ultimately the productivity of manufacturing SMEs is affected.

Furthermore, Guo and Wang (2014:262–3) state that competitiveness is critical to the productivity of manufacturing SMEs. Manufacturing SMEs, by participating in international markets via export, can access potential sources for innovation. As a result, this type of innovation helps these SMEs to acquire knowledge that is more external, which enables their businesses to sustain their productivity. These manufacturing SMEs become innovative by benchmarking with other innovative manufacturing SMEs, hiring employees with experience and training coming from innovative firms, and collaborating with universities on research and development and/or with other research agents and businesses or university exhibitions.

In terms of intangible assets such as knowledge and exposure to innovation, costs are reduced by manufacturing SMEs and contribute to the productivity of their businesses. So, the absence of new knowledge hinders productivity among manufacturing SMEs (Guo & Wang, 2014:263; José Rodríguez-Gutiérrez et al., 2015:196).

9.3.5 Networking

As pointed out by Panizzolo et al. (2012:776), networks such as electronic data interchange, internet, extranet, intranet, websites, barcoding, and fax are used to ensure productivity enhancement in manufacturing SMEs. Panizzolo et al. (2012:776) emphasize the importance of teamwork through networking for decision making in manufacturing SMEs. The literature shows that working as a team in manufacturing SMEs, from top to bottom, improves the degree of integration within the business as well as that between the buyer and the supplier, encourages innovation through brainstorming, and cuts down manufacturing costs by examining the manufacturing operations, all of which influence the productivity of manufacturing SMEs.

One of the components of innovation influencing growth in manufacturing SMEs is networking. When knowledge is shared among the members of these SMEs as well as stakeholders in the growth of manufacturing SMEs, these SMEs manage to sustain their productivity (Brink, 2015:264).

May, Stahl, and Taisch (2016:633–6) suggest that knowledge transfer plays a pivotal role in the manufacturing process of manufacturing SMEs. The transfer of knowledge through networking among the members of the manufacturing SMEs becomes critical to the productivity of their businesses.

9.3.6 Technological Skills

Cantamessa et al. (2012:209–10) identify technological capital factors such as education in information technology (IT). Findings from the literature studied show that employees exposed to IT skills make provision for the productivity of manufacturing SMEs. As a result, the productivity of manufacturing SMEs continues to improve.

According to Wu et al. (2012:164), technological skills in manufacturing SMEs influence the productivity of their businesses. When technical knowledge or managerial skills are available, manufacturing SMEs add value to their productivity.

Kostka et al. (2013:59–61) endorse that training in technical know-how and skills in energy efficiency is essential to employees of manufacturing SMEs. However, manufacturing SMEs that have acquired specialized knowledge in energy efficiency add more value to the productivity of their businesses than those that do not.

Al-Somali et al. (2015:9–10) examine the importance of the internet in the productivity of manufacturing SMEs. The research indicates that the internet is a useful network tool that manufacturing SMEs can use to excel in improving their productivity through creating a bulletin board for brochures, employee telephone directories, and other documents such as catalogues and price lists. This is a simple way that manufacturing SMEs can facilitate the flow of information to customers. May et al. (2016:633–6) advise that technological skills play an important role in the manufacturing process of manufacturing SMEs. These skills are embedded by training employees to prepare them to be engaged in achieving the results of manufacturing SMEs by contributing to productivity.

9.4 Case Study Findings and Analysis for South African SMEs

This section examines the results of various case studies focused on Companies A, B, C, and D in Gauteng, South Africa. This research focused on finding productivity problems facing Companies A, B, C, and D attributed to manufacturing capabilities and on adding-value activities applied in terms of technological factors and capital factors. The interventions necessary to resolve these problems are then introduced, with the intention of improving productivity in the companies addressed in the case studies.

During the manufacturing process of all the companies targeted in this study, in terms of technological capital, tangible assets such as computers, telephone, and printers are used in order to facilitate information flow from management at the top toward employees and between management for communication. The computer was used for recording information regarding inventory ordered, sales, orders, and pricing. The telephone was used as a communication tool for ordering and expediting. Finally, the printer was used to print meeting rosters, job cards, flow charts, production reports based on achievement, and standard working procedures. With regard to intangible assets, employees were free to use their own creativity based on the on-the-job training offered by the company. Secondly, emails were used to communicate with suppliers, management, and customers in the company. Only Companies A, B, and D used the scanner to track inventory for inventory management.

9.5 Comparative Analysis and Discussion

Based on the case study results discussed and analyzed in terms of technological capital factors used in South African manufacturing SMEs such as Companies A, B, C, and D, a comparative analysis and discussion of challenges facing these companies is shown in Table 9.1 in order to highlight discrepancies between these companies. Furthermore, a comparative analysis of the application of WS is also made based on these companies to determine differences on the manner in which they contribute to the productivity of their businesses; see the outline of Table 9.1.

Based on the background of these companies, Companies A, B, and C are exposed to a number of WS tools to ensure productivity progress. These

Table 9.1 Comparative Analysis and Discussion of Challenges Facing Companies A, B, C, and D

Background of Manufacturing	*Company A*	*Company B*	*Company C*	*Company D*
Challenges facing manufacturing capabilities based on input resource factors utilized in companies	**Technology:** **Tangible:** Computers, telephone, printer, scanner **Intangible:** Innovation and creative skills Mass and batch systems	**Technology:** **Tangible:** Computers, telephone, printer, scanner **Intangible:** Integrated assembly system	**Technology:** **Tangible:** Computers, telephone, printer **Intangible:** Batch system on packaging	**Technology:** **Tangible:** Computers, telephone, printer, scanner **Intangible:** Research and development for innovation assembly system
WS tools available in these companies	Standards such as Quality ISO 9001 certified	Quality plan SABS certified standards	Quality plan neither SABS nor ISO 9001 certified standards	Quality ISO 9001 certified standards
WS: Method Study	Benchmark with Productivity SA, brainstorming, work flow charts, flow diagrams, cause-and-effect charts, filming techniques	Benchmark with Productivity SA, brainstorming, work flow charts, flow diagrams, cause-and-effect charts, filming techniques	Benchmark with other companies, work flow charts	Benchmark with other companies, brainstorming, work flow charts, flow diagrams, cause-and-effect charts, filming techniques
WS: Work measurement	Measure time from start to end for cycle time	Use WS expert for measuring time using time study	N/A	Measuring of time using computer recording

tools are exercised through method study and involve benchmarking, brainstorming, work flow charts, flow diagrams, cause-and-effect diagrams, and filming techniques, whereas with work measurement, these companies focus on measuring time. The difference between Companies A, B, and C is that Companies A and C have their supervisors conduct their own time measurement, whereas Company B makes use of the WS intern to conduct a time study for the job being carried out by employees.

As indicated in Table 9.1, Companies A, B, C, and D used WS tools such as method study and work measurement mainly on technological capital factors such as computers, telephones, and printers for productivity improvement in their businesses.

9.6 Summary

Chapter 9 focused on the impact of WS on technological capital for productivity improvement in manufacturing SMEs as well as addressing productivity challenges facing companies in Gauteng, South Africa. WS was used to measure technological capital factors such as tangible assets and intangible assets for productivity improvement in manufacturing SMEs. The aim is to determine whether there is a relationship between WS and technological capital for productivity of manufacturing SMEs. Tangible assets included equipment such as hardware, trackers, and scanners, whereas intangible assets consisted of software, technological tools, systems, innovation, networking, and technological skills.

The impact of WS into management for productivity in manufacturing SMEs is discussed in the next chapter to be presented.

References

Al-Somali, S.A., Gholami, R., and Clegg, B. 2015. A stage-oriented model (SOM) for ecommerce adoption: A study of Saudi Arabian organisations. *Journal of Manufacturing Technology Management*, 26(1): 2–35.

Akinboade, O.A. 2015. Determinants of SMEs growth and performance in Cameroon's central and littoral provinces' manufacturing and retail sectors. *African Journal of Economic and Management Studies*, 6(2): 183–196.

Akroush, M.N. 2012. Organisational capabilities and new product performance: The role of new product competitive advantage. *Competitive Review: An International Business Journal*, 22(4): 343–365.

Asrofah, T., Zailani, S., and Fernando, Y. 2010. Best practices for the effectiveness of benchmarking in the Indonesian manufacturing companies. *Benchmarking: An International Journal*, 17(1): 115–143.

Ball, P. 2015. Low energy production impact on lean flow. *Journal of Manufacturing Technology Management*, 26(3): 412–428.

Barnes, D. and Hinton, C.M. 2012. Reconceptualising e-business performance measurement using an innovation adoption framework. *International Journal of Productivity and Performance Management*, 61(5): 502–517.

Benkraiem, R. and Gurau, C. 2013. How do corporate characteristics affect capital structure decisions of French SMEs? *International Journal of Entrepreneurial Behaviour and Research*, 19(2): 149–164.

Brennan, L., Ferdows, K., Godsell, J., Ruggero, G., Keegan, R., Kinkel, S., Srai, S., and Taylor, M. 2015. Manufacturing in the world: Where next? *International Journal of Operations and Production Management*, 35(9): 1–35.

Brink, T. 2015. Passion and compassion represent dualities for growth. *International Journal of Organizational Analysis*, 23(1): 41–60.

Cantamessa, M., Montagna, F., and Neirotti, P. 2012. Understanding the organizational impact of PLM systems: Evidence from an aerospace company. *International Journal of Operations and Production Management*, 32(2): 191–215.

Chaplin, L., Heap, J., and O'Rourke, M.T.J. 2016. Could "Lean Lite" be the cost effective solution to applying lean manufacturing in developing economies? *International Journal of Productivity and Performance Management*, 65(1): 126–136.

Cuerva, M.C., Triguero-Cano, A., and Córcoles, D. 2014. Drivers of green and non-green innovation: Empirical evidence in Low-Tech SMEs. *Journal of Cleaner Production*, 68: 104–113.

Damoa, O.B.O. 2013. Strategic factors and firm performance in an emerging economy. *African Journal of Economic and Management Studies*, 4 (2): 267–287.

Esan, A.O., Khan, M.K., Qi, H.S., and Craig Naylor, C. 2013. Integrated manufacturing strategy for deployment of CADCAM methodology in a SMME. *Journal of Manufacturing Technology Management*, 24(2): 257–273.

Gao, J., Yao, Y., Zhu, V.C.Y., Sun, L., and Lin, L. 2011. Service-oriented manufacturing: A new product pattern and manufacturing paradigm. *Journal of Intelligent Manufacturing*, 22: 435–446.

Granly, B.M. and Welo, T. 2014. EMS and sustainability: Experiences with ISO 14001 and Eco-Lighthouse in Norwegian metal processing SMEs. *Journal of Cleaner Production*, 64: 194–204.

Grigoriev, S.N., Yeleneva, J.Y., Golovenchenko, A.A., and Andreev, V.N. 2014. Technological capital: A criterion of innovative development and an object of transfer in the modern economy. *Procedia CIRP*, 20: 56–61.

Guo, B. and Wang, Y. 2014. Environmental turbulence, absorptive capacity and external knowledge search among Chinese SMEs. *Chinese Management Studies*, 8(2): 258–272.

Gupta, S., Acharya, P., and Patwardhan, M. 2013. A strategic and operational approach to assess the lean performance in radial tyre manufacturing in India: A case based study. *International Journal of Productivity and Performance Management*, 62(6): 634–651.

Heidrich, O. and Tiwary, A. 2013. Environmental appraisal of green production systems: Challenges faced by small companies using life cycle assessment. *International Journal of Production Research*, 51(19): 5884–5896.

Huang, Y. and Handfield, R.B. 2015. Measuring the benefits of ERP on supply management maturity model: A "big data" method. *International Journal of Operations and Production Management*, 35(1): 2–25.

Jain, S.K. and Ahuja, I.S. 2012. An evaluation of ISO 9000 initiatives in Indian industry for enhanced manufacturing performance. *International Journal of Productivity and Performance Management*, 61(7): 778–804.

Jardon, C.M. and Martos, M.S. 2012. Intellectual capital as competitive advantage in emerging clusters in Latin America. *Journal of Intellectual Capital*, 13(4): 462–481.

Johnson, M. and Templar, S. 2011. The relationships between supply chain and firm performance: The development and testing of unified proxy. *International Journal of Physical Distribution and Logistics Management*, 41(2): 88–103.

José Rodríguez-Gutiérrez, M.J., Moreno, P., and Tejada, P. 2015. Entrepreneurial orientation and performance of SMEs in the services SME. *Journal of Organizational Change Management*, 28(2): 194–212.

Jun, S., Seo, J.H., and Son, J. 2013. A study of the SME technology road mapping program to strengthen the R&D planning capability of Korean SMEs. *Technological Forecasting and Social Change*, 80: 1002–1014.

Kostka, G., Moslener, U., and Andreas, J. 2013. Barriers to increasing energy efficiency: Evidence from small- and medium-sized enterprises in China. *Journal of Cleaner Production*, 57: 59–68

Kumar, S., Heustis, D., and Graham, J.M. 2015. The future of traceability within the U.S. food industry supply chain: A business case. *International Journal of Productivity and Performance Management*, 64(1): 129–146.

Lourens, A.S. and Jonker, J.A. 2013. An integrated approach for developing a technology strategy framework for small-to-medium sized furniture manufacturers to improve competitiveness. *South African Journal of Industrial Engineering*, 23(2): 144–153.

Mandal, S.K. and Madheswaran, S. 2011. Energy use efficiency of Indian cement companies: A data envelopment analysis. *Energy Efficiency*, 4: 57–73.

Manderson, E. and Kneller, R. 2012. Environmental regulations, outward FDI and heterogeneous firms: Are countries used as pollution havens? *Environmental Resource Economics*, 51: 17–352.

May, G., Stahl, B., and Taisch, M. 2016. Energy management in manufacturing: Toward eco-factories of the future—A focus group study. *Applied Energy*, 164: 628–638.

McDermott, C.M. and Prajogo, D.I. 2012. Service innovation and performance in SMEs. *International Journal of Operations and Production Management*, 32(2): 216–237.

Mensah, M.S.B. 2012. Access to market of a manufacturing small business sector in Ghana. *International Journal of Business and Management*, 7(12): 36–46.

Mitchelmore, S., Rowley, J., and Shiu, E. 2014. Competencies associated with growth of women-led SMEs. *Journal of Small Business and Enterprise Development*, 21(4): 588–601.

Naude, W. and Matthee, M. 2011. The impact of transport costs on new venture internalisation. *Journal of International Entrepreneurship*, 9: 62–89.

Oliveira, T., Thomas, M., and Espadanal, M. 2014. Assessing the determinants of cloud computing adoption: An analysis of the manufacturing and services sectors. *Information and Management*, 51: 497–510.

Panizzolo, R., Garengo, P., Sharma, M.K., and Gore, A. 2012. Lean manufacturing in developing countries: Evidence from Indian SMEs. *Production Planning and Control*, 23(10–11): 769–788.

Poba-Nzaou, P., Raymond, L., and Fabi, B. 2014. Risk of adopting mission-critical OSS applications: An interpretive case study. *International Journal of Operations and Production Management*, 34(4): 477–512.

Radziwon, A., Bilberg, A., Bogers, M., and Madsen, E.S. 2014. The smart factory: Exploring adaptive and flexible manufacturing solutions. *Procedia Engineering*, 69: 1184–1190.

Ramdass, K. and Pretorius, L. 2011. Implementation of modular manufacturing in the clothing industry in Kwazulu Natal: A case study. *South African Journal of Industrial Engineering*, 22(1): 167–181.

Ren, L., Zhang, L., Tao, F., Zhao, C., Chai, X., and Zhao, X. 2015. Cloud manufacturing: From concept to practice. *Enterprise Information Systems*, 9(2): 186–209.

Saxer, M., de Beer, N., and Dimitrov, D.M. 2012. High speed 5-axis machining for tools applications. *South African Journal of Industrial Engineering*, 23(2): 144–153.

Sethi, S.P., Veral, E.A., Shapiro, H.J., and Emelainova, O. 2011. Mattel, Inc.: Global manufacturing principles (GMP)—A life-cycle analysis of a company-based code of conduct in the toy industry. *Journal of Business Ethics*, 99: 483–517.

Sharma, C. and Sehgal, S. 2010. Impact of infrastructure on output, productivity and efficiency: Evidence from the Indian manufacturing industry. *Indian Growth and Development Review*, 3(2): 100–121.

Singh, D., Oberoi, J.S., and Ahuja, I.S. 2013. An empirical investigation of dynamic capabilities in managing strategic flexibility in manufacturing organizations. *Management Decisions*, 51(7): 1442–1461.

Tan, C.S.L., Smyrnios, K.X., and Xiong, L. 2014. What drives learning orientation in fast growth SMEs? *International Journal of Entrepreneurial Behavior and Research*, 20(4): 324–350.

Tarí, J.J., Molina-Azorín, J.F., and Heras, I. 2012. Benefits of the ISO 9001 and ISO 14001 standards: A literature review. *Journal of Industrial Engineering and Management*, 5(2): 297–322.

Teeratansirikool, L., Siengthai, S., Badir, Y., and Charoenngam, C. 2013. Competitive strategies and firm performance: The mediating role of performance measurement. *International Journal of Productivity and Performance Management*, 62(2): 168–184.

Vuong, Q.H. and Napier, N.K. 2014. Resource cure or destructive creation in transition: Evidence from Vietnam's corporate sector. *Management Research Review*, 37(7): 642–657.
Wu, W., Yu, B., and Wu, C. 2012. How China's equipment manufacturing firms achieve successful independent innovation. *Chinese Management Studies*, 6(1): 160–183.
Yazdanfar, D. and Öhman, P. 2015. The growth–profitability nexus among Swedish SMEs. *International Journal of Managerial Finance*, 11(4): 1–26.

Chapter 10

The Impact of Work Study on Management in Relation to Productivity of Manufacturing SMEs

10.1 Introduction

In this chapter, the impact of work study (WS) into management on the productivity of manufacturing small and medium enterprises (SMEs) is studied and addressed with the aim of revealing the relationship between the two concepts. The research indicates that the relationship of these concepts can be divided into *method study* and *work measurement.* For WS to be effective in term of method study in the manufacturing process, WS experts should encourage management to use method study tools such as benchmarking, brainstorming, preliminary surveys, charts, flow diagrams, filming techniques, value stream mapping (VSM), cause-and-effect diagrams, and business process reengineering (BPR). Furthermore, work measurement techniques should also be considered, such as benchmarking, brainstorming, preliminary surveys, time study, activity sampling, synthesis, PMTS, standards, VSM, statistical process control (SPC), and BPR. When method study and work measurement are encouraged, manufacturing capabilities exercised by operation managers, including management, on the technical side will effectively contribute to the productivity of manufacturing SMEs. These techniques are indicated in Figure 10.1.

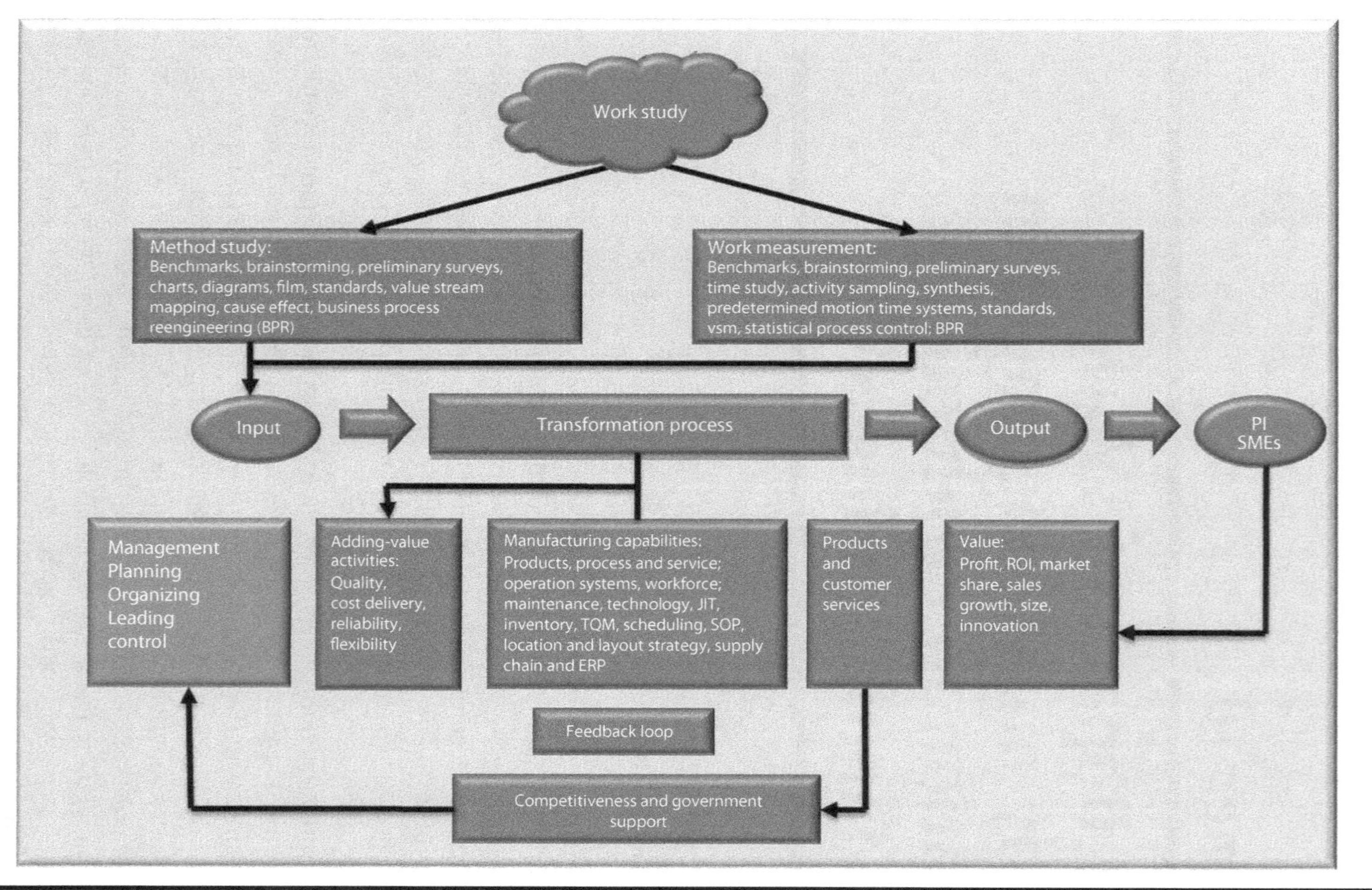

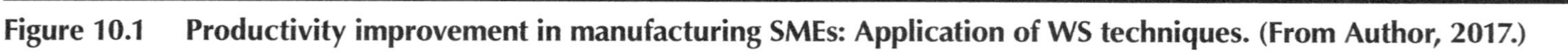
Figure 10.1 Productivity improvement in manufacturing SMEs: Application of WS techniques. (From Author, 2017.)

10.2 Method Study on Management in Relation to Productivity of Manufacturing SMEs

In this section, method study on management in relation to the productivity of manufacturing SMEs is studied and discussed on the basis of various literature sources. The purpose was to provide an understanding of how method study influences management on the productivity of manufacturing SMEs. Various method study techniques are used that impact on management to ensure that the productivity of manufacturing SMEs is influenced.

10.2.1 Benchmarking

Based on the literature, WS plays an important role in management operational planning in relation to the productivity of manufacturing SMEs. The outcome of results in the literature shows that the use of benchmarking by management through their work plan, as guided by WS specialists, influences manufacturing capability areas such as just-in-time (JIT) and existing technology to reduce lead time and inventory and eliminate bottlenecks in manufacturing SMEs. As a result, the productivity of manufacturing SMEs continues to improve (Saboo et al., 2014:42–6).

As found by Garza-Reyes, Ates, and Kumar (2015:1097–105), benchmarking plays a vital role among managers in ensuring that they interact with other sections in manufacturing or other organizations of similar status in order to be competitive locally and globally in terms of productivity. Findings from the literature studied show that benchmarking is based on planning, delegation, motivation, and control to continuously improve the productivity of manufacturing SMEs.

Sanjog et al. (2016:33–47) investigate benchmarking by management by comparing one main production process to the other and focusing on employee capabilities, inventory, machine utilization, location, layout, and technology within the manufacturing SME. This research finds that when benchmarking is encouraged by management in terms of the appropriate use of the input resource factors indicated earlier, as compared to the success of other production process sections, the productivity of manufacturing SMEs tends to improve.

10.2.2 Brainstorming Used in Method Study for Employees by Management in Manufacturing SMEs

Mathur, Mittal, and Dangayach (2012:756–7) and Dora, Kumar, and Gellynck (2016:11) investigate brainstorming through method study on the behavior of management in relation to productivity in manufacturing SMEs. Findings from the literature reviewed show that the influence of WS experts on management brainstorming helps management to be creative in terms of the utilization of the cause-and-effect diagram in order to ensure that employee working methods, material specification, machinery capacity, methods used, procedures followed, and environment improve employee performance and, in turn, advance productivity in manufacturing SMEs.

Gupta, Acharya, and Patwardhan (2013:640); Gupta and Jain (2014:23–27); and Chompu-inwai, Jaimjit, and Premsuriyanunt (2015:1354) investigate WS expert encouragement of brainstorming by management in relation to the productivity of manufacturing SMEs. Based on the results of the literature reviewed, management commitment toward brainstorming creates a positive relationship in engaging the effort to achieve good performance through control of inventory and elimination of defects in manufacturing SMEs. As a result, productivity improvement is projected.

Sharma and Shah (2016:582–99) assess brainstorming in management attitudes to the productivity of manufacturing SMEs. Findings from the literature reviewed report that brainstorming in manufacturing SMEs takes place during meetings or seminars on problem issues requiring clarification and knowledge sharing and discussing problems with employee abilities, with ideas coming from employees, supervisors, and top management working as a team. Training is assured and provided to employees by managers after the brainstorming session to improve coordination, motivation, and trust among employees with the intention of ensuring that they are dedicated to accomplishing results that will enhance productivity.

10.2.3 Preliminary Survey

Gupta et al. (2013:634) investigate how WS experts encourage the management of manufacturing SMEs to hold interviews with employees in the interests of improving productivity. The research recommends that commitment by management to the use of method study surveys carried out by WS experts, such as interview practices on the background of work activities, allows WS experts to easily notice problem areas affecting manufacturing SMEs and to guide management as to where productivity can be improved.

Pandey et al. (2014:114–5) explore the importance of the use of observation by WS experts in manufacturing SMEs. Based on the literature studied, the research shows that when method study tools, such as observation of problem areas that do not add value to the manufacturing operations, are made known through commitment from management, management is able to anticipate the risks represented by those areas and to make corrective action so that productivity can improve. The research emphasizes the importance of introducing method study surveys, such as questionnaires, for the productivity of manufacturing SMEs. The outcome of the findings in the literature studied indicates that when the information from the questionnaires, noted down by WS experts, on the risks facing employees in manufacturing SMEs is introduced to management, management is able to direct the respective operations by supervisors to ensure that these risk factors are addressed so that operational processes can be improved. When operations supervisors are aware of the problems facing manufacturing SMEs, remedial actions are undertaken that enhance productivity in the business (Marodin & Saurin, 2015:69–71).

Duran, Cetindere, and Aksu (2015:110–1) analyze the layout of manufacturing SMEs through the use of flow diagrams. Findings from the literature studied show that the use of flow diagrams helps management to plan efficient methods of reducing excess work and eliminating wasteful use of resources in manufacturing SMEs. Simplifying the way activities are carried out in the workplace will enhance the productivity of manufacturing SMEs.

10.2.4 Method Study Process Charts

Method study process charts consist of charts such as flow process, outline, and two-handed process charts.

10.2.4.1 Process Charts in Manufacturing SMEs

Wajanawichakon and Srimitee (2012:956–7) investigate how WS specialists guide management by using the flow process chart in the work place to collect information on the behavior of an employee doing their job. The outcome of the literature reviewed shows that the use of the flow process to make observations of the work activities being carried out by employees reveal problem areas to management, such as waste, unnecessary motion, and bottlenecks. By eliminating waste, bottlenecks, combining, rearranging, and simplifying work process activities in the workstation, the productivity of manufacturing SMEs is improved.

Burchart-Korol (2013:233–46); Alkaya and Demirer (2014:597–601); and Silva et al. (2015:175–6) consider a flow process chart used by a WS specialist to guide management on the operations taking place in their working environment, including transport, delay, and storage, that impact on the activities being carried out by employees. Findings from the literature studied show that when WS specialists encourage managers to ensure that delays are eliminated in the flow of raw materials from the starting point in the storeroom to the transformation process where this material is transformed into a product, the symptoms of productivity decline or are avoided.

Sanjog et al. (2016:34–44) investigate how the WS specialist directs management by using a flow process chart or other chart on the activities taking place via employees, material, machinery, location, and the environment in manufacturing SMEs. Based on the literature read, the research reports that lack of using process charts guided by the work study specialist to management limits the insight of operations managers to realize the activities occurring in the working environment in manufacturing SMEs. If these process chart are not utilized, management hardly notice inappropriate work procedures taking place, which results in productivity regression.

10.2.4.2 Flow Diagram

As advised by Burchart-Korol (2013:233–46); Alkaya and Demirer (2014:597–601); and Silva, Delai, de Castro et al. (2015:175–6), the use by WS specialists of flow diagrams to enlighten management on the activities taking place in the workplace is vital to the productivity of manufacturing SMEs. By using process chart symbols such as operation, inspection, transport, delay, and storage, WS specialists facilitate the manufacturing process and allow management to take corrective measures for productivity improvement.

Sanjog et al. (2016:34–44) investigate how the WS specialist leads management by using the flow diagram to analyze the activities carried out by employees who are expending material in poor locations and exploiting machines in a poorly functioning workplace. Based on the literature studied, it can be shown that if WS is not used to reveal excessive transport, delay, and employee ill-health due to poor location and an inappropriate workplace environment, productivity will be negatively affected.

10.2.4.3 Filming Techniques

Dai et al. (2012:53) explore the use of images by managers to study problem areas experienced by all parties involved through the guidance of WS specialists in the workplace. These problem areas involve a poor working environment, material damage, poor maintenance, and high inventory. Findings from the literature examined indicate that if managers do not use images as guided by WS specialists, manufacturing SMEs will go in the wrong direction. As a result, wrong decisions are made in terms of employee safety, wasted material elimination, scheduled maintenance, and minimum inventory. Ultimately, the productivity of manufacturing SMEs declines.

As explained by Heidrich and Tiwary (2013:5884–91) and Muruganantham, Krishnan, and Arun (2014:420–6), film techniques are essential to the manufacturing environment of manufacturing SMEs. The outcome of the results from the literature reviewed show that, by encouraging management to monitor the workplace through photos taken of employees as well as taking an operational direction on the planning of standards, WS specialists enable management to improve the workplace. Furthermore, images of employees' behavior in terms of compliance to standards in the working environment also positively contribute to the productivity of manufacturing SMEs.

As found by Sanjog et al. (2016:35–46), the availability of workplace pictures taken by WS specialists to provide evidence of work activities carried out by employees, machinery operation, material flow, location for inventory control, and material environmental safety are critical to management's attitude in relation to the productivity of manufacturing SMEs. On the basis of the literature reviewed, the research shows that if WS specialists do not take pictures to guide management on the input resource factors indicated earlier, there will be shortcomings in management performance in manufacturing SMEs. As a result, symptoms of productivity decline are foreseen.

Garza-Reyes, Ates, and Kumar (2015:1097) study the utilization of cause-and-effect diagrams by management in relation to the productivity of manufacturing SMEs. When these managers use cause-and-effect diagrams as a guideline for problem solving, a breakdown structure is provided for input resource factors such as poor process, lack of knowledge, insufficient inventory, poor maintenance, a disorganized workplace, poor planning, and inappropriate standards. Using this structure, managers are then able to attend to these input resource factors for problem solving in order to ensure waste elimination and, in turn, the improvement of productivity.

10.3 Work Measurement on Management in Relation to Productivity of Manufacturing SMEs

In this section, various literature sources are read and addressed in terms of the impact of work measurement on management in relation to the productivity of manufacturing SMEs. The focus was on the literature to provide an understanding of how work measurement influences management's approach to productivity. Various work measurement techniques used in this way are presented next.

10.3.1 Time Study

Garza-Reyes et al. (2012:179–84); Das, Ghosh, and Gangopadhyay (2013:247–9); and Vinodh, Kumar, and Vimal (2014:291–300) explore how time study is used to guide operations managers to ensure effective planning and monitoring of the workplace for better employee utilization, improved material workflow, maintenance for easy access to machinery, better working conditions, and improved systems. The outcome of the effective use of these input resource factors, as indicated by the literature studied, shows an understanding of how management benefits in ensuring improved productivity.

Duran, Cetindere, and Aksu (2015:110) state that it is not only effective planning and monitoring of the workplace for better employee utilization, improved material workflow, maintenance for easy access to machinery, better working conditions, and improved systems as indicated by WS specialists that are considered in improving the productivity of manufacturing SMEs; these specialists also advise operations managers to consider the forecasting of energy management, such as the effective use of electricity in the workplace, in order to improve productivity. Based on the literature read, time study plays a vital role in ensuring that operations managers follow the correct procedure for delegating employees to use the appropriate operating standards in producing a quality product and avoiding waste material. In situations where standard operating procedures are not complied with during the current work flow, time study assists operations managers in identifying ineffective time and delays resulting in waste and, in turn, in productivity decline in manufacturing SMEs (Srinivasan et al., 2016:364–76).

10.3.2 Work (Activity) Sampling

Garza-Reyes et al. (2012:186–93) and Gupta et al. (2013:636–41) explore the use of the work sampling technique by WS specialists in directing operations managers and supervisors in relation to the productivity of manufacturing SMEs. WS specialists suggest work sampling as one of the most straightforward and direct work measurement techniques, which can be used on any work activity performed in any environment as well as any material or tools used in any location for progress in the improvement of manufacturing processes. The recommendation made by WS specialists to operations managers and supervisors as indicated in the literature reviewed is that when work activities are not normally distributed to employees, the workload in workstations builds up and the output of the manufacturing SMEs becomes delayed, which results in waste and high costs. The consequences of waste and high costs are that manufacturing SMEs experience low production and ultimately productivity decline in their businesses.

Based on the results of the literature reviewed, work sampling plays a pivotal role in enabling operations managers to discover non-value-adding activities caused by employees or processes in manufacturing SMEs. By revealing what is taking place in the manufacturing environment, this technique then contributes to driving management's effort to ensure productive work from employees, adding value for manufacturing SMEs in improving productivity in their businesses and, in turn, good customer service (Hasle, 2014:40–43; Dinis-Carvalho et al., 2015:235–41).

Duran, Cetindere, and Aksu (2015:110) investigate the impact of time utilized on work sampling on the work activities done by employees in manufacturing SMEs. The results of the literature studied show that the use of work sampling as part of management's plan to identify the appropriate standard time for carrying out work activities by employees improves efficiency by eliminating ineffective time spent and avoiding delays. As a result, the productivity of manufacturing SMEs improves. Czumanski and Lödding (2016:2938–48) examine the effect of work sampling on the attitude of operations managers in relation to the productivity of manufacturing SMEs. As a result, operations managers have the confidence to eliminate waste and costs for productivity improvement in manufacturing SMEs. The literature studied indicates that when more observations are made through the application of work sampling on employees, machines, processes, and activities, operations managers can determine the accuracy of the work being carried out.

10.3.3 Predetermined Motion Time Systems

According to Kuo and Wang (2012:6520–27) and Larsson (2013:891–901), predetermined motion time systems (PMTS) are critical to operations managers in addressing the productivity of manufacturing SMEs. Findings from the literature studied show that the utilization of PMTS by WS specialists encourages operations managers to ensure that employees know how accuracy, operational standards, skills gained, and efficiency affect the activities carried out. These motions need to be done by taking into consideration standards in place, such as employee posture, in terms of safety, health, and well-being. When these standards are complied with, the productivity of manufacturing SMEs improves.

Suwittayaruk, Van Goubergen, and Lockhart (2014:725) highlight the importance of PMTS in operations managers' efforts to influence employee attitudes in the manufacturing process in relation to the productivity of manufacturing SMEs. Based on the results of the literature researched, if management is not influenced by WS specialists in stimulating employees, if employees are operating machinery without proper training, or if they are demotivated and unqualified to carry out work activities under specified conditions, the application of PMTS by WS experts will not provide adequate results from the data collected during the time study investigation, and there will be indications of productivity deterioration.

As reported by Battini et al. (2015:825–7) and Wu et al. (2016:292), PMTS plays a fundamental role in assisting operations managers and engineers in ensuring that ergonomics such as human movement contributes to the manufacturing process. Therefore, when these managers encourage employees to follow standard operating procedures in their workplace when doing work activities involving motions, these employees are able to adhere to safety measures and properly execute their work activities. Consequently, employee abilities and willingness through the compliance of safety measures in the workplace enriches the productivity of manufacturing SMEs.

10.3.4 Analytical Estimating

As indicated by Mathur et al. (2012:755–61); and Sharma and Sharma (2014:751–55), the literature studied examines analytical estimating as a technique used to guide operations managers on the activities done by previous method and time studies in relation to the productivity of manufacturing SMEs. Findings from the literature report that when WS specialists are skilled, knowledgeable, and experienced from these studies to assist operations managers in the effective running of their manufacturing process through the use of employees abilities focusing on productive time using quality, delivery, flexibility and cost of the product produced in the manufacturing SMEs, productivity of manufacturing SMEs progresses.

Padhi et al. (2013:1887–94) and Colledania, Rattia, and Senanayake (2015:357–58) investigate the application of analytical estimating as a guide to management in facilitating the manufacturing process line layout in relation to productivity. This tool is applied by WS specialists to assist management in ensuring that there is sufficient time for the sequence of manufacturing processes and inspection of the operational frequency and skill of the employees as well as the replacement of tools during operations and machine operations in manufacturing SMEs. By focusing on the time used in the sequence of manufacturing processes, inspection of the operational frequency and skill of the employees, and the changing of tools during operations and machine operations in the workplace, these specialists guide operations managers to ensure that employees are able to perform well in their work activities and thus improve productivity.

Marodin and Saurin (2015:69–71) investigate the use of analytical estimating by specialists in clarifying operations managers in relation to productivity of manufacturing SMEs. Findings from the literature studied show that analytical estimating studies conducted by WS specialists based on reports of the job done for previous studies show projected improved standards for the new work activity. By ensuring that analytical estimating is conducted with the support of operations managers, the productivity of manufacturing SMEs can be projected and improved.

Based on the literature reviewed, analytical estimating is one of the indirect work measurement techniques influencing managers in ensuring that employees are engaged in their effort to influence the productivity of manufacturing SMEs. When WS specialists apply knowledge of this tool in the manufacturing process, they can enlighten operations managers in ensuring that employees not only use their competencies such as skills, knowledge, training, and experience but also have the power to make their own decisions to cope with complex situations in manufacturing processes, especially in cases when there is no one superior present during their operation. By making these decisions, employees becoming more confident, which enables them to engage in improving manufacturing processes. As a result, these improved processes drive the improvement of productivity in manufacturing SMEs (Małachows & Korytkowski, 2016:166–70).

10.3.5 Comparative Estimating (Benchmarking)

The literature reviewed indicates that comparative estimating plays a vital role in ensuring that operations managers plan and monitor the work methods selected to ensure improved process flow, efficiency, and line balancing in the manufacturing process. The proposed method is compared with the traditional manufacturing process, whereby operations managers are guided by WS specialists on the proposed measuring of time whereby current workload distribution and the process flow performance takes place. However, by comparing tasks, this type of estimating helps operations managers to reduce the throughput times in the operation environment with the intention of foreseeing an improvement in productivity (Garza-Reyes et al., 2012:178–94; Boulter, Bendell, & Dahlgaard, 2013:201–3).

Wang et al. (2015:221–26); Dora, Kumar, and Gellynck (2016:3–17); and Randhawa and Ahuja (2017:335–53) make an observation of comparative estimating used by a WS specialist to show the contrast between the proposed results of work activities and the standard of the previous activities in order to guide operations managers on the productivity of manufacturing SMEs. The research finds that when operations managers comply with WS guidelines in terms of proposed standard operation procedures and remuneration systems used against the number of employees, machines, and/or process production hours for the activities being carried out by the employees, it becomes easier for operations managers to anticipate improvements in productivity. These managers can only project this type of productivity if there is a better understanding of the morale, safety, and health of employees in the working environment.

10.3.6 Synthesis

As reported by Santosa, Vidal, and Moreira (2012:1657–62), synthesis created by WS specialists plays a vital role on the attitude of operations and planning managers in influencing the productivity of manufacturing SMEs. Based on the findings of the literature studied, when synthesis is exercised as a standard work measurement approach for building up cycle time for a task carried out by a qualified employee through training provided by operations and planning managers, guided by WS specialists, an appropriate standard of performance is attained for achieving the goals of manufacturing SMEs. Consequently, the productivity of manufacturing SMEs is enhanced.

Doltsinis, Ratchev, and Lohse (2013:85–87); and Beyer (2014:1–6) provide a report of the application of synthesis by a WS specialist through the authority of operations managers in their operations section in manufacturing SMEs. According to the research conducted, when synthesis is taken seriously by operations managers and qualified employees carry out the work activities appropriately and within the efficient standard time, delays and high costs are eliminated and the productivity of the manufacturing SME improves.

Glock and Grosse (2015:66437–46) examine the application of synthesis by WS specialists in guiding operations managers on the build-up of time for the work activities done by the employees in association with the productivity of manufacturing SMEs. The outcome of the findings in the literature indicate that when synthesis is implemented as a standard work measurement approach for building up cycle time for a task carried out by a qualified employee through training provided by operations and planning managers, guided by WS specialists, an appropriate standard of performance is attained for achieving the goals of manufacturing SMEs. Pedersen and Slepniov (2016:45–49) testify that the application of synthesis by WS specialists plays an important role for plant and operations managers in influencing the productivity of manufacturing SMEs. The research reports that when the application of synthesis is introduced to the operations section, managers are able to ensure that employees have qualities such as training, knowledge, and experience to carry out the work activities at standard performance. WS specialists, by encouraging plant and operation managers to allow employees to gain relevant training, knowledge, and experience, identify this gap using synthesis to show how the activities are built up considering costs in order for manufacturing SMEs to generate the final products and, in turn, improve the productivity of their businesses.

10.3.7 Statistical Process Control

Mathur, Mittal, and Dangayach (2012:756) study the capability of operations managers in using control charts in relation to productivity of manufacturing SMEs. The study reports that operations managers, by applying statistical tools as guided by WS specialists, are able to acquire an understanding of the operational systems and in turn to make knowledgeable decisions in improving the same systems for continuous productivity improvement in manufacturing SMEs. These statistical tools are statistical process control (SPC) techniques, process capability analysis, and Pareto diagrams.

Heidrich and Tiwary (2013:5891–2) and Du and Lv (2013:377–86) examine the importance of the use of control charts by operations management in the manufacturing environment of manufacturing SMEs. The literature reviewed finds that operations managers monitor product variations in the workplace to identify problem areas that need to be resolved, as guided by WS specialists, to improve productivity. These control charts are illustrated in the in the form of pie charts. As found by Mishra and Sharma (2014:524–41), SPC plays an important role in introducing control charts to assist operations managers to project the progress of productivity in their businesses. By enforcing statistical process, operations managers are then able to detect failures before they happen, and corrective actions are undertaken to ensure minimal cost and product variability, thus improving productivity. Ahmed, Ramadan, and Saghbini (2015:26); and Zhu, Zhang, and Deng (2016:1804–6) declare that SPC guides operations managers in identifying necessary design and process modifications, which take the form of projects for achieving customer satisfaction and productivity improvement. Operations managers, as guided by WS specialists, can only realize the achievement of business goals through process monitoring using Pareto and control charts. The aim of process monitoring is to ensure quality, cost, reliability, and delivery with the intention of sustaining and improving productivity.

10.3.8 Value Stream Mapping

Chiarini (2012:681–90) investigates the behavior of management as guided by VSM through WS specialists in relation to the productivity of manufacturing SMEs. Findings from the literature studied show that the unavailability of VSM results in management failing to recognize waste incurred in manufacturing SMEs. As a result, productivity decreases.

Dora et al. (2013:125–36) and Matt (2014:333–45) examine how WS specialists use VSM to enlighten managers in relation to productivity. The research shows that that absence of VSM to identify waste along the supply chain with the aim of reducing costs and increasing customer satisfaction impacts on the productivity of manufacturing SMEs. In addition, failure to visualize additional capabilities such as just-in-time (JIT), total productivity maintenance (TPM), human resource training and employee work involvement, supply chain, quality management, location, working environment, scheduling and lean processes such as good housekeeping, 5S, kaizen, and quick changeover through VSM results in managers being unable to improve the manufacturing processes of manufacturing SMEs. This failure becomes a risk to productivity performance.

Saboo et al. (2014:42–46) examine WS on management operational planning for productivity of manufacturing SMEs. The results found from the literature reviewed indicate that when management uses VSM, as guided by WS specialists, to follow their work plan on manufacturing capability areas such JIT and existing technology in the manufacturing processes, problem areas such as increased lead time, high inventory, and bottlenecks can be anticipated. By identifying the expected problem areas, management, through the guidance of VSM, will be able to use JIT efficiently and introduce new technology to replace the existing technology so as to achieve continuous improvement of productivity in manufacturing SMEs.

Based on the results of the literature reviewed, VSM on physical capital factors contributes to the inspection of waste, focusing on various capabilities such as employee abilities, material flow, costs involved, location, layout, inventory control, supply chain control, quality improvement, scheduling, time management, production planning and control, and product, process, and service design in order to ensure that manufacturing SMEs do not incur waste, which may negatively impact on the productivity of their businesses. Adding-value activities are also taken into consideration since they impact on the effective use of capabilities applied by manufacturing SMEs in their businesses. These activities are quality, cost, delivery, reliability, dependability, and flexibility (Tyagi et al., 2015:202–13).

10.3.9 Statistical Process Control

Mathur, Mittal, and Dangayach (2012:756) study management capability on the application of control charts in relation to the productivity of manufacturing SMEs. The study reports that managers, by applying statistical tools as guided by WS experts, are able to acquire an understanding of the operational systems and, in turn, to make knowledgeable decisions on how to improve the same systems for continuous productivity improvement. These statistical tools are statistical process control (SPC) techniques, process capability analysis, and Pareto diagrams.

Heidrich and Tiwary (2013:5891–2) examine the importance of using control charts by management in the manufacturing environment of manufacturing SMEs. The study of the literature finds that management monitors product variations in the workplace to identify problem areas that need to be resolved and to improve productivity. These control charts are illustrated in the in the form of pie charts.

10.3.10 Business Process Reengineering as a Technique for Both Method Study and Work Measurement on Management in Manufacturing SMEs

BPR is the last tool of WS for both method study and work measurement applied to operations management through the authority of operations managers, and it is the ultimate WS tool to be used to measure other qualitative tools such as benchmarking, brainstorming, preliminary surveys, charts, diagrams, film techniques, VSM, and cause-and-effect diagrams (Mathur, Mittal, & Dangayach, 2012:755–7; Gupta, Acharya, & Patwardhan, 2013:635–45; Panwar et al., 2013:778; Gupta & Jain 2014:22–34; Silva et al., 2015:176; Chay et al., 2015:1033–39; Prasad, Khanduja, & Sharma, 2016:409–12; Srinivasan et al., 2016:817–27) as well as quantitative tools such as benchmarking, brainstorming, preliminary surveys, time study, work sampling, PMTS, analytical estimating, comparative estimating, synthesis, VSM, and SPC (Garza-Reyes et al., 2012:185; Gupta et al. 2013:636–41; Suwittayaruk, Van Goubergen, & Lockhart, 2014:727; Sharma & Sharma, 2014:751–3; Dushyanth-Kumar, Shivashankar, & Rajeshwar, 2015:10; Wang et al., 2015:221–26; Pedersen & Slepniov, 2016:44–56; Zhu, Zhang, & Deng, 2016:1804–6), in enabling manufacturing SMEs to measure and improve productivity.

These qualitative and quantitative tools, which are identified and used to guide operations managers, are exercised on manufacturing areas such as input resource factors that are converted through manufacturing capabilities and adding-value activities in order to achieve the goals of manufacturing SMEs.

As found by Poudelet et al. (2012:192–200); Olhager (2013:6837–40); Low, Kamaruddin, and Azid (2015:703); and Al-Sa'di, Abdallah, and Dahiyat (2017:351–4), a verdict is made on how BPR, as used by a WS specialist, impacts on operations managers' behavior in relation to productivity. The aim of using BPR is to assist in ensuring that operations managers detect the current decision-making process in the manufacturing process, based on those resource factors, capabilities, and activities indicated earlier, in order to rethink and carry out a radical redesign of the current process of the manufacturing operations for continuous improvement and, in turn, to improve productivity.

On the contrary, Böhme et al. (2014:6518) and Sunder (2016:133) warn that if there are no best practices in terms of BPR, manufacturing SMEs fail in their product development and will have insufficient financial and employee resources, poor manufacturing systems, and a lack of innovation due to the shortcomings of operations managers. As a result, productivity declines.

10.4 Case Study Results and Analysis for South African SMEs

This section examines the results originating from various case studies based on Companies A, B, C, and D in Gauteng, South Africa. The objective of the research was to ascertain productivity problems facing Companies A, B, C, and D in terms of manufacturing capabilities and adding-value activities exercised in management challenges. The interventions necessary to resolve these problems were then initiated, with the aim of improving productivity in the companies tackled in the case studies.

10.4.1 Company A (Small Company)

The research conducted commenced with Company A, the vision of which was to manufacture its product focusing on precise quality and exercising ISO 9001 standards and employee commitment through the delegation of management in order to meet the needs of its customers.

Since the company's assignment was to become one of the most competitive companies in the manufacturing industrial sector in the Gauteng region, the aim of management, including the managing director, the operations manager, and the supervisors, was to ensure that employees are motivated to achieve the goals of company as requested by the company. For Company A to be competitive, it was guided by Productivity SA on the use of toolkits in order to enhance productivity.

This company provided services to customers involved mining, gas, and electrical industries as well as households. The intentions and target goals of Company A were to measure actual results against targeted results in order to ensure productivity improvement in the company. The plan target, as indicated in Chapter 1, was 10,000 components to be produced by eight employees in an 8-hour shift, but the challenge facing Company A was that the actual results had become less than the expected results. The actual results averaged 60% of the products produced, which represents a 40% decrease in productivity.

Even though injuries and damages were stringently avoided over and above inappropriate suppliers' delivery, customer returns, and the elimination of complaints, Company A did not comply with ISO 14001 in terms of the environment or with ISO 18001 with regard to occupational safety and health in the manufacturing process. Furthermore, employee training was limited. In addition, stock accuracy was also a serious situation facing the company. Company A competed with other components engineering manufacturing companies in terms of pricing, flexibility, product quality, and delivery. Despite its competitors, Company A was still going strong by exercising ISO 9001 quality management system to a certain extent in attempting to improve the productivity of its business locally and internationally.

Company A only utilized technology focusing on intangible components such as emails and telephone in the ordering of steel components. Printers were also used to print the documents required for records of ordered and manufactured products. During the manufacturing process in Company A, the transformation of physical capital and technological capital through the processes of cutting, computer numeric control, drilling, and tapping involved various capabilities such as production planning and control; product, process, service and capacity design; scheduling; employee ability and job rotation; control of components produced; machine maintenance; layout strategy for employee safety and health; location strategy for inventory risk management; supply chain management; and, lastly, JIT flow and delivery. These capabilities were applied along with adding-value activities such as quality, cost, and flexibility in Company A, but the reliability and speed of the operation to ensure less cycle time and early and on-time delivery for the customers was disregarded. The lack of reliability and speed resulted in the company experiencing a 35% rate of late delivery.

Other challenges facing the manufacturing process were unpacked by Company A. The first challenge was the late delivery of inventory, incorrect sizes, and, at times, the unavailability of stock to be delivered. Inventory handling was often done using machines such as the pallet jack and carrying material manually.

10.4.2 Company B (Medium Company)

The second research project was conducted at Company B, the vision of which was to provide the best quality products through the authority of the

South African Bureau of Standards. The mission of Company B was to supply the best possible variety of headwear for officers in the police and armed forces, as well as security firms, in Southern Africa and other Asian countries, including islands off the east coast of South Africa. The types of material used by Company B to produce headwear included wool, fabric, and wire. The suppliers of Company B were located both within the country and abroad. For the company to gain a competitive edge, it also took a decision to liaise with Productivity SA to assist on the use of toolkits in order to boost the productivity of the business.

Through toolkits, this company empowered employees with on-the-job training and job enrichment, followed job descriptions for every task, and put standards in place using a time study carried out by a WS specialist. In addition, filming techniques were applied in the form of pictures of employees and of the environment where employees carry out activities. The remaining tools used, as advised by Productivity SA, included cause-and-effect analysis to identify areas for improvement in the company.

The operations managing director introduced the use of an integrated system to ensure continuous improvement in the manufacturing process, but this system did not influence employees to change their mind-set. In order for Company B to ensure improvement in productivity, the operations managing director used various operational capabilities such as total quality improvement, supervisory and employee training, supply chain and procurement, inventory control, three-monthly machine maintenance, design, process, and service along with adding-value activities such as quality, pricing, and reliability.

The assembly process of Company B consists of stores, sample department, cutting room, preparation, embroidery, quality check, and, finally, packaging. The average standard time for the whole process is 64.17 minutes. The company produces 1000 caps per day. Company B competes with other headwear companies in Malaysia as well as other East Asian countries. Company B was still battling in its attempt to improve labor productivity in its businesses. These problems resulted from challenges such as the supply side, management, the human element, and manufacturing.

The first problem experienced stems from the very low lead times for the delivery of material to Company B by suppliers, due to imports done through shipping.

Secondly, the company is faced with poor planning on the part of the operations managing director, whereby 41 operators, many of them elderly, are not properly assigned to their respective workstations. As a result, minimum production was taking place, hindering the productivity of Company B. For instance, 1000 caps are produced by 41 operators, with each employee taking an hour to produce three caps. This represented a serious labor productivity problem, resulting in high costs and waste for the company.

The challenge of the human element is that most employees in this company are aging and face a challenge when it comes to carrying heavy boxes; this is demotivating and results in poor employee performance. Incentives such as benefits and medical aids are not provided to employees. Communication between management and supervisors and employees is also a major problem in terms of group decision making. Employee training was provided internally and growth in terms of formal training was ineffective.

Management lacked an understanding of the role of integrated systems in the manufacturing process. As a result, more people were assigned to the assembly line without appropriate planning to control the system, resulting in high costs for the company. There was also limited space for the storage of inventory and finished goods. Based on the results reported, 20% of the time available was wasted on late delivery due to inflexibility as a result of the company's dependence on imports and the lack of management innovation in terms of research and development.

10.4.3 Company C (Small Company)

The third company was Company C, and its vision was to pack colored cocky pens through precise quality provision as well as management commitment to the customer. The company's aim was to become one of the most competitive companies to carry out a precise and efficient packaging operation in the manufacturing industrial sector around the central part of Gauteng, South Africa. The concept of Productivity SA as an agency in assisting on the use of toolkits was not applicable to this company due to the managing director's shortcomings and unawareness of the existence of the agency.

The target of Company C centered on the process of packaging various types of boxes of colored cocky pens that are packed locally daily for export. The packaging process starts when pens are packed and ends with the wrapping of boxes of pens in plastic, incorporating corners to protect the finished products from damage. Company C packaged 2250 cocky pens per hour on three parallel lines, with 12 pens being packed in each envelope by 26 packers. This meant that each packer packed approximately 60 envelopes with 12 pens in each per day in the company. The concept of productivity measurement is not applicable in this company since it has only been in business for four months and thus is still regarded as a start-up small business. In the light of this statement, Company C focuses more on production than productivity improvement, which is a serious labor productivity challenge facing the company.

Company C needed to consider other avenues of the market to target other customers for business productivity growth. The challenge faced by Company C is that management lacked the knowledge of standards in place such as ISO 9001 for quality packaging; ISO 14001 for cleanliness and congestion of the work place, including storage and warehousing; and ISO 18001 for safety precautions. The supply process was done locally.

The various challenges experienced by Company C included inappropriate SOPs and job rotation, resulting in employee redundancy.

The operations managing director in Company C recruits candidates and provides on-the-job training in the workplace rather than improving employee career paths based on formal education. This company depended on outsourcing forklifts for material handling.

The physical conveyor belt was also used by employees to wrap boxes rather than being automated. In terms of the location, Company C uses the same door for both receiving material and issuing packaged products.

For Company C to function in the manufacturing process, capabilities such as process and service design, technology, employee involvement and job rotation, supply chain, location, and transport along with adding values such as pricing, flexibility, product quality, and delivery are exercised.

10.4.4 Company D (Medium Company)

The final research was conducted at Company D, the vision of which was to develop innovative solutions for the local and international market. The mission of Company D was to provide the best quality product and to offer a high level of service to valued customers. For this company to gain a competitive edge, it also took the decision to liaise with Productivity SA to assist with the use of toolkits in order to boost productivity. With Company D, information signals were converted into specific tones in order to diffuse signals over wire, wireless, and cables to a target point. This company produced components in the form of batches. Company D supplies a wide range of radio communication systems for mobile communication based on sites and controls centers locally and internationally. Company D generates 2500 radio security communication components per hour with a workforce of 102 employees.

The competition in Company D is based on adding-value activities such as price and good service driving capabilities such as supplier–customer negotiation; product, process, and service design; maintenance; new technology; and JIT. The strategy of Company D is based on the niche market. Employees in Company D were motivated by receiving extra payment from overtime and bonuses when they exceeded the company's expectations. All these incentives were paid according to what is planned by the manufacturing department at Company D.

Company D was producing a single product. Due to the growth of the business, the company moved to producing a multiple product, which forced the company to initiate changes through the cellular network in order to ensure that the system done by Company D is working efficiently. Before and during the manufacturing process as well as at the completion of the manufacturing process, some WS tools were applied by Company D, such as cause-and-effect diagrams and filming techniques, with the intention of identifying problem areas such as poor planning.

The delivery of items produced by Company D was done by company transport or through the use of couriers to reduce costs. Even though the challenges facing this company as indicated by an electronic engineer were based on product quality, planning, logistics, and absenteeism due to chronic ill-health, research and development was encouraged to improve product design and allow the firm to remain competitive.

Cycle time variation impacted on the efficiency of the manufacturing process in Company D. Waste and damages were experienced due to poor quality systems.

Company D was also faced with a major change whereby international supply needed to be considered in order to gain a competitive edge. The company was failing to control inventory due to the lack of a computerized system. The company began to become more complex due to the multiple products produced by Company D. Logistic problems are experienced where the supply chain is not attended in a timely manner from the first stage to the last stage. This problem may lead to late delivery and loss of customers. Safety issues in terms of employee well-being and risk factors in terms of material damage are a major problem in Company D. This may be due to poor planning or to the shortcomings of management in Company D. Material handling is very poor, whereby employees have to carry some of the material physically, causing severe health problems among these employees. For Company D to manage inventory, senior officials foresee the use of a software package to track any component coming through the company, through the completion of the product, till it reaches the customer. The challenge facing Company D was a 30% rate of late delivery to the customer.

10.5 Comparative Analysis and Discussion

Based on the case study results discussed and analysis made for South African SMEs such as Companies A, B, C, and D, a comparative analysis and discussion of challenges facing these companies is shown Table 10.1 in order to highlight discrepancies between these companies. Furthermore, a comparative analysis of the application of WS is also made based on these companies to determine differences in the manner in which they contribute to the productivity of their businesses; see the outline in Table 10.1.

Regarding the manufacturing process in Table 10.1, operations directors and managers in Companies A, B, C, and D fail to integrate all the necessary capabilities that will efficiently improve the performance of their businesses. Furthermore, without the driving force behind the integration of adding-value activities to reinforce these manufacturing capabilities, these companies will find it difficult to compete locally and abroad.

Despite the use of these capabilities and adding-value activities implemented by management, the unavailability of WS in the manufacturing process will delay the development of productivity in these companies. Even though these individual companies use some WS tools, the limited application of these tools means that these companies will continue to fail to reach the productivity targets of their businesses. Based on research studied and discussed in the literature, the absence of WS techniques delays improvements in the productivity of manufacturing SMEs, and Companies A, B, C, and D may fall into the same trap in their businesses.

Since WS is an exciting area of operations management and industrial engineering, having a considerable effect on productivity, as advised by various academic scholars in the literature reviewed, the improvement of productivity can be achieved through the establishment of an enabling environment through the application of WS.

Table 10.1 Comparative Analysis and Discussion of Challenges Facing Companies A, B, C, and D

Background of Manufacturing	*Company A*	*Company B*	*Company C*	*Company D*
Challenges facing manufacturing capabilities based on input resource factors utilized in companies	**Management shortcomings** Poor inventory control, poor planning, poor employee safety control, poor sharing of responsibility **Transport** 35% delivery is used inefficiently, whereby products are delivered late **Supply chain** Late deliveries, incorrect sizes, stock not available in other instances	**Management shortcomings** Poor planning **Transport** 20% late delivery **Supply chain** Lead times very low, depend on suppliers for quantity of material required and no enough suppliers	**Management shortcomings** Poor planning, poor forecast on buying, poor inventory control **Transport** Products delivered when required **Supply chain** Late delivery, and unavailability of stock	**Management shortcomings** Poor planning **Transport** 30% late delivery **Supply chain** Major change with international supply, poor supply chain
Adding-value activities exercised by these companies	Quality, costs, and flexibility	Quality, pricing and reliability	Quality; cost and reliability	Price and service
WS tools available in these companies	Standards such as Quality ISO 9001 certified	Quality plan SABS certified standards	Quality plan neither SABS nor ISO 9001 certified standards	Quality ISO 9001 certified standards

(*Continued*)

Table 10.1 (Continued) Comparative Analysis and Discussion of Challenges Facing Companies A, B, C, and D

Background of Manufacturing	*Company A*	*Company B*	*Company C*	*Company D*
WS: Method study	Benchmarking with Productivity SA, brainstorming, work flow charts, flow diagrams, cause-and-effect charts, filming techniques	Benchmarking with Productivity SA, brainstorming, work flow charts, flow diagrams, cause-and-effect charts, filming techniques	Benchmarking with other companies, work flow charts	Benchmarking with other companies, brainstorming, work flow charts, flow diagrams, cause-and-effect charts, filming techniques
WS: Work measurement	Measure time from start to end for cycle time	Use WS expert for measuring time using time study	N/A	Measure of time using computer recording

10.6 Summary

Chapter 10 centered on the impact of WS into management on productivity improvement in manufacturing SMEs as well as addressing productivity challenges facing companies in Gauteng, South Africa.

The types of WS techniques used to evaluate management in terms of the productivity performance of manufacturing SMEs were method study and work measurement, respectively. The impact of WS into management on the productivity of manufacturing SMEs was studied and discussed with the intention of determining the relationship between the two concepts. The research reported that, for WS to be useful in terms of method study in the manufacturing process of manufacturing SMEs, WS experts should persuade management to use method study tools such as benchmarking, brainstorming, preliminary surveys, charts, flow diagrams, filming techniques, VSM, cause-and-effect charts, and BPR. Furthermore, work measurement techniques should also be considered, such as benchmarking, brainstorming, preliminary surveys, time study, activity sampling, synthesis, PMTS, standards, VSM, SPC, and BPR. The research emphasized that when method study and work measurement are encouraged and manufacturing capabilities exercised by operation managers, including management from the technical side, the productivity of manufacturing SMEs will continue to improve.

References

Alkaya, E. and Demirer, G.N. 2014. Sustainable textile production: A case study from a woven fabric manufacturing mill in Turkey. *Journal of Cleaner Production*, 65: 595–603.

Al-Sa'di, A.F., Abdallah, A.B., and Dahiyat, S.E. 2017. The mediating role of product and process innovations on the relationship between knowledge management and operational performance in manufacturing companies in Jordan. *Business Process Management Journal*, 23(2): 349–376.

Bateman, N., Hines, P., and Davidson, P. 2014. Wider applications for Lean: An examination of the fundamental principles within public sector organisations. *International Journal of Productivity and Performance Management*, 63(5): 550–568.

Battini, D., Delorme, X., Dolgui, A., Persona, A., and Sgarbossa, F. 2015. Ergonomics in assembly line balancing based on energy expenditure: A multi-objective model. *International Journal of Production Research*, 54(3): 824–845.

Bechar, A. and Eben-Chaime, M. 2014. Hand-held computers to increase accuracy and productivity in agricultural work study. *International Journal of Productivity and Performance Management*, 63(2): 194–208.

Beyer, C. 2014. Strategic implications of current trends in additive manufacturing. *Journal of Manufacturing Science and Engineering*, 136: 1–8.

Böhme, T., Deakins, E., Pepper, M., and Towill, D. 2014. Systems engineering effective supply chain innovations. *International Journal of Production Research*, 52(21): 6518–6537.

Boulter, L., Bendell, T., and Dahlgaard, J. 2013. Total quality beyond North America. *International Journal of Operations and Production Management*, 33(2): 197–215.

Burchart-Korol, D. 2013. Life cycle assessment of steel production in Poland: A case study. *Journal of Cleaner Production*, 54: 235–243.

Chay, T., Xu, Y., Tiwari, A., and Chay, F. 2015. Towards lean transformation: The analysis of lean implementation frameworks. *Journal of Manufacturing Technology Management*, 26(7): 1031–1052.

Chompu-inwai, B., Jaimjit, B., and Premsuriyanunt, P. 2015. A combination of material flow cost accounting and design of experiments techniques in an SME: The case of a wood products manufacturing company in northern Thailand. *Journal of Cleaner Production*, 108: 1352–1364.

Colledania, M., Rattia, A., and Senanayake, C. 2015. An approximate analytical method to evaluate the performance of multi-product assembly manufacturing systems. *Procedia CIRP*, 33: 357–363.

Czumanski, T. and Lödding, H. 2016. State-based analysis of labour productivity. *International Journal of Production Research*, 54(10): 2934–2950.

Dai, Q., Zhong, R., Huang, G.Q, Qu, T., Zhang, T., and Luo, T.Y. 2012. Radio frequency identification-enabled real-time manufacturing execution system: A case study in an automotive part manufacturer. *International Journal of Computer Integrated Manufacturing*, 25(1): 51–65.

Das, B., Ghosh, T., and Gangopadhyay, S. 2013. Child work in agriculture in West Bengal, India: Assessment of musculoskeletal disorders and occupational health problems. *Journal of Occupational Health*, 55(4): 244–258.

Dinis-Carvalho, J., Moreira, F., Bragança, S, Costa, E., Alves, A., and Sousa, R. 2015. Waste identification diagrams. *Production Planning and Control*, 26(3): 235–247.

Doltsinis, S.C., Ratchev, S., and Lohse, N. 2013. A framework for performance measurement during production ramp-up of assembly stations. *European Journal of Operational Research*, 229: 85–94.

Dora, M., Kumar, M., and Gellynck, X. 2016. Determinants and barriers to lean implementation in food-processing SMEs: A multiple case analysis. *Production Planning & Control*, 27(1): 1–23.

Duran, C., Cetindere, A., and Aksu, Y.E. 2015. Productivity improvement by work and time study technique for earth energy-glass manufacturing company. *Procedia Economics and Finance*, 26: 109–113.

Dushyanth-Kumar, K.K.R., Shivashankar, G.S., and Rajeshwar S.K. 2015. Application of value stream mapping in pump assembly process: A case study. *Industrial Engineering and Management*, 4(3): 1–11.

Garza-Reyes, J.A., Ates, E.M., and Kumar, V. 2015. Measuring lean readiness through the understanding of quality practices in the Turkish automotive suppliers industry. *International Journal of Productivity and Performance Management*, 64(8): 1092–1112.

Garza-Reyes, J.A., Oraifige, L., Soriano-Meier, H., Forrester, P.L., and Harmanto, D. 2012. The development of a lean park homes production process using process flow and simulation methods. *Journal of Manufacturing Technology Management*, 23(2): 178–197.

Glock, C.H. and Grosse, E.H. 2015. Decision support models for production ramp-up: A systematic literature review. *International Journal of Production Research*, 53(21): 6637–6651.

Gupta, S., Acharya, P., and Patwardhan, M. 2013. A strategic and operational approach to assess the lean performance in radial tyre manufacturing in India: A case based study. *International Journal of Productivity and Performance Management*, 62(6): 634–651.

Gupta, S. and Jain, S.K. 2014. The 5S and kaizen concept for overall improvement of the organisation: A case study. *International Journal of Lean Enterprise Research*, 1(1): 22–40.

Hasle, P. 2014. Lean production: An evaluation of the possibilities for an employee supportive Lean practice. *Human Factors and Ergonomics in Manufacturing and Service Industries*, 24(1): 40–53.

Heidrich, O. and Tiwary, A. 2013. Environmental appraisal of green production systems: Challenges faced by small companies using life cycle assessment. *International Journal of Production Research*, 51(19): 5884–5896.

Herman, F. 2011. Textile disputes and two-level games: The case of China and South Africa. *Asian Politics & Policy*, 3(1): 115–130.

Ingvaldsen, J.A., Holtskog, H., and Ringen, G. 2013. Unlocking work standards through systematic work observation: Implications for team supervision. *Team Performance Management: An International Journal*, 19(5/6): 279–291.

Jagoda, K., Lonseth, R., and Lonseth, A. 2013. A bottom-up approach for productivity measurement and improvement. *International Journal of Productivity and Performance Management*, 62(4): 387–406.

Jain, A., Bhatti, R.S., and Singh, H. 2015. OEE enhancement in SMEs through mobile maintenance: A TPM concept. *International Journal of Quality & Reliability Management*, 32(5): 503–516.

Jain, R., Gupta, S., Meena, M.L., and Dangayach, G.S. 2016. Optimisation of labour productivity using work measurement techniques. *International Journal of Productivity and Quality Management*, 19(4): 485–510.

James, R. and Jones, R. 2014. Transferring the Toyota lean cultural paradigm into India: Implications for human resource management. *The International Journal of Human Resource Management*, 25(15): 2174–2191.

Kafetzopoulos, D.M., Gotzamani, K.D., and Psomas, E.L. 2014. The impact of employees' attributes on the quality of food products. *International Journal of Quality & Reliability Management*, 31(5): 500–521.

Kuo, C. and Wang, M.J. 2012. Motion generation and virtual simulation in a digital environment. *International Journal of Production Research*, 50(22): 6519–6529.

Larsson, A. 2013. The accuracy of surgery time estimations. *Production Planning and Control*, 24(10–11): 891–902.

Lasrado, F., Arif, M., and Rizvi, A. 2015. Employee suggestion scheme sustainability excellence model and linking organizational learning: Cases in United Arab Emirates. *International Journal of Organizational Analysis*, 23(3): 425–455.

Low, S., Kamaruddin, S., and Azid, I.A. 2015. Improvement process selection framework for the formation of improvement solution alternatives. *International Journal of Productivity and Performance Management*, 64(5): 702–722.

Magu, P., Khanna, K., and Seetharaman, P. 2015. Path Process Chart: A technique for conducting time and motion study. *Procedia Manufacturing*, 3: 6475–6482.

Malachows, B. and Korytkowski, P. 2016. Competence-based performance model of multi-skilled workers. *Computers and Industrial Engineering*, 91: 165–177.

Marodin, G.A. and Saurin, T.A. 2015. Managing barriers to lean production implementation: Context matters. *International Journal of Production Research*, 53(13): 3947–3962.

Maskaly, J. and Jennings, W. 2016. A question of style: Replicating and extending Engel's supervisory styles with new agencies and new measures. *Policing: An International Journal of Police Strategies & Management*, 39(4): 620–634.

Mathur, A., Mittal, M.L., and Dangayach, G.S. 2012. Improving productivity in Indian SMEs. *Production Planning and Control: The Management of Operations*, 23(10–11): 754–768.

Mathur, A., Mittal, M.L., and Dangayach, G.S. 2012. Improving productivity in Indian SMEs. *Production Planning and Control: The Management of Operations*, 23(10–11): 754–768.

Mehralian, G., Nazari, J.A., Nooriparto, G., and Rasekh, H.R. 2017. TQM and organizational performance using the balanced scorecard approach. *International Journal of Productivity and Performance Management*, 66(1): 111–125.

Muruganantham, V.R., Krishnan, P.N., and Arun, K.K. 2014. Integrated application of TRIZ with Lean in the manufacturing process in machine shop for productivity improvement. *International Journal of Productivity and Quality Management*, 13(4): 414–429.

Oeij, P.R.A., De Looze, M.P., Have, K.T., Van Rhijn, J.W., and Kuijt-Evers, L.F.M. 2012. Developing the organisation's productivity strategy in various sectors of industry. *International Journal of Productivity and Performance Management*, 61(1): 93–109.

Olhager, J. 2013. Evolution of operations planning and control: From production to supply chains. *International Journal of Production Research*, 51(23–24): 6836–6843.

Ongkunaruk, P. and Wongsatit, W. 2014. An ECRS-based line balancing concept: A case study of a frozen chicken producer. *Business Process Management Journal*, 20(5): 678–692.

Padhi, S.S., Wagner, S.M., Niranjan, T.T., and Aggarwal, V. 2013. A simulation-based methodology to analyse production line disruptions. *International Journal of Production Research*, 51(6): 1885–1897.

Pandey, A., Singh, M., Soni, N., and Pachorkar, P. 2014. Process layout on advance CNG cylinder manufacturing. *International Journal of Application or Innovation in Engineering and Management (IJAIEM)*, 3(12): 113–116.

Panwar, A. Nepal, B., Jain, R., and Yadav, O.P. 2013. Implementation of benchmarking concepts in Indian automobile industry: An empirical study. *Benchmarking: An International Journal*, 20(6): 777–804.

Pedersen, P. and Slepniov, D. 2016. Management of the learning curve: A case of overseas production capacity expansion. *International Journal of Operations and Production Management*, 36(1): 42–60.

Poudelet, V., Chayer, J., Margni, M., Pellerin, R., and Samson, R. 2012. A process-based approach to operationalize life cycle assessment through the development of an eco-design decision-support system. *Journal of Cleaner Production*, 33: 192–201.

Prasad, S., Khanduja, D., and Sharma, S.K. 2016. An empirical study on applicability of lean and green practices in the foundry industry. *Journal of Manufacturing Technology Management*, 27(3): 408–426.

Ramish, A. and Aslam, H. 2016. Measuring supply chain knowledge management (SCKM) performance based on double/triple loop learning principle. *International Journal of Productivity and Performance Management*, 65(5): 704–722.

Randhawa, J.S. and Ahuja, I.S. 2017. 5s: A quality improvement tool for sustainable performance: Literature review and directions. *International Journal of Quality & Reliability Management*, 34(3): 334–361.

Richardson, M., Evans, C., and Gbadamosi, G. 2014. The work-study nexus: The challenges of balancing full-time business degree study with a part-time job. *Research in Post-Compulsory Education*, 19(3): 302–309.

Saboo, A., Garza-Reyes, J., Er, A., and Kumar, V. 2014. A VSM improvement-based approach for Lean operations in an Indian manufacturing SME. *International Journal of Lean Enterprise Research*, 1(1): 41–58.

Sanjog, J., Patnaik, B., Patel, T., and Karmakar, S. 2016. Context-specific design interventions in blending workstation: An ergonomics perspective. *Journal of Industrial and Production Engineering*, 33(1): 32–50.

Santosa, M.S., Vidal, M.C.R., and Moreira, S.B. 2012. The RFad method: A new fatigue recovery time assessment for industrial activities. *Work*, 41: 1656–1663.

Sharma, S. and Shah, B. 2016. Towards lean warehouse: Transformation and assessment using RTD and ANP. *International Journal of Productivity and Performance Management*, 65(4): 571–599.

Silva, D.A.S., Delai, I., de Castro, M.A.S., and Ometto, A.R. 2015. Quality tools applied to cleaner production programs: A first approach toward a new methodology. *Journal of Cleaner Production*, 47: 174–187.

Srinivasan, K., Muthu, S., Devadasan, S.R., and Sugumaran, C. 2016. Enhancement of sigma level in the manufacturing of furnace nozzle through DMAIC approach of Six Sigma: A case study. *Production Planning and Control: The Management of Operations*, 27(10): 810–822.

Srinivasan, S., Ikuma, L.H., Shakouri, M., Nahmens, I., and Harvey, C. 2016. 5S impact on safety climate of manufacturing workers. *Journal of Manufacturing Technology Management*, 27(3): 364–378.

Sunder, M.V. 2016. Lean six sigma project management: A stakeholder management perspective. *The TQM Journal*, 28(1): 132–150.

Suwittayaruk, P., Van Goubergen, D., and Lockhart, T.E. 2014. A preliminary study on pace rating using video technology. *Human Factors and Ergonomics in Manufacturing and Service Industries*, 24(6): 725–738.

Tanvir, S.I. and Ahmed, S. 2013. Work study might be the paramount methodology to improve productivity in the apparel industry of Bangladesh. *Industrial Engineering Letters*, 3(7): 51–60.

Vijai, J.P., Somayaji, G.S.R., Swamy, R.J.R., and Aital, P. 2017. Relevance of F.W. Taylor's *Principles to Modern Shop-Floor Practices: A Benchmarking Work Study. Benchmarking: An International Journal*, 24(2): 445–466.

Vinodh, S., Vasanth Kumar, S.V., and Vimal, K.E.K. 2014. Implementing lean sigma in an Indian rotary switches manufacturing organisation. *Production Planning and Control*, 25(4): 288–302.

Wajanawichakon, K. and Srimitee, C. 2012. ECRS's principles for a drinking water production plant. *International Organization of Scientific Research (IOSR) Journal of Engineering*, 2(5): 956–960.

Wang, Q. Huang, P., Li, J., and Ke, Y. 2015. Uncertainty evaluation and optimization of INS installation measurement using Monte Carlo Method. *Assembly Automation*, 35(3): 221–233.

Wu, S., Wang, Y., BolaBola, J.Z., Qin, H., Ding, W., Wen, W., and Niu, J. 2016. Incorporating motion analysis technology into modular arrangement of predetermined time standard (MODAPTS). *International Journal of Industrial Ergonomics*, 53: 291–298.

Zhu, H., Zhang, C., and Deng, Y. 2016. Optimisation design of attribute control charts for multi-station manufacturing system subjected to quality shifts. *International Journal of Production Research*, 54(6): 1804–1821.

Chapter 11

Presentation and Report Writing in Manufacturing SMEs

11.1 Introduction

The purpose of this chapter is to examine presentation and report writing in manufacturing small and medium enterprises (SMEs). The nature of presentations provided by work study (WS) specialists in manufacturing SMEs can be broken down into the kinds of presentation given, guidelines to follow when presenting, and presentation procedures in manufacturing SMEs. The discussion of report writing, meanwhile, focuses on the purpose of report writing by WS specialists in manufacturing SMEs, the kinds of reports used, and the format of report writing. Report writing is then discussed based on case study investigations done among manufacturing SMEs in South Africa. The first section to be discussed is the nature of presentations provided by WS specialists in manufacturing SMEs.

11.2 The Nature of Presentations Provided by Work Study Specialists in Manufacturing SMEs

The nature of presentations provided by WS specialists in manufacturing SMEs is addressed next with the intention of providing a detailed background to such presentations. Various types of presentation can be used, such as oral presentations, informative talks, investigative reports, and visual aids (such as information, graphs, and tables showing operational information about the company). The presentations can be used in manufacturing SMEs depending on the nature of results needed in the business (Bocken, Morgan, & Evans, 2013:857–62). The next section will address the kinds of presentations given in manufacturing SMEs.

11.2.1 Kinds of Presentations in Manufacturing SMEs

The presentations provided by WS specialists in manufacturing SMEs can be of various kinds, such as oral presentations, informative talks, discussions, demonstrations, meetings, and progress reports on operational functioning. As stated by Prajogo and Sohal (2013:1533–4), oral presentations, also referred to as *oral instructions* in manufacturing SMEs, involve passing on a message to all parties across departments or units or to an audience in order to ensure that the message has efficiently reached the parties involved. An informative talk is information passed on to a group of people regarding the manufacturing operations facing manufacturing SMEs (Wells, 2016:42–6).

According to Dulange, Pundir, and Ganapathy (2014:218–9), a *discussion* refers to the activity in which WS experts talk collectively about challenges that impact on the performance of the business, such as change, competition, technology, poor-quality material, employee incapability, absenteeism, unavailability of training facilities, high volume orders, inappropriate standard time, and bottlenecks found through cause-and-effect diagrams, and these experts share their ideas on the productivity of manufacturing SMEs.

Demonstrations differ in the manner in which they are addressed by manufacturing SMEs. Various academic scholars define the term *demonstration.* For instance, a demonstration involves a practical exhibition of issues such as profit, the size of the business, competition, market share, and assets against the planned target in the form of visual aids and the explanation of difference in order to show the position of the business and areas for improvement (Pakdil & Leonard, 2015:728–31). As is indicated by Rostam (2015:252–6), visual aids are presented in the form of organization structure, graphs, diagrams, charts, and tables showing the productivity progress of manufacturing SMEs.

Ciarapica, Bevilacqua, and Mazzuto (2016:183–200) mention that meetings involve an assembly of WS experts in their unit, and selected individual experts, with the operations management department for a particular purpose, especially for formal discussion on challenges and solutions in the form of progress reports, the aim being to improve productivity in manufacturing SMEs.

A progress report examines problem areas experienced, such as a poor-quality product, waste, and high costs using cause-and-effect diagrams and statistical process control attributed to value stream mapping as one of the WS techniques. Where value stream mapping has already used to identify these problem areas in the manufacturing process, WS experts introduce business process reengineering to resolve these problems.

The next section discusses guidelines to follow when presenting in manufacturing SMEs.

11.2.2 Guidelines to Follow When Presenting in Manufacturing SMEs

In this section, various literature sources that deal with guidelines to follow when presenting in manufacturing SMEs are read and discussed to provide readers with a detailed insight into how WS experts can ensure a successful presentation for the business.

As pointed out by Hu et al. (2015:991–1001) and Lim et al. (2017:688–93), the following guidelines need to be unpacked.

The presentation should firstly be realistic, objective, and relevant to the current situation facing manufacturing SMEs. This will involve graphs and tables to show the manner in which manufacturing SMEs are operating. Graphs show areas such as the structure of the positions of managers, WS tools employed, and manufacturing capabilities, along with adding-value activities that are exercised in manufacturing SMEs. The facts need to be selected according to the parties attending the meeting in manufacturing SMEs. After selection of the facts, the information gathered needs to be integrated and interpreted for management and other parties in the business. WS experts need to use simple and understandable language that will cater for all parties in manufacturing SMEs. In the next section, presentation procedures are examined.

11.2.3 Presentation Procedures in Manufacturing SMEs

Hu et al. (2015:999) identified the following presentation procedures to be undertaken by WS specialists in manufacturing SMEs. The following presentation procedures are listed and addressed. First, WS specialists need to determine the topic and research material for the presentation. This material needs to be addressed to all parties, and the benefits of the material addressed for the future of the business should be considered.

WS specialists are advised to take into consideration all the parties involved during a presentation in a manufacturing SME. The specialist should avoid using unfamiliar language and should present simple and brief information for the parties attending. WS specialists need to be objective and to use simple sentences. The presentation should be logical and systematic in terms of the order of importance of the information. WS specialists should avoid using new material that will confuse the participants. The delivery of the presentation should include nonverbal communication. Finally, the WS practitioner should conclude the presentation on a strong and encouraging positive note, allowing parties in the presentation venue to comment and ask questions. Report writing is the next section to be discussed.

11.3 Report Writing Provided by Manufacturing SMEs

As explained by Movahedi, Miri-Lavassani, and Kumar (2016:470), report writing involves writing a document with the specific intention of conveying information to management about the challenges facing the manufacturing process and making recommendations that will contribute to the efficient functioning of the manufacturing SME. Report writing involves the purpose of the report, the kinds of reports used, and the format of report writing. The purpose of report writing in manufacturing SMEs is the next topic to be discussed.

11.3.1 The Purpose of Report Writing in Manufacturing SMEs

The purpose of report writing is, first, to identify deficiencies such as poor quality practices, shortage of employees, absence of new technology (Garza-Reyes, Ates, & Kumar, 2015:1094–1108), and financial and nonfinancial measures (Mehralian et al., 2017:116–17). Second, report writing is done to keep all parties in the manufacturing SMEs informed of the progress being made in the business (Lee & Wong, 2015:2014–5; Siddiqui et al., 2016:183; Govindan, 2017:7–12). Reports are essential tools in management decision making in manufacturing SMEs (Wiesner, Chadee, & Best, 2017:3–12).

Finally, reports provide management with an efficient method of conveying a complex message to all parties in the business, including top management, supervisors, and employees (Hu et al., 2015:987).

11.3.2 Kinds of Reports Used in Manufacturing SMEs

Different kind of reports are used in manufacturing SMEs, namely *investigative* reports and *evaluative* reports (Carrillo-Castrillo et al., 2016:70–4). An investigative report is a report conducted by a WS specialist to identify problem areas such as a lack of production planning and control; accidents due to poor layout; poor maintenance; poor product, process, and service design; infrequent schedule; poor employee training; and time wasted with the intention of improving them for productivity progress in manufacturing SMEs.

An evaluative report in manufacturing is a systematic report and assessment of ongoing or completed projects, the aim of which is to improve efficiency, effectiveness, and the competitive edge of the business (McLean, Antony, & Dahlgaard, 2017:222–8).

Based on the reports that are conducted, an investigative report is the main report benefiting manufacturing SMEs in identifying problem areas in the manufacturing process. When the information is gathered by the WS expert based on challenges resulting from inefficient use of capabilities, leading to waste, high costs, and poor-quality products, this information is structured in the form of a report. This investigative report is finalized by drawing conclusions and making recommendations that will assist management in improving the productivity of the manufacturing SME (Carrillo-Castrillo et al., 2016:70–4). The next topic to be discussed is the format of report writing in manufacturing SMEs.

11.3.3 The Format of Report Writing in Manufacturing SMEs

The format of report writing entails an executive summary, an introduction, the method followed to conduct the report, the results of the findings, and discussion of the conclusions, and recommendations, as well as annexure and appendices at the end of the report (Fonseca, McAllister, & Fitzpatrick, 2014:70–7). As advised by Helo et al. (2014:646–51), the executive summary is a report by management or WS specialists about the background of the current operational running of the business, for example, its challenges in terms of the operational systems used, the performance of employees, scheduling, production queues, management production planning and control of the operational line and supply chain, work instructions, and reporting with the aim of achieving the goals of manufacturing SMEs.

As stated by Antwerpen and Curtis (2016:239–41), an introduction in report writing outlines the background to the current situation, such as threads, environment, operational factors, employee behavior, timelines, and the availability of facilities influencing the performance of the manufacturing SME. The method followed, meanwhile, explains the techniques and procedures followed by management on the capabilities to achieve goals.

As clarified by Klewitz and Hansen (2014:64–5), the results of the findings reported by any senior official from management, consultants, or WS specialists regarding the operational functioning of the manufacturing SME involves the challenges SMEs face in terms of operational redesign within the value chain and best practice recommended, such as innovative practices to sustain productivity progress. In the conclusion, meanwhile, management or WS specialists make an inference as to the types of best practice that could be used to improve on the results addressed and make recommendations.

The recommendations involve a suggestion made or an action plan devised at the top according to the guidelines provided by WS specialists to come up with best practices for the improvement of the operational process to achieving the goals of manufacturing SMEs (Klewitz & Hansen, 2014:64–5; Marodin & Saurin, 2015: 3951–2).

11.4 Report Writing on Case Studies Investigations Conducted in Manufacturing SMEs in South Africa

During the interview conducted, questions were probed further by the book authors to Companies A, B, C, and D, with regard to the application of report writing on the progress of their businesses. Out of these four companies, only Companies C and D had work study investigations conducted. Since Company C had a work study intern specialist, the aim was to develop that person toward becoming a competent work study specialist. Even though this intern specialist was conducting time studies to provide reports to the operations manager and the operations director, this specialist was not yet in the position of writing formal reports in terms of the productivity results of the company, which needed to be measured from period to period to ensure the company's productivity progress. In Company D, the operations manager did the duties of both managing operations in the units as well as conducting time studies, which impacts on the workload of the operations manager. Even though the operations manager conducted time studies, this manager was not in the position of writing formal work study reports for the company, which meant that the company could not foresee the progress of the company's productivity. Companies A and B as well, by not using formal report writing, are in danger of not foreseeing progress with regard to the productivity of their businesses. The importance of literature sources used regarding report writing on the results of the companies' productivity levels is that, when reports are not written as indicated in the report writing literature sources used, companies that do not use formal report writing are in the dilemma of not foreseeing the future of the business in relation to the productivity improvement of their company.

11.5 Comparative Analysis and Discussion

Based on the reports done for Companies A, B, C, and D, the differences in Figure 1.8 and Tables 8.2, 9.1, and 10.1 and similarities in Figure 1.8 were recognized. Interviews about the background of these companies found that these companies write reports, but not formal ones that guide them on the challenges faced by their businesses and suggest corrective measures to take to improve their operational processes and sustain their businesses.

11.6 Summary

This chapter focused on presentations and report writing in manufacturing SMEs. The nature of presentations provided by WS specialists in manufacturing SMEs can be broken down into the kinds of presentation, guidelines to follow when presenting, and presentation procedures in manufacturing SMEs. Report writing focuses on the purpose of report writing by WS specialists in manufacturing SMEs, the kinds of reports used, and the format of report writing. Report writing was discussed based on case study investigations done in manufacturing SMEs in South Africa. The results found regarding the writing of reports on the background of their operational process is that these companies need training on report writing if they are to see their business performance improve for sustainability in terms of productivity and competitiveness worldwide. The next chapter will present conclusions and discuss further research studies.

References

Antwerpen, C. and Curtis, N.J. 2016. A data collection and presentation methodology for decision support: A case study of hand-held mine detection devices. *European Journal of Operational Research*, 251: 237–251.

Bocken, N., Morgan, D., and Evans, S. 2013. Understanding environmental performance variation in manufacturing companies. *International Journal of Productivity and Performance Management*, 62(8): 856–870.

Carrillo-Castrillo, J.A., Rubio-Romero, J.C. Onieva, L., and Lopez-Arquillos, A. 2016. The causes of severe accidents in the Andalusian manufacturing sector: The role of human factors in official accident investigations. *Human Factors and Ergonomics in Manufacturing and Service Industries*, 26(1): 68–83.

Ciarapica, F.E., Bevilacqua, M., and Mazzuto, G. 2016. Performance analysis of new product development projects: An approach based on value stream mapping. *International Journal of Productivity and Performance Management*, 65(2): 177–206.

Dulange, S.R., Pundir, A.K., and Ganapathy, L. 2014. Prioritization of factors impacting on performance of power looms using AHP. *Journal of Industrial Engineering International*, 10: 217–227.

Fonseca, A., McAllister M.L., and Fitzpatrick, P. 2014. Sustainability reporting among mining corporations: A constructive critique of the GRI approach. *Journal of Cleaner Production* 84, 70–83.

Garza-Reyes, J.A., Ates, E.M., and Kumar, V. 2015. Measuring lean readiness through the understanding of quality practices in the Turkish automotive suppliers industry. *International Journal of Productivity and Performance Management*, 64(8): 1092–1112.

Govindan, K. 2017. Sustainable consumption and production in the food supply chain: A conceptual framework. *International Journal of Production Economics*: 1–14.

Helo, P., Suorsa, M., Hao, Y., and Anussornnitisarn, P. 2014. Toward a cloud-based manufacturing execution system for distributed manufacturing. *Computers in Industry*, 65: 646–656.

Hu, Q., Mason, R., Sharon J., and Found, W.P. 2015. Lean implementation within SMEs: A literature review. *Journal of Manufacturing Technology Management*, 26(7): 980–1012.

Klewitz, J. and Hansen, E.G. 2014. Sustainability-oriented innovation of SMEs: A systematic review. *Journal of Cleaner Production*, 65: 57–75.

Lee, C.S. and Wong, K.Y. 2015. Development and validation of knowledge management performance measurement constructs for small and medium enterprises. *Journal of Knowledge Management*, 19(4): 711–734.

Lim, S.A.H., Antony, J., He, Z., and Arshed, N. 2017. Critical observations on the statistical process control implementation in the UK food industry: A survey. *International Journal of Quality and Reliability Management*, 34(5): 684–700.

Marodin, G.A. and Saurin, T.A. 2015. Managing barriers to lean production implementation: Context matters. *International Journal of Production Research*, 53(13): 3947–3962.

McLean, R.S., Antony, J., and Dahlgaard, J.J. 2017. Failure of continuous improvement initiatives in manufacturing environments: A systematic review of the evidence. *Total Quality Management and Business Excellence*, 28(3–4): 219–237.

Mehralian, G., Nazari, J.A., Nooriparto, G., and Rasekh, H.R. 2017. TQM and organizational performance using the balanced scorecard approach. *International Journal of Productivity and Performance Management*, 66(1): 111–125.

Movahedi, B., Miri-Lavassani, K., and Kumar, U. 2016. Operational excellence through business process orientation: An intra- and inter-organizational analysis. *The TQM Journal*, 28(3): 467–495.

Pakdil, F. and Leonard, K.M. 2015. The effect of organizational culture on implementing and sustaining lean processes. *Journal of Manufacturing Technology Management*, 26(5): 725–743.

Prajogo, D. and Sohal, A. 2013. Supply chain professionals: A study of competencies, use of technologies, and future challenges. *International Journal of Operations and Production Management*, 5(33): 1532–1554.

Rostam, S. 2015. Decision making model based on PROMETHEE for manufacturing strategy direction and performance improvement in manufacturing SMEs. *American Journal of Science and Technology*, 2(5): 251–257.

Siddiqui, S.Q., Ullah, F., Thaheem, M.J., and Gabriel, H.F. 2016. Six Sigma in construction: A review of critical success factors. *International Journal of Lean Six Sigma*, 7(2): 171–186.

Wells, P. 2016. Economies of scale versus small is beautiful: A business model approach based on architecture, principles and components in the beer industry. *Organization and Environment*, 29(1): 36–52.

Wiesner, R., Chadee, D., and Best, P. 2017. Managing change toward environmental sustainability: A conceptual model in small and medium enterprises. *Organization and Environment*, 1–8.

Chapter 12

Conclusion and Further Research Studies

12.1 Summary

Chapters 1 through 11 in this research book presented the real-world challenges facing manufacturing small and medium enterprises (SMEs), as well as revealing advances in work study, in terms of the productivity level of these SMEs. Work study was regarded as one of the key tools contributing to manufacturing SMEs worldwide, including the companies used as case studies in this book.

Various work study tools were discussed, such as method study and work measurement, which contribute to the productivity of manufacturing SMEs. Method study involved various elements, such as benchmarking, brainstorming, preliminary surveys, charts, diagrams, filming techniques, value stream mapping, cause-and-effect diagrams, and business process reengineering. Work measurement tools also comprised benchmarking, brainstorming, preliminary surveys, value stream mapping, and business process reengineering, which are the same as in method study, but work measurement differs from method study in including tools such as time study, work sampling, predetermined motion time systems, analytical systems, comparative estimating, and synthesis. Both of these tools were used to identify problem areas facing these SMEs, but they differed in the sense that method study mainly focused on simplifying employees' working activities and making improvements to the method of carrying out activities for good performance, whereas with work measurement, the focus was on eliminating ineffective time spent and determining the appropriate standard time for the work activities being carried out by qualified employees at a defined rate of performance.

Furthermore, the input resource factors for physical, technological, and management were investigated and related to the competitive priorities and strategic capabilities required by manufacturing SMEs to ensure productivity optimization. The reality of the case study observations and the literature studied is that the concept of work study tools was unfamiliar to the companies visited as compared with the challenges facing other SMEs abroad with regard to the literature read.

Most of these tools were unknown to the companies visited, making it difficult for them to realize the role that work study can play in enabling their businesses to compete effectively and efficiently on a global scale.

The research outcome indicated that the manufacturing SMEs in Gauteng, South Africa, focused mainly on product quality and also considered integrating other capabilities that would help them eradicate the waste, defects, and late deliveries hindering productivity growth in their businesses. There was a lack of implementation of other standards such as ISO 14001 and 18001 for the regulation of employee environment and safety, respectively, in these manufacturing SMEs. When waste and defects are eliminated and cycle time reduced, quality products are generated and cost savings are realized, thus improving productivity in South African businesses.

One challenge facing these companies is the issue of the human element, whereby employees are resistant to change. The issue of employee resistance to change was attributed to poor communication between management and employees due to a lack of disclosure concerning other issues impacting on the attitudes of employees in the workplace.

An element of communication and employee motivation might be necessary from top to bottom, as well as through a bottom-up approach whereby management involve employees in decision making and in communication workshops in order to build trust and openness toward employees.

In view of the challenges experienced in manufacturing SMEs, one of which is communication, the intervention of a work study expert is crucial in explaining and guiding management, supervisors, unions, and employees on the information that can be presented on productivity improvement, which will build a healthy atmosphere among all parties in manufacturing SMEs.

This research book reflects on the future trend of introducing work study specialists to encourage productivity improvement in manufacturing, in particular in SMEs. The cases presented are significant in assisting management/supervisors in making decisions that will promote productivity in their businesses. Academic scholars, operations management, work study practitioners, industrial engineers, and other successful manufacturing industries already engaged in work study can broaden SMEs' knowledge in relation to productivity improvement through the application of work study.

12.2 Concluding Remarks

A review of the literature on the importance of productivity and work study has been carried out to provide an extensive background to the concepts involved. An inference is also made that input resource factors such as physical capital technological capital, and management contribute to the manufacturing operations of these manufacturing SMEs. In addition, the availability of management capabilities and adding-value activities reinforce these input resource factors to contribute to the effective and efficient running of these SMEs. Despite the presence of these capabilities and adding-value activities, the ignorance of management on standard operating procedures in areas such as ergonomics and good housekeeping, which aim to ensure a safe, healthy, and comfortable environment for employees, hinders productivity in their businesses.

Furthermore, based on the literature studied and discussed as well as companies visited, this research book concludes that when work study is applied in the manufacturing processes, considering capabilities and adding-value activities in SMEs, the productivity of these SMEs will continue to improve.

The challenges facing manufacturing SMEs were mapped and discussed in detail in the South African context. Despite the challenges and variations in relation to the literature and existing knowledge, there is potential in the use of work study (work method and measurement) to improve productivity in manufacturing SMEs. Improving productivity in manufacturing SMEs ensures business competitiveness in terms of quality, cost, and delivery. This is achieved by ensuring input resource factors are effectively and efficiently utilized and minimized throughout the manufacturing process in manufacturing SMEs. The knowledge from work study will be extended in equipping manufacturing SMEs.

The application of work study tools by management can be a challenge in the absence of a work study specialist to assist management and supervisors, with the support of the union. Resistance to change and lack of knowledge and skills in manufacturing SMEs is still a concern if productivity optimization is to become a reality. The elements of communication, formal training, and employee motivation are necessary to drive manufacturing SMEs' objectives. Based on further studies highlighted, it will be necessary to investigate results emanating from the effect of application of work study in manufacturing SMEs.

Despite the fact that there is sufficient consensus on work study and the productivity of manufacturing SMEs globally, work study remains questionable as a methodology in measuring the productivity of manufacturing SMEs in South Africa. This research book attempts to shed light on these emerging perspectives.

12.3 Further Research Studies

In view of the challenges identified within the case studies, future research studies should concentrate on various aspects relating to productivity improvement in South Africa. First, research on the efficient use of input resource factors in improving productivity in the country needs to be conducted among manufacturing SMEs. By improving productivity, these resources can be efficient through the application of work study tools such as method study and work measurement in the manufacturing process of these SMEs. Second, the literature on the integration of these tools needs to be carried out by focusing on physical capital, technological capital, and management. In terms of management, this literature should provide guidance on the use of various manufacturing capabilities and adding-value activities that contribute to the productivity of manufacturing SMEs processes in South Africa. Finally, longitudinal studies based on the measurement of productivity among manufacturing SMEs need to be conducted in order to provide productivity performance results before, during, and after the performance of these SMEs.

Index

A

B

C

D

E

W